Spacecraft Power Technologies

Published by
Imperial College Press
57 Shelton Street
Covent Garden
London WC2H 9HE

Distributed by
World Scientific Publishing Co. Pte. Ltd.
P O Box 128, Farrer Road, Singapore 912805
USA office: Suite 1B, 1060 Main Street, River Edge, NJ 07661
UK office: 57 Shelton Street, Covent Garden, London WC2H 9HE

British Library Cataloguing-in-Publication Data
A catalogue record for this book is available from the British Library.

SPACECRAFT POWER TECHNOLOGIES
Copyright © 2000 by Imperial College Press
All rights reserved. *This book, or parts thereof, may not be reproduced in any form or by any means, electronic or mechanical, including photocopying, recording or any information storage and retrieval system now known or to be invented, without written permission from the Publisher.*

For photocopying of material in this volume, please pay a copying fee through the Copyright Clearance Center, Inc., 222 Rosewood Drive, Danvers, MA 01923, USA. In this case permission to photocopy is not required from the publisher.

ISBN 1-86094-117-6

Printed in Singapore by Regal Press (S) Pte. Ltd.

Space Technology — Volume 1

Spacecraft Power Technologies

A.K. Hyder
University of Notre Dame

R.L. Wiley
Booz, Allen & Hamilton Inc.

G. Halpert
Jet Propulsion Laboratory

D.J. Flood
NASA Glenn Research Center

S. Sabripour
Lockheed Martin CPC

Imperial College Press

CONTENTS

PREFACE   *xi*
ACKNOWLEDGEMENTS   *xiii*

1 INTRODUCTION   1
1. The beginnings   1
    1.1 The increasing demand for spacecraft electrical power   3
    1.2 The architecture of a spacecraft   5
2. The electrical power system   7
    2.1 An overview of electrical power systems   7
    2.2 Electrical power system designs   10
    2.3 Examples of missions and their electrical power systems   12
    2.4 Spacecraft electrical power technologies   17
    2.5 An overview of the book   18
3. References   20

2 ENVIRONMENTAL FACTORS   23
1. Introduction   23
2. Orbital considerations   25
    2.1 Orbital elements   27
    2.2 Eclipse times   29
3. The near-Earth space environment   33
    3.1 The neutral environment   33
    3.2 The plasma environment   41
    3.3 The radiation environment   46
    3.4 The particulate environment   62
4. References   66

3 SOLAR ENERGY CONVERSION 71
1. Introduction 71
   1.1 Space photovoltaic power systems 72
   1.2 Space power system applications and requirements 73
   1.3 Space solar cell and array technology drivers 74
2. Solar cell fundamentals 75
   2.1 Introduction 75
   2.2 Basic theory 77
3. Solar cell calibration and performance measurements 80
   3.1 Calibration techniques 81
   3.2 Laboratory measurement techniques 84
4. Silicon space solar cells 88
   4.1 Advanced silicon solar cells 89
   4.2 Radiation damage in silicon solar cells 90
5. III-V compound semiconductor solar cells 95
   5.1 Single junction cells 96
   5.2 Multiple junction cells 108
6. Thin film solar cells 117
7. Space solar cell arrays 119
   7.1 Space solar array evolution 120
   7.2 Rigid panel planar solar arrays 120
   7.3 Flexible, flat panel arrays 122
   7.4 Concentrator arrays 126
   7.5 Array environmental interactions 129
   7.6 Power system design and array sizing 136
8. Space thermophotovoltaic power systems 140
   8.1 TPV system efficiency 141
   8.2 Solar thermophotovoltaic space power systems 144
9. Conclusion 147
10. References 149

4 CHEMICAL STORAGE AND GENERATION SYSTEMS 157
1. Introduction 157
2. Inventions 158
3. Evolution of batteries in space 159

4. Fundamentals of electrochemistry	163
4.1 Standard electrode potential and free energy	164
4.2 The Nernst equation	165
4.3 Capacity and the Faraday relationship	166
5. Cell and battery mechanical design	166
5.1 Cell design	166
5.2 Battery design	170
6. Performance metrics	171
6.1 Voltage	172
6.2 Capacity and energy	174
6.3 Specific energy and energy density	175
6.4 Life and performance limitations	176
6.5 Charge control	180
6.6 Efficiency and thermal properties	181
7. Electrochemical cell types	185
7.1 Primary cells	185
7.2 Rechargeable cells and batteries	203
8. Fuel cell systems	223
8.1 History	224
8.2 Fuel cell system basics	226
8.3 Alkaline fuel cells	229
8.4 Proton exchange membrane fuel cells	230
8.5 Regenerative fuel cells	231
8.6 Direct methanol liquid-feed fuel cell/PEM	232
9. Definitions and terminology	235
10. References	239

5 NUCLEAR SYSTEMS	**241**
1. Introduction	241
2. History of the U.S. space nuclear program	242
2.1 Radioisotope space power development	244
2.2 Space reactor power development	250
2.3 The future	253
3. History of the Russian space nuclear program	255
4. Radioisotope systems	258

5. Reactors	264
6. Safety	273
6.1 U.S. safety	274
6.2 Russian space nuclear safety experience	280
6.3 International developments in space nuclear safety	282
7. References	282

6 STATIC ENERGY CONVERSION — 287

1. Introduction	287
2. Thermoelectrics	288
3. Thermionics	294
4. AMTEC	307
5. Thermophotovoltaics	311
6. References	318

7 DYNAMIC ENERGY CONVERSION — 323

1. Introduction	323
2. Stirling cycle	324
3. Closed Brayton cycle	332
4. Rankine cycle	340
5. References	349

8 POWER MANAGEMENT AND DISTRIBUTION — 353

1. Introduction	353
1.1 The ideal power system	353
1.2 Power subsystem overview	357
1.3 Electrical power system options	361
2. Functions of PMAD	363
2.1 Power management and control	363
2.2 Power distribution	372
2.3 Fault management and telemetry	374
2.4 Point-of-load DC-DC converters	375

CONTENTS

3. Components and packaging	397
3.1 High-reliability space-grade parts	398
3.2 Packaging technologies	404
4. System examples	404
4.1 The Lockheed Martin A2100	405
4.2 Global positioning system block IIR	407
4.3 The International Space Station	407
4.4 The Modular Power System	409
5. References	412

9 THERMAL MANAGEMENT — 415

1. Introduction	415
1.1 Definition and purpose of a TCS	416
1.2 Characterization and design of the thermal control process	417
2. The thermal environment	421
2.1 Solar radiation	422
2.2 Planetary radiation	424
2.3 Spacecraft-generated heat	426
3. Heat transfer mechanisms	429
3.1 Heat transfer by conduction	429
3.2 Heat transfer by radiation	430
3.3 Absorptivity and emissivity	432
4. The basics of thermal analysis	435
5. Thermal management techniques	437
5.1 Passive thermal management	438
5.2 Active thermal management	445
6. References	449

Appendix: MAGNETIC MATERIALS	453
INDEX	467

PREFACE

It is always an interesting exercise, after a book is written, to justify the need for the just-completed manuscript. The situation with this text is no different. The genesis for this book began well over a decade ago when one of us (AKH) began thinking about a book on spacecraft electrical power, with a focus on electrical power systems. The immediate need for such a book was made clear when, as a principal in an emerging academically-based research institute focusing on space power, I searched for a generic reference text for use by undergraduate and graduate students entering this interdisciplinary research arena. It was apparent then that the information was available, but dispersed among a large number of industrial technical memoranda, NASA and ESA technical reports, and proceedings of conferences. Although there were texts (liberally referenced in this work) that included space power systems as a topic, there was no comprehensive text treating the subject in a global sense.

The need for such a book, beyond its use in an academic setting, was evident for a number of other reasons: the uniqueness of space as an operating medium, the increasing demand for electrical power aboard newer spacecraft, the emergence of new power technologies that made higher power systems more feasible, the realization that power system design was a pacing factor in future space operations, and, as mentioned earlier, the absence of such a reference text on the subject.

In the Fall of 1997, Imperial College Press presented an opportunity to further develop my thoughts on just how such a text might be organized. The first step was to seek the advice of several colleagues more knowledgeable than I in the subject of space power– colleagues who appear as coauthors of this book. In the initial attempt to outline the book, the title chosen was *Spacecraft Power Systems*. That title was short-lived, however. After struggling for quite some time, it finally appeared to us that, while such a text could prove to be a useful addition to the technical literature on spacecraft design, any approach based on a *systems* concept would necessarily exclude a full discussion of the breadth and richness of the technologies upon which those systems are built.

The distinction between 'systems' and 'technologies' is not unique to spacecraft electrical power. It does, however, present an interesting challenge in the context of this particular subject of power.

Spacecraft electrical power systems are designed to address specific mission needs. The mission requirements would, for example, dictate a variety of design parameters such as operating lifetime, constraints imposed by launch vehicle and orbit choices, average and peak power levels required by the payloads, the degree of reliability and redundancy appropriate for the mission, the operating temperature limits, the total project cost, *etc.* And since there is an almost limitless number of missions that can be performed in space, the number of spacecraft systems that are possible can quickly grow to a very large number also. A book based on power systems would necessarily lead to redundancies or gaps in the presentation.

Consider, for example, a photovoltaic system coupled to a battery reserve to serve as the power source during eclipse, the most commonly found power system in space. Even within this relatively simple system, the options for various photovoltaic cell materials, concentrator designs, and battery couples are numerous. Similarly, the description of a nuclear reactor power system would include a full discussion of a specific conversion process but could overlook the fact that the source of heat could equally well be a chemical or solar source, or that the conversion process could be one of several other static or dynamic conversion options.

We were thus drawn to the present organization of the book, with, as the title reflects, an emphasis on the technologies enabling the power systems rather than systems themselves. While we recognize that this, too, falls short of optimum in that it does not allow a full discussion of the integration of the various technologies into operating systems, it does provide a comprehensive basis on which that integration can proceed.

Under the premise that the three energy sources possible for space application are solar, chemical, and nuclear, the book attempts to explore each from several aspects. The two larger chapters are devoted to solar conversion and chemical storage/conversion, appropriately since those, by far, constitute the most mature space electrical power systems. Accomplishments in the Russian and U.S. nuclear power programs are presented in a series of chapters devoted to nuclear reactors and radioisotopes as heat sources interfaced to either static or dynamic conversion methods. A detailed discussion is also offered on the techniques and technologies of power management and distribution aboard spacecraft. Finally, two other chapters discuss topics which, while not directly related to power technologies, are critical in the design of spacecraft electrical power systems: the space environment within which the spacecraft operates and the thermal environment within the spacecraft.

We do hope that, in spite of all of its shortcomings, the present effort may prove to be of some value to those designers, engineers, scientists, and students for whom space is not just a place, but a profession.

A. K. Hyder
Notre Dame, Indiana U.S.A

ACKNOWLEDGEMENTS

Many people, other than the authors, contributed significantly to the preparation of this book. It is our pleasure to acknowledge their efforts.

There are a number of space power experts whose kindness and willingness to contribute to the work we gratefully acknowledge:

Ned Rasor (consultant), James Dudenhoefer and Richard Shaltens (NASA Lewis Research Center), Mark Morgan (Edtek, Inc.), Subbarao Surampudi (Jet Propulsion Laboratory), Joseph Sholtis (consultant), Robert Sievers and Thomas Hunt (Advanced Modular Power Systems, Inc.), and Richard Hemler and C. Edward Kelly (Lockheed Martin).

Once the words are finally on paper, the real work begins! There are a number of people who assisted in the editing of the text and without whose contributions this effort would not have been completed. First and foremost, our thanks to Nancy Hanson, the principal editor to whom we went with a thousand questions and who responded a thousand times with patience and guidance. Thank you, Nancy, for all that you did to make this a reality.

Our thanks go also to Andy Deliyannides who was responsible for the conversion of the text into Adobe™ Pagemaker ™ format, and for the countless hours he spent in formatting the manuscript. Again, Andy, thank you.

We also acknowledge the many contributions of Jaime Raba in the final stages of the manuscript preparation, especially in the generation of the book's index.

Our thanks to John Navas, commissioning editor for Imperial College Press, for the opportunity to bring this book to press and for his patience and support during its preparation, and to Caroline Ching and K. N. Hong of World Scientific Publishing for all of their assistance .

And, finally, we thank our families for their support during what may have seemed for them an eternity.

CHAPTER 1

INTRODUCTION

1. The beginnings

The first artificial satellite, the 184-pound Sputnik I (Figure 1.1), was launched on October 4, 1957, and carried a silver-zinc primary battery as its only power source. The battery provided one watt to power the two transmitters which ceased broadcasting three weeks later. The satellite reentered the atmosphere in January, 1958, but not before marking the dawn of the space age (Walls, 1995). The primary battery (i.e., not a rechargeable one) effectively defined the useful life of the spacecraft since the spacecraft itself did not re-enter the Earth's atmosphere until some weeks after the batteries were spent. This initial satellite was followed soon thereafter by the launch of Vanguard I, the first satellite to carry solar cells coupled to secondary (i.e., rechargeable) batteries. The batteries were included to provide electrical power during periods of eclipse. Since then, the sophistication of artificial satellites and the attendant demands for electrical power to make them functional have increased by many orders of magnitude. What was once a scientific curiosity has become an indispensable tool of modern communications, meteorology, observation, navigation, geodesy, national defense, and entertainment, as well as scientific discovery.

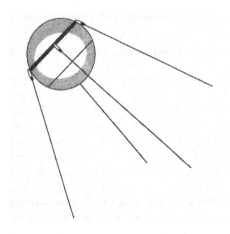

Figure 1.1 Sputnik I, the Earth's First Artificial Satellite

Since those early days, the frequency of satellite launches has made the event commonplace. Figure 1.2 shows this growth in the number of spacecraft launched worldwide over the past 40 years (Curtis, 1994; Thompson, 1994). This growth has occurred not only in the number of satellites launched, but in their size also. While the first Sputnik was only a few kilograms, the size of present-day satellites can be judged by the capabilities of several current launch vehicles shown in Table 1.1.

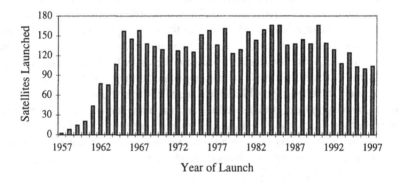

Figure 1.2 The Satellite Launch Rate, Worldwide, Since 1957

Table 1.1 Representative Capabilities of Current Launch Vehicles

Launch Vehicle	Payload to LEO (kg)	Payload to GEO(kg)	GTO (kg)
Delta II- 7925	5,000	1,800	
Titan IV	17,700	4,450	
Ariane 5			6,800
Proton K	20,100	2,100	4,615
Shuttle	24,400	5,900	

Although enormous payloads can be placed in orbit with relative ease using these and other modern launch systems, the cost of launch still remains very high (typically several thousands of dollars (U.S.) per kilogram into low Earth orbit). This cost places a premium on minimum mass and high system reliability, especially for the bus systems

INTRODUCTION 3

(e.g., guidance and control, telemetry, power, etc.) that are often assumed by the mission planners.

1.1 The increasing demand for spacecraft electrical power

The increases in satellite sophistication, and with it increases in payload size, have been accompanied by an ever-growing requirement for electrical power aboard the spacecraft. Figure 1.3 shows the growth in electrical power needed for specific spacecraft over the past 40 years. In some sense, the communications satellite demands for electrical power have diverged along two tracks based on orbits: geosynchronous communications satellites which often require ten to twenty kilowatts of power versus the lower-orbit, smaller communications spacecraft which typically require only tens to hundreds of watts. For many other applications, the trend has generally been for more power, although not exclusively. The NASA program to develop less-expensive, lighter satellites has also increased demands for less, rather than more power. While the demand for spacecraft electrical power extends across a broad range of values from several hundred watts to many tens of kilowatts, in a real sense, the more challenging task may be at the lower end of the power range.

Regardless of the power levels, in all space applications there is the need to improve the system specific power. This has placed great demands on the engineering skills of spacecraft power designers, and the response has been to develop new technologies and to refine existing technologies especially to enable the missions which require higher power levels. This has been done in the face of a rather limited menu of options available for generating electrical power in space.

Prior to 1957, most design experience related to the engineering of electrical power systems was rooted in terrestrial systems or power systems designed for use on aircraft. Most of the terrestrial power design guidelines were not relevant to operations in space

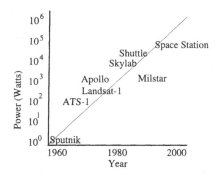

Figure 1.3 The Growth in Requirements for Spacecraft Electrical Power

because their design assumes certain elements that simply are not available in space. These have been identified, with some humor, as Earth, air, fire, and water. The Earth provides a convenient grounding mechanism (grounding to the space plasma is problematic), air offers efficient convective cooling (radiative cooling from the spacecraft is the only realistic option), fire is an inexpensive heat source for conversion to electricity (not a space option), and water is used universally in thermal management (the control of waste heat, especially in conjunction with power generation, is a serious design constraint). On a less whimsical basis, Walls (1995) points to several differences between space-based and Earth-based power systems that seriously limit technology interchange. These differences are contrasted in Table 1.2.

It is not surprising, therefore, that early-on aircraft engineering practices were adapted for spacecraft. But even that was insufficient to accommodate the more stringent constraints imposed on operations in the space environment. While mass, reliability, and cost are considerations in aircraft power systems, those systems have the advantages of very large prime power sources, the aircraft engines that develop many times more power than any electrical power that might be drawn as an incidental load. Further, flight times are measured in hours rather than years.

No constraint is more demanding on space operations, however, than reliability.

Table 1.2 Comparison of Terrestrial and Space-Based Power Systems

Attribute	Terrestrial System	Space-Based System
Scale	Tens to hundreds of megawatts	One to ten kilowatts, typically
Sources	Many options: hydro, coal, nuclear, large rotating machines, etc. or chemical	Few options with premium on mass: solar, nuclear,
Transmission	High voltage operations, AC is standard, switching and voltage conversion are simplified	High voltage not compatible with space plasmas, DC is standard, DC-DC inverters are needed
Costs	Addressed through scale	Includes cost of delivery to orbit, premium paid for high reliability and mass and volume reduction
Energy management	Providers adjust to customers' need	Energy budget is fixed by the power system and all power management is load management
Operations	Large, interconnected grids to provide redundancy	Autonomous operation, no interconnectedness for redundancy

INTRODUCTION 5

While reliability is taken seriously in the design of terrestrial and airborne power systems, the appearance of a problem in either of these applications can be addressed with relative ease. In the case of space-based systems, maintainability is a prohibitively expensive option, if it is available at all. It takes only ten minutes to get into space, but at $50 million a minute, everything must be totally reliable (Kinesix, 1998). Reliability must be engineered from the beginning since failure in a non-redundant electrical power system can mean the end of the mission.

1.2 The architecture of a spacecraft

The complexity and cost of building and launching satellites have also increased during this same period. In this evolution, spacecraft operators have worked with innovation and great engineering skill to improve the lifetime, efficiency, reliability, and compactness of each of the subsystems aboard the spacecraft. What began as a simple design centered on a power source and a transmitter has become a complicated interrelationship among a number of subsystems, each requiring electrical power. The general architecture of a spacecraft is shown in Figure 1.4. The satellite can be viewed as being comprised of two major parts: the mission payloads and the support subsystems. The payloads, specific to each satellite, are the reasons for the mission and to a large extent will define the overall satellite design. As Stark *et al.* (1995) point out, however, the

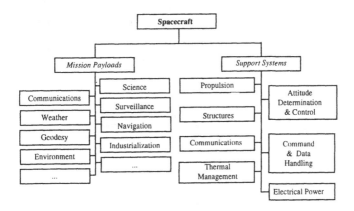

Figure 1.4 Spacecraft Systems

final design goes beyond the nature of the payloads and is often a reflection of the design philosophy of the contractor or the nation responsible for the satellite.

Support systems, sometimes referred to as bus systems, are generic and tend to be the same, functionally, from satellite-to-satellite. In an effort to reduce costs, much effort has been devoted in recent years to the modular design of these bus systems. Virtually every system onboard a satellite, payload or bus, will require electrical power, usually at differing peak and average power levels, voltages, and duty cycles. To illustrate the complexity of this load, consider the attitude control system (ACS) which is designed to maintain the satellite pointing in the proper direction. It was the failure of the ACS on Galaxy IV in May, 1998, which caused the loss of that communications satellite carrying 90% of the electronic-pager traffic in the United States. Figure 1.5, adapted from Barter (1992), shows the many parts of a modular attitude-control system (ACS), and it is clear that each component will require electrical power. Several of the ACS subsystems such as the accelerometers, sensors, and computers for data manipulation require low voltages and currents, while the drives and electromagnetic actuators for solar arrays require high peak powers. This situation is repeated in each of the mission and bus systems shown in Figure 1.4. In the overall satellite design, a power budget for each of these systems is an important part of the process of sizing the power system. A more complete discussion of the architecture of spacecraft and the functions and power requirements of the bus systems can be found in several recent books in the

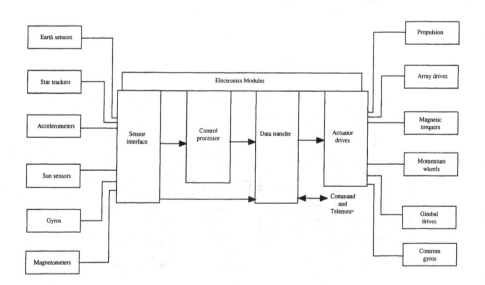

Figure 1.5 A Modular Attitude-Control System

INTRODUCTION

field, among them Fortescue and Stark (1995), Griffin and French (1991), Larson and Wertz (1992), Pisacane and Moore (1994), and DeWitt, Duston, and Hyder (1993).

2. The electrical power system

2.1 An overview of electrical power systems

The enabling system aboard any satellite is the electrical power system (EPS). In its simplest form, a satellite electrical power system consists of four major components as shown in Figure 1.6. The prime power source will provide energy for conversion into electricity. As an intermediate step, in certain cases (e.g., thermophotovoltaic), all or part of that prime energy may be stored before conversion takes place. Conversion into electricity then occurs through a variety of methods, depending on the nature of the prime source and the spacecraft electrical loads. The electricity that is generated will need to be managed, regulated, monitored, and conditioned to match the electrical needs of the spacecraft systems.

The conversion of one form of energy into another involves technologies that are both old and new. Table 1.3 lists several forms of energy that can be considered as potential input energy for conversion to another form, electricity in the case of spacecraft power. While the technologies listed do not form an exhaustive list of options, their great number reflects the richness of possibilities. For example, nuclear sources, primarily viewed as sources of heat, are not included but are certainly important sources of energy for space operations. Entries on the diagonal of Table 1.3 may be viewed as storage options rather than conversion mechanisms.

The choices available as prime power sources in space are limited to three: nuclear, chemical, or solar. As shown in Figure 1.7, the duration of the mission is a key factor in the selection of the prime power source. For short-duration missions, or to supply the

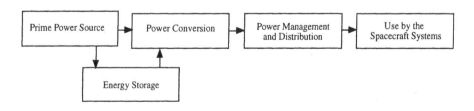

Figure 1.6 Elements of the Electrical Power System

Table 1.3 Energy Conversion and Storage Options

Output Energy / Input Energy	Electricity	Heat	Chemical	Photons	Kinetic
Electricity	Batteries Fuel cells Superconducting magnets Inductors	Ohmic heaters Heat pumps	Electrolysis Ionization and recombination	LEDs Discharges Light bulbs Lasers & transistors Microwaves	Flywheels JxB thrusters Motors
Heat	Thermoelectrics Thermionics Generators Fuel cells	Phase-change materials Chemical reactions High C_p materials	Thermochemical electrolysis	Radiators	All thermodynamic cycles
Chemical	Fuel cells Capacitors Batteries	Combustors	Propellants Explosives	Chemical lasers	Rocket exhaust Gas turbines
Photons	Photovoltaic cells	Thermal concentrators Thermal absorbers	Photolysis electrolysis	Resonant cavities Metastable atoms	Radiometers
Kinetic	MHD Generators Homopolar devices Compulsators	Friction	Impact ionization	Triboluminescence	Flywheels

INTRODUCTION

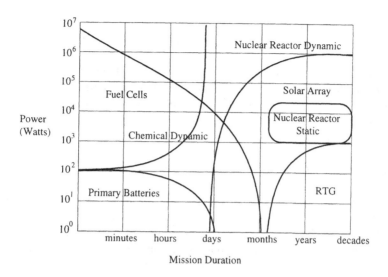

Figure 1.7 Options For Various Mission Power Needs and Durations

power for activities that will be completed relatively quickly within the framework of a longer mission, chemical systems such as primary batteries, fuel cells, or chemical dynamic conversion, may be the appropriate choice, depending on the total power required. Often, primary batteries are used in meeting the high power and high energy demands of the launch vehicle itself as well as in the activation of pyrotechnic devices related to explosive stage separation. For longer duration missions, the choices are restricted to solar arrays in conjunction with secondary batteries or regenerative fuel cells, or to nuclear systems, either reactors or radioisotope thermoelectric generators. Other operational issues may certainly influence the choice of prime power sources. For example, the survivability of solar arrays in certain orbits could exclude their choice in spite of their ability to provide the necessary power within limitations of mass, cost, etc. The restricted maneuverability of large solar arrays, an unacceptable level of the infrared signature of nuclear systems, or compatibility with mission-related sensors can also eliminate certain prime power options which otherwise would have been logical choices.

In some applications, the demand can be for a very large peak power, but for a short enough duration that the total energy can be surprisingly small. In other cases, such as housekeeping power requirements aboard an operational satellite, the average power requirements can again be modest but the extended time over which the power is needed can create the need for large amounts of total energy.

As an example, a military mission that requires 100 MW for ten minutes will demand about the same total energy as ten kW missions that must remain on orbit for 10 years. In this example, while solar cells can be sized to provide 100 MW, since the duration is so short, an expendable fuel option makes more sense from a total mass argument. A space-based radar which requires one MW of power in a one ms burst needs only 10^3 J per pulse, an application that could be met with the use of capacitors and a few kW baseload power system. Figure 1.7, therefore, presents options based on total energy that is required for the mission as well as the rate at which that energy can be delivered.

2.2 Electrical power system designs

The electrical power system (EPS) is designed and configured to perform several key functions: it must be a continuous and reliable source of peak and average electrical power for the life of the mission; it must control, distribute, regulate, and condition the power provided to the various loads; it must be capable of providing data regarding the health and status of its operation; and it must protect itself and its loads from electrical faults anywhere within the spacecraft (McDermott, 1992). Many factors contribute to the final design and the choice of technologies that must be integrated. This process (shown schematically in Figure 1.8) starts with the mission and its requirements. The mission payloads will define the peak and average power needed, together with the lifetime of the satellite, the orbit, and the overall configuration of the spacecraft. Each of these constraints will carry implications for the design of the EPS, such as the end-of-life power needs, the degree of redundancy needed for an acceptable level of reliability, the environmental factors against which the system must be protected, and options for the thermal management (TM) system.

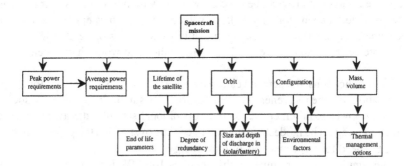

Figure 1.8 The EPS Selection Process

INTRODUCTION

The selection process will focus not only on the prime power source but each of the other subsystems in the EPS (Figure 1.9) as well. In many cases, the prime power source serves only as a source of heat to be converted into electrical power. All three sources– nuclear, solar, and chemical– are capable of producing heat for conversion into electricity through either static or dynamic processes. As the name suggests, the static processes– thermoelectric, thermionic, AMTEC, and others– do the conversion without benefit of moving parts. This is often demanded by the pointing-and-tracking requirements of the payload. Dynamic conversions involve the thermodynamic-cycle processes such as Rankine, Stirling, and Brayton. The most common EPS in use is the photovoltaic array, involving solar energy and a static conversion process, the photovoltaic cell. In these cases, the energy must be stored, usually through a chemical process (mostly batteries, but sometimes regenerative fuel cells), so that the spacecraft can be powered during the eclipse periods or when load demands exceed solar array output. Regardless of the prime power source, energy storage is an option using thermal, chemical, or mechanical means. Mechanical storage mechanisms (e.g., high-rpm flywheels) are not in use but offer very large storage potential, again addressing the possible need for large peak powers simultaneous with modest average power. Following the conversion process, the unregulated electrical power is delivered to the Power Management and Distribution (PMAD) subsystem. The PMAD links the generation process to the storage elements and the spacecraft loads. Although PMAD is indicated as a subsystem interfacing with the spacecraft loads, in reality it is distributed throughout the EPS, and functional elements can be found virtually everywhere in the electrical system.

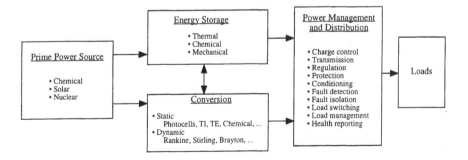

Figure 1.9 Functional Breakdown of the EPS

2.3 Examples of missions and their electrical power systems

To provide examples of the impact of mission requirements on EPS design, we conclude this introduction with a brief discussion of several satellites recently launched or in the late planning stages.

Spartan

The Spartan (Figure 1.10a), a free-flying platform for scientific experiments, is released and recovered from the Shuttle. It uses a common service module containing the ACS, electronics, batteries, the TM system, data handling electronics, and a cold plate. In a throwback to Sputnik I, since the unit has an operational lifetime of only 40 to 50 hours, silver-zinc primary batteries supply the electrical power. The batteries have a capacity of 30 kWh and deliver power at 28 VDC. Since it will remain in a relatively benign low-Earth orbit, there are no special constraints imposed by the orbital environment.

Cassini

At the other extreme, the mission of spacecraft Cassini (Figure 1.10b) is to explore the Saturnian system. It is designed to carry 12 instruments on the 2,100-kg orbiter (remaining in orbit around Saturn for four years) and six on the 350-kg probe (which will explore the moon Titan in situ). The satellite was launched in October, 1997, and is due to arrive at Saturn in June, 2004. Several requirements distinguish Cassini from other interplanetary missions: the distances to the Earth and the Sun, the extended length of the mission, the number and complexity of the scientific experiments, and the four gravity-assists enroute to Saturn. It has a design life of 13 years, much of which will be at such great distances from the Sun that solar arrays are impractical. These factors, even with relatively low power requirements (about 750 W at beginning–of–life and 628 W at end-of-mission), make a nuclear prime power system mandatory. In the case of Cassini, three radioisotope thermoelectric generators (RTGs) are employed. Lithium sulfur-dioxide (LiSO2) primary batteries provide power for the Huygens probe.

Magellan

Magellan (Figure 1.11) was designed to study the geological structure of Venus. The primary payloads included a synthetic aperture radar that imaged 98 percent of the Venetian surface with a resolution of 100 m and an S-band radio-tracking package to measure the planet's gravitational field. The satellite was lost in October, 1994. Because of

INTRODUCTION

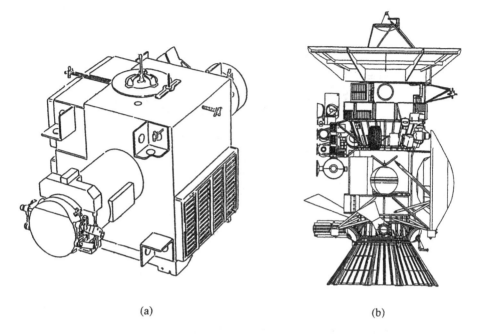

(a)   (b)

Figure 1.10 Schematic Representations of the (a) Spartan and (b) Cassini Spacecraft

the relatively short mission duration and the proximity to the Sun, power for the four-year mission was provided by a 12.5 m^2 solar array used in conjunction with two 30-Ah nickel-cadmium batteries. The system provided 1029 W of power at the end-of-life.

The International Space Station

The basic element of the ISS power system is the photovoltaic power module PVPM (Baraona, 1990). It consists of five major components: (1) two solar array assemblies and associated sequential shunt units; (2) the beta gimbals; (3) the integrated equipment assembly (IEA); (4) the thermal control system and radiator; and (5) the truss structural elements that enable the PVPM to be attached to the main ISS structure. The IEA holds several different types of boxes called orbital replacement units, or ORUs. Each ORU is dedicated to a specific subsystem of the full power system. There is an ORU for the batteries, battery charge/discharge units, direct current switching units, dc-dc converters, power distribution and control units, junction boxes for fluid and electrical services, and thermal control system pump units.

There are to be 4 PVPMs with a total of eight solar array wings, each of which will produce 32 kilowatts at the beginning of full ISS operation. The net power delivered to

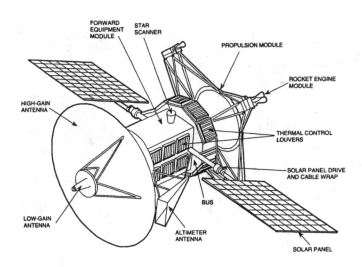

Figure 1.11 The Magellan Spacecraft

the ISS will be about 75 kilowatts of the 250 kilowatts available at the arrays. The remaining power is used to charge the batteries and to account for various system inefficiencies. The array operating voltage is 160V dc, and the distribution system voltage is 120V dc. The solar array power is regulated to the primary distribution voltage by the sequential shunt units in the PVPM and transmitted through the beta gimbals by roll rings to dc switching units. The power management and distribution (PMAD) system is designed so that any combination of two power system failures will not cause a loss of all electrical power to the ISS. This is accomplished by the use of redundant switching and controlling units and multiple independent cables to each critical user, such as the pressurized modules and the experimental pallets on the station trusses. Power management and distribution control will be through the use of semiautonomous local controllers linked to a central controller. The control system is designed to monitor and detect faults, isolate malfunctioning circuits, and reconfigure and recover system performance. The same semiautonomous controllers will schedule power use to a certain extent to help prevent overloads and to assure full battery charge at the start of each eclipse period. The batteries are 81 amp-hr nickel-hydrogen cells arranged to provide 120V dc output over a 35% depth of discharge. Power system thermal control is accomplished via a pumped loop cooling system which regulates the temperature of the batteries and other critical electrical control system hardware. A radiator assembly completes the system. Battery design life is five years, and array design life is 15 years.

INTRODUCTION

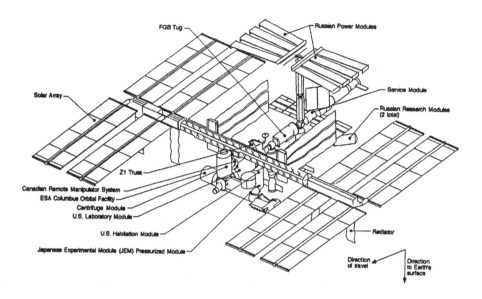

Figure 1.12 The International Space Station

Galileo

Galileo (Figure 1.13) was designed to study Jupiter's atmosphere, moons, and surrounding magnetosphere. The spacecraft also deployed a probe into Jupiter's atmosphere in December, 1995. In spite of a failure in the spacecraft's high-gain antenna, most of the scientific objectives were accomplished. Its eight-year design life and it distance from the Sun favored the choice of ^{238}Pu RTGs as the power source. Two were used, and each provided 570 W at the beginning-of-life decreasing to 485 W at the end-of-life. Had solar arrays been used for the power source, over 150 m^2 of solar panels would have been needed. Eighteen-Ahr primary lithium-sulfur batteries powered the probe, which was released into the planet's atmosphere. Galileo was deployed in October, 1989, and entered orbit around Jupiter in July, 1995. The long shelf life of the primary batteries was a factor in their selection.

16 SPACECRAFT POWER TECHNOLOGIES

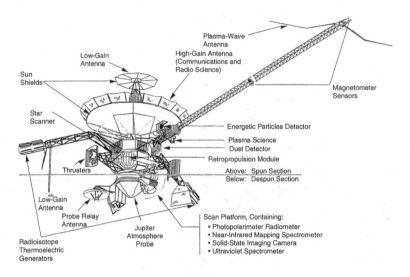

Figure 1.13 The Galileo Spacecraft

TOPEX/Poseidon

The TOPEX/Poseidon mission, launched in August, 1992, is a remote-sensing scientific program undertaken jointly by the Centre National d'Estudes Spatiales (CNES) and the National Aeronautics and Space Administration (NASA). The TOPEX (Ocean Topography Experiment) satellite monitors the Earth's oceans from an altitude of about 1300 km to better understand the ocean climate, weather, and surface features, as well as to enhance coastal storm warnings and safety at sea. TOPEX is one of a series of satellites to use the multimission modular spacecraft bus (MMS) to support the essential subsystems such as attitude control, power, command and data handling, and propulsion. The payloads for TOPEX include radar altimeters, a Doppler-tracking receiver, a microwave radiometer, and a laser retro-reflector array. During its first three years in orbit, the spacecraft has measured sea heights to within 4 cm.

Figure 1.14 shows the fully deployed TOPEX/Poseidon satellite. The spacecraft electrical system is powered by a solar array, providing about 3400 W at the beginning-of-life and designed for maximum energy transfer using a non-dissipative, unregulated main power bus. The modest power requirement and the relatively benign LEO orbit naturally lead to a photovoltaic power option. The MMS and its Modular Power Subsystem (MPS) was also used on the Solar Max Mission, Landsat 4 and 5, the Upper Atmosphere Research Satellite (UARS), and the Gamma-Ray Observatory, among others. (See Chapter 8 for an extended discussion of the TOPEX power system.)

INTRODUCTION

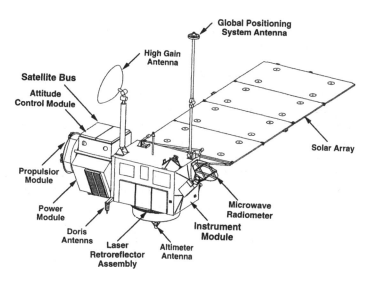

Figure 1.14 The Fully-Deployed TOPEX Satellite

Envisat

The Envisat system, scheduled to deploy in late 1999, is in some ways similar to the TOPEX/Poseidon program. The purpose is to observe the Earth's oceans and ice fields and their interactions with the atmosphere and specific aspects of atmospheric chemistry. The spacecraft will operate in a sun-synchronous orbit (see Chapter 2) at an altitude of about 800 km. The complex mission payloads include radars, laser reflectors, spectrometers, radiometers, altimeters, interferometers, and occultation optics. The low-Earth orbit, four-year lifetime, and large electrical power budget led to a design using a 6.5 kW (end-of-life) solar array with eight 40Ahr NiCd batteries. The power will be distributed on a voltage bus at 23 to 37 volts. Much of the EPS design is driven by the demands of the synthetic aperture radar which will draw a peak power of 1.2 kW and an average power of 750 W.

2.4 Spacecraft electrical power technologies

The improvements that have taken place in the technologies which enable spacecraft power systems have been achieved in spite of two serious constraints on power system development. The first is the issue of ownership of power technology development. Often the payload mission office is reluctant to endorse a spacecraft design that demands more power than is readily available from the existing technology. This is understandable since to do otherwise would impose a development burden on the mission that might be considered tangential to the primary purpose of the flight. This recurring theme limits the resources to promote dramatic technological breakthroughs in power system design. As a result, the improvements in spacecraft electrical power systems that have come about over the past three decades are more properly characterized as evolutionary rather than revolutionary. The second constraint, as we have seen, is related to the limited options available for generating electrical power in space. In the case of chemical or nuclear systems, the source of the power must be transported into orbit at great expense in cost and mass. In the case of solar-battery systems, the total power available is limited by both nature (the solar constant at one A.U. is 1.4 kW/m^2) and technology (still relatively low conversion efficiencies). There have been improvements, however, and they will continue. Table 1.4 offers a comparison of the state-of-the-art in several key technology areas from the mid-1980s to the end of this decade. Progress will continue and with it expanded options for powering satellites and their payloads.

2.5 An overview of the book

This book attempts to present, in a systematic way, a discussion of the evolving technologies that make today's spacecraft power systems the reliable energy systems they have proven to be. The focus is intentionally on the technologies rather than systems because of the large number of candidate systems that can be designed by mixing-and-matching the subsystems as shown in Figure 1.9. The solar array-battery system is the most common and will be emphasized throughout the book. Unfortunately, a full discussion of all possible system configurations is not feasible. To demonstrate this point, consider the system shown in Figure 1.15, a solar-thermal dynamic Brayton-cycle conversion system. Here, solar energy is concentrated and collected into a storage element and is then used to heat a gas in a closed-Brayton-cycle turbine connected to an alternator. This represents one of many combinations that can be considered as candidate systems for large power level electrical systems in space. Hopefully, with the discussions of the underlying technologies presented in this book, the reader will be better prepared to address spacecraft electrical power issues at the systems level.

The book begins with a discussion of the near-Earth environment (Chapter 2) and the limitations that operations in that arena will impose on the EPS. Solar energy is the most common method of generating electricity on satellites, and Chapter 3 presents the theory and practical implications of that form of prime power. Coupled closely with solar power is the use of batteries, and Chapter 4 discusses, from basic chemistry to space-qualified components, the array of chemical storage technologies including both batteries and fuel cells. Although nuclear power, other than RTGs, has been scarcely used in space, it already has a rich heritage, and the evolution of that technology is

INTRODUCTION

Table 1.4 The Evolution of Selected Power Technologies

System or Component	Parameter	Circa 1985	Estimated 2000
Solar-Battery Systems	Power Output	5 kW	100 kW
	Specific Power	10 W/kg	50 W/kg
	Solar Array-Battery Costs	$3000/W	$1000/W
Solar Cells and Arrays	Cell Power Output	5 kW	100 kW
	Cell Efficiency (in space)	14%	25%
	Array Specific Power	35 W/kg	150W/kg
	Array Design Life (LEO/GEO)	5yr/7yr	10yr/15yr
	Array Specific Cost	$1500/W	$500/W
Batteries			
Primary			
AgZn	Energy Density	150W-hr/kg	
	Design Life	2 yr	
$LiSOCl_2$	Energy Density	200W-hr/kg	700 W-hr/kg
	Design Life	3 yr	5 yr
Secondary			
NiCd (LEO)	Energy Density	10W-hr/kg	
NiCd (GEO)	Energy Density	15 W-hr/kg	
NiCd (LEO/GEO)	Design Life	5yr/10yr	
NiH_2 (LEO)	Energy Density	25 W-hr/kg	
NiH_2 (GEO)	Energy Density	30 W-hr/kg	
NiH_2 (LEO/GEO)	Design Life	2yr/3yr	
Primary Fuel Cells	Power Load	7 kW	50kW
	Specific Power	100 W/kg	150W/kg
	Specific Cost	$40/W	$25/W
	Design Life	~2000 hrs	4000 hrs
Nuclear Power			
Reactors	Power Level	10kW	10kW
	Specific Power	10W/kg	10W/kg
	Efficiency	10%	10%
RTG	Power Level	2 kW	2 kW
	Specific Power	6 W/kg	10W/kg
	Efficiency	8%	12%
Typical Overall System Parameters			
	Power	12 kW	25kW
	Voltage	28 V	50V
	Frequency	DC	DC/AC
	Cost –on–Orbit	~$1000/kW-hr	
	Radiator Specific Mass	20kg/kW	

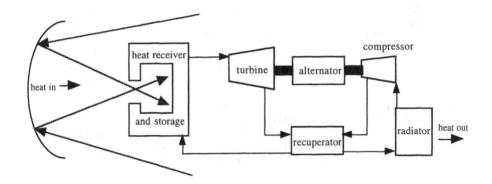

Figure 1.15 A Solar-Thermal Dynamic System

presented in Chapter 5. Chapters 6 and 7 discuss a number of static and dynamic conversion processes that have been used or are being proposed for use in space. While these conversion technologies are presented within the context of a nuclear-powered heat source, any heat source, including solar energy, can be applied. The PMAD subsystem is the subject of Chapter 8, which also presents descriptions of components and techniques for achieving the various PMAD functions and examples of several PMAD configurations. The generation and use of electrical energy on satellites present a difficult design problem in thermal management (TM) and the interactions between power production and thermal management is discussed in the final chapter. Also presented in Chapter 9 are the theory and practical engineering aspects of maintaining the proper thermal environment onboard satellites.

3. References

Agrawal, B. N., *Design of Geosynchronous Spacecraft*, (Prentice-Hall, Inc., Englewood Cliffs, NJ, 1986), 1-53.

Baraona, C. R., *Photovoltaic Power for Space Station Freedom,* Proceedings of the 21st Photovoltaic Specialists Conference, New York, (1990), 28-30.

Barter, N. J. (ed.), *TRW Space Data Book,* (The TRW Company, Redondo Beach, California, 1992).

Curtis, A. R. (ed.), *Space Satellite Handbook* (Third Edition), (Gulf Publishing Co., Houston, Texas, 1994), 190-331.

DeWitt, R. N., Duston, D., and Hyder, A. K. (eds.), *The Behavior of Systems in the Space Environment*, (Kluwer Academic Publishers, Dordrecht, 1993).

Fortescue, P. and Stark, J., *Spacecraft Systems Engineering*, (Wiley and Sons, Chichester, 1995).

Griffin, Michael D. and French, James R., *Space Vehicle Design*, (AIAA Education Series, Prezemieniecki, J. R., ed., 1991).

Kinesix Corporation, Houston, Texas, (personal communication, 1998).

Larson, W. J. and Wertz, J. R. (eds.), *Space Mission Analysis and Design*, (Microcosm, Inc., Torrance, California, and Kluwer Academic Publishers, Dordrecht, 1992).

McDermott, J. K., *Power*, in *Space Mission Analysis and Design* (Second Edition), Larson, W. J. and Wertz, J. R. (eds.), (Microcosm, Inc., Torrance, California, and Kluwer Academic Publishers, Dordrecht, 1992), 391-409.

Pisacane, V. L. and Moore, R. C., (eds.), *Fundamentals of Space Systems*, (Oxford University Press, New York, 1994).

Reeves, E., *Spacecraft Design and Sizing*, in *Space Mission Analysis and Design* (Second Edition), Larson, W. J. and Wertz, J. R. (eds.), (Microcosm, Inc., Torrance, California, and Kluwer Academic Publishers, Dordrecht, 1992), 285-337.

Stark, J. P. W., Fortescue, P. W., and Tatnall, A. R. L., *Introduction*, in *Spacecraft Systems Engineering* (Second Edition), Fortescue, P. and Stark, J. (eds.), (Wiley and Sons, Chichester, 1995), 2-9.

Thompson, T., D. (ed.), Space Log 1994, (TRW, Redondo Beach, California, 1994).

Walls, B., *Utility Aspects of Space Power: Load Management versus Source Management* (NASA Technical Memorandum 108496, 1995).

CHAPTER 2

ENVIRONMENTAL FACTORS

1. Introduction

The near-Earth space environment is complex and dynamic and its interaction with spacecraft has provided satellite designers with challenging design problems from the beginning of the space age. In his study of environmentally-induced spacecraft anomalies, Vampola (1994) points out that, because of our inability to simulate the full range of space environmental conditions, spacecraft designed and tested on the ground cannot be expected to operate in the same manner when placed in orbit. Even in a dormant mode, a spacecraft in orbit will experience degradation from its environment.

Virtually all elements of the environment will affect the design and operation of spacecraft power systems. In this discussion, the environmental factors will be divided into four groups: the neutral environment, radiation, plasma, and micrometeoroids. These categories are somewhat arbitrary but are chosen to reflect the primary effects influencing power system design that can be expected at various orbital altitudes. For example, as Tribble (1995) suggests, at altitudes of 300 km or less, the interactions with the neutral environment will dominate. As the altitude increases to geosynchronous orbit, the primary environmental interactions become those with the high-energy plasma and the outer radiation belts. In between is an array of environments and interactions between the spacecraft and its environment that is the subject of this chapter.

Figure 2.1 depicts the structure of the space environment. As implied by the figure, not all of the effects are present at all altitudes, but neither is there a sharp delineation of the various effects with altitude. As we shall see later in this chapter, the near-Earth environment is dominated by the Sun and the interactions between the Sun's various forms of radiation and the Earth's magnetic field. The sun's influence is enormous: it radiates about 10^{27} watts and emits about 10^6 kilograms per second of protons and electrons. The solar electromagnetic radiation extends from wavelengths of 0.2 μm to about 4 μm with varying degrees of intensity and variability, with the region of the extreme ultraviolet (EUV) potentially doing the most damage to materials. In addition, cosmic rays, very energetic particles from outside the solar system, will interact with all spacecraft at all altitudes, with the most serious effects occurring in electronics components of satellites at the higher altitudes, in polar orbits, or in the region of the South Atlantic Anomaly (SAA). The micrometeoroid environment can be either natural or man-made, with the more serious interactions being the result of collisions with man-made debris.

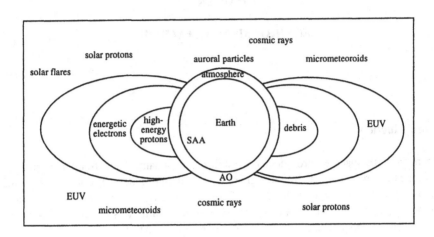

Figure 2.1 The Near-Earth Environment (after Vampola, 1994)

The concern from man-made debris becomes less serious at altitudes above about 2000 km.

Each of these environmental conditions will produce interactions with spacecraft systems, often with deleterious effects. As shown in Figure 2.2, these interactions often combine to produce synergetic effects that must be anticipated in the design. For example, chemical interactions of materials with atomic species will change the surface properties of most materials. If that surface is part of a radiator, the overall thermal balance of the spacecraft will be modified and will continue to change throughout its operational life. This, in turn, will force the thermal engineer to overdesign radiator surfaces at the beginning-of-life (BOL), thereby decreasing the operating temperatures within the satellite at BOL, and increasing the need for electrical power to provide heat to sensitive components during the early years of operation.

Figure 2.3 provides a more detailed example of the types of interactions that might influence one aspect of power systems, in this case regions of high electric fields, and several of the effects that those interactions could produce. These high fields can be found on high-voltage power-system components such as ion thrusters or the plasma contactors used to ground the spacecraft to the external plasma. In this example, several elements of the environment, such as atomic oxygen, the plasma, or micrometeoroids, will affect both the bulk and surface properties dielectrics and conductors, increasing the likelihood of an electrical breakdown in the bulk dielectric, surface flashover of the insulator, or arcing at the triple junction of the dielectric-conductor-vacuum.

The environment and the effects it produces on the operation of the overall spacecraft and specifically on the electrical power system are dynamic. They are functions of

ENVIRONMENTAL FACTORS

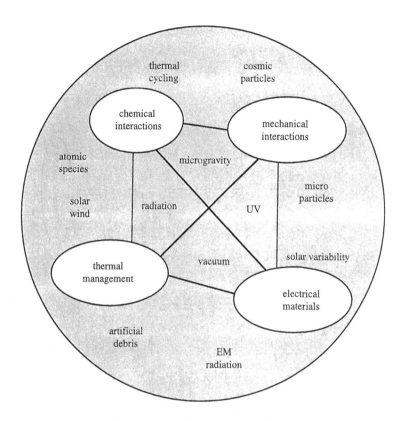

Figure 2.2 Synergetic Interactions with the Environment

many variables: the season, diurnal variations, solar activity, and the satellite orbit, to suggest just a few. In this next section, several of the orbital aspects of spacecraft-environment interactions are presented, followed by a description of the four components of the environment itself.

2. Orbital considerations

The orbit of the satellite is an important factor in determining the parameters of the electrical power system and the thermal management system. This is true principally because the type of orbit will determine the frequency and duration of eclipse periods, and these periods will place special demands on the power system, especially photovoltaic systems. In this section, we explore a few basic concepts necessary to define the

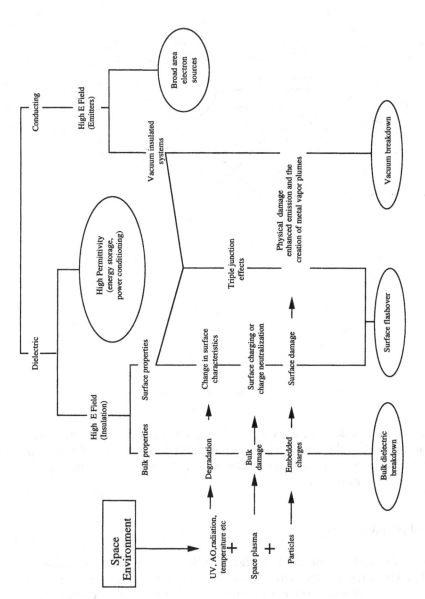

Figure 2.3 Interactions of the Environment with High-Electric-Field Components

ENVIRONMENTAL FACTORS 27

characteristics of several common near-Earth orbits and the impact of orbit selection on the design of the spacecraft power system.

2.1 Orbital elements

Orbits are characterized by several parameters which uniquely define the satellite motion. Rather than attempt any comprehensive survey of orbital mechanics, we present a brief review of a number of the key parameters which will be useful in the discussions that follow.

The motion of a spacecraft in near-Earth orbit can be defined by the use of six orbital elements that refer the spacecraft to a frame of reference fixed with respect to the stars (Fortescue and Stark, 1995). The frame of reference often used is Cartesian with the x-y plane coincident with the equatorial plane of the Earth. The x-axis direction is in the direction of the Sun at the vernal equinox (March 21). This direction is often referred to as the first point of Aries because at the time of vernal equinox several centuries ago the Sun was in the direction of the constellation Aries. The z-axis is aligned

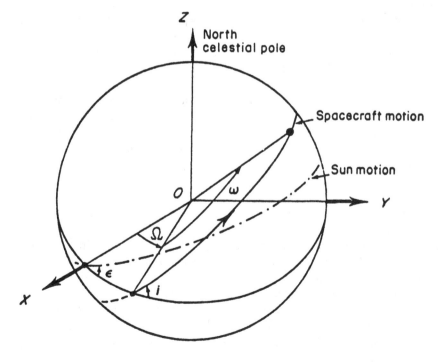

Figure 2.4 Orbital Elements (Fortescue and Stark, 1995, used with permission)

along the spin axis of the Earth in the northerly direction. Within this frame of reference, the six orbital elements can be defined.

i, the inclination, is the angle between the orbital plane and the equatorial plane.

Ω is the angle between the intersection of the x-axis and line defined by the intersection of the orbital and equatorial planes.

ω, the argument of perigee, is the angle between the line of nodes (the x-axis) and the perigee of the orbit.

Together, the first two elements define the plane of the orbit and the third defines the orientation of the orbit within the plane (Figure 2.4).

To these three are added three additional elements to define the trajectory of the spacecraft. These can be the semimajor axis, a, which defines the size of the orbit, the eccentricity, e, which defines the shape of the orbit, and a sixth which defines the position of the satellite in the orbit. This last element is often the time of last passage through perigee (the point of highest altitude of the satellite), or equivalently, the true anomaly, or the mean anomaly. These six elements can be, and often are, replaced by an alternate set of elements chosen to reflect the idiosyncrasies of a particular orbit such as a very small eccentricity or an ill-defined perigee. For our purposes, the inclination and apogee/perigee will be seen to have the most impact on defining the environment. Orbits whose planes are coincident with the equatorial plane have a zero degree inclination while those in polar orbits have a 90 degree inclination. Circular orbits are those with zero eccentricity, or equivalently, whose apogee and perigee are equal.

There are five easily definable orbits that are common to near-Earth space operations:

LEO, the low-Earth orbits, occur at altitudes below about 1000 km. At this altitude the spacecraft is below the radiation belts but does have to contend with several atmospheric effects, especially those related to reactions with atomic oxygen and atmospheric drag. A special case of the LEO is the polar orbit (~90 degree inclination) which places the satellite in the high-radiation environment of the auroral zones around the north and south poles.

MEO, the mid-Earth orbits, occur at altitudes between 1000 km and 10,000 km. This is one of the least-used regions of space, in large part because of the severe radiation environments of the Van Allen belts. The extended Earth coverage afforded by these orbits, at lower power levels than geosynchronous satellites require, has sparked renewed interest for communications satellites at these altitudes.

GEO, the geosynchronous orbit, occurs at an altitude of about 36,000 km and is one of the most-populated regions of near-Earth space, principally by communications satellites. This altitude places the satellite above the intense radiation belts. In geosynchronous orbit a satellite is near the equatorial plane of the Earth and has an orbital period of 24 hours. With this period the spacecraft appears to remain fixed over a single point on the Earth.

ENVIRONMENTAL FACTORS

SSO, the Sun-synchronous orbit, has the satellite orbital plane fixed with respect to the Sun. In these orbits, the satellite's inclination is greater than 90 degrees (i.e., the rotation is retrograde) and is chosen for a given altitude to insure that the plane precedes at the same rate as the Earth orbits the Sun. In an SSO, a satellite passes over a fixed point of the Earth at approximately the same time each day.

Molniya, the highly eccentric orbits, have apogees of several tens of thousands of kilometers and perigees of several hundreds of kilometers, and inclinations correlated with the altitude to eliminate changes in perigee. Satellites in these orbits provide extended (~11 hours per orbit) coverage of the higher latitudes, something that GEO orbits fail to do well. The periods of these orbits will vary but are generally about 12 hours during which time they will traverse the full range of the Van Allen belts and may extend as well into the regions where atmospheric effects are important. A preferred inclination is 63 degrees, the angle at which, to first order, there is no precession of the major axis of the orbital ellipse.

Each of these near-Earth orbits carry different implications for the power system design, as well as for other systems of the spacecraft. In LEO, the satellite will interact with the decreasing density of the neutral atmosphere, orbital debris, the ionospheric plasma, atomic oxygen corrosion, increased exposure to charged particles around the South Atlantic Anomaly, and solar electromagnetic radiation (ultraviolet). In MEO, the primary concern will be with solar ultraviolet radiation and the belts of trapped protons and electrons. In GEO, the environmental factors of most concern are the outer electron belts, the solar ultraviolet radiation, cosmic rays, and solar particulate radiation. These environmental effects will cause degradation of solar arrays and may induce arc discharges which can be energetic enough to adversely affect not just the power system but structural components as well. Satellites in all orbits are subject to the effects of cosmic rays and solar flares.

2.2 Eclipse times

Eclipses present several problems for both the spacecraft power system and the thermal management system. Depending on the design, during eclipse the overall equilibrium temperature of the spacecraft will drop, perhaps significantly, and solar cells will cease to function. Upon entering eclipse, batteries will be brought on-line, heaters started, and some energy-intensive operations interrupted. The Earth's IR emission and albedo, generally small compared to the direct solar radiation (see Chapter 9), become the dominant external heat sources at eclipse.

Unfortunately, it is not possible to offer a closed-form solution to the problem of calculating eclipse times for a general orbit. The reasons are detailed by Wolverton (1961) who demonstrates that the eclipse profile will depend on such variables as the time and date of launch, the orbit eccentricity, the longitude of nodes and argument of perigee (which vary constantly due to the Earth's oblateness), the semimajor axis of the orbit (which will change with atmospheric drag), and the variations in orbital parameters introduced by lunar and solar perturbations.

There are, however, several points regarding eclipse seasons, frequencies, and duration that can be made for specialized orbits. In general, LEO satellites will undergo an eclipse during each orbit with a duration somewhat less than about one-half of the period. This contrasts with spacecraft in GEO which will enter eclipse season only twice during the year, around autumnal and vernal equinox. Each geosynchronous eclipse season lasts about 46 days and the maximum duration of the eclipse in each season is about 72 minutes.

At winter and summer solstice, the Earth's equatorial plane never intersects the shadow of the Earth so there can be no eclipse (Figure 2.5). As the equinox seasons approach, the shadow of the Earth intersects the geosynchronous orbits for increasingly longer duration, beginning about February 26 and again about August 31. When the inclination of the orbit is not zero, the time of the season and the length of the season will change, but the maximum duration will remain unchanged (Soop, 1994).

Not all eclipse times are as easily visualized as that for the geosynchronous orbit.

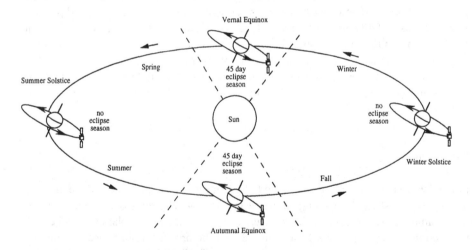

Figure 2.5 GEO Eclipse Seasons

The reader is referred to more comprehensive texts (for example, Wolverton, 1961) on orbital operations for a more complete discussion of the calculations of eclipse periods for general orbits.

Wise (1993) offers a compact expression for the maximum fractional suntime that a satellite sees in a circular orbit as a function of the orbit radius :

ENVIRONMENTAL FACTORS

$$f = \frac{1}{2} + \frac{1}{\pi}\sin^{-1}\sqrt{\frac{1-\left(\frac{R}{h+R}\right)^2}{\sin\theta}}$$

where f is the maximum fraction of the orbit in sunlight, R is the radius of the Earth, h is the altitude of the circular orbit measured from the surface of the Earth, and θ is the angle between the normal to the orbit and the direction of the Sun. These variables are shown in Figure 2.6.

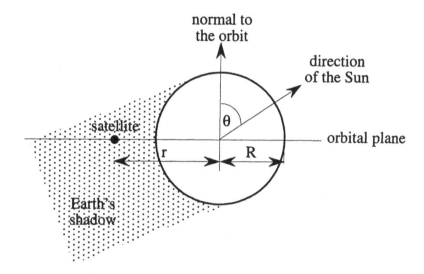

Figure 2.6 Fractional Sunlight Calculation Model

This fraction of the orbit in sunlight can easily be converted to actual time in sunlight by recalling that the period (in minutes) of a satellite in Earth orbit is given by the expression

$$P \cong 84.489 \sqrt[3]{(R+h)/R}$$

The maximum fraction of the period in eclipse, together with the maximum eclipse time and the orbital period, are plotted against the altitude and shown in Figure 2.7 for circular orbits. Note also that the maximum eclipse time begins to increase sharply for altitudes greater than about 3,000 nmi.

SPACECRAFT POWER TECHNOLOGIES

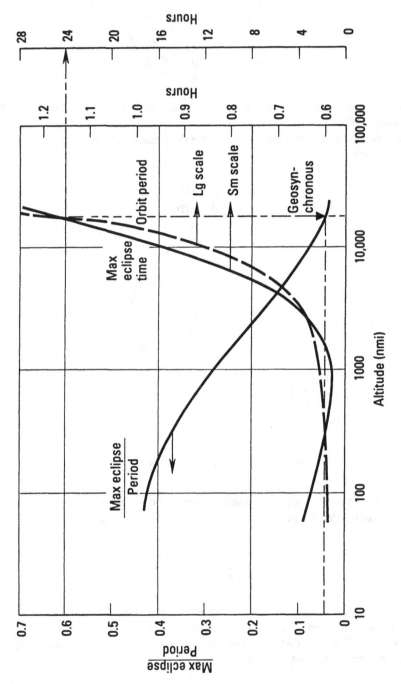

Figure 2.7 Altitude Effects on Eclipses (Barter, 1992, used with permission)

ENVIRONMENTAL FACTORS

The direct effect of the eclipse period on the performance of a solar array power system is easily understood with respect to parameters such as the sizing and depth of discharge of the battery subsystem and the sophistication needed in the power management and distribution system. The eclipses also introduce thermal cycling which may be as often as every 45 minutes or as infrequent as every several months, and these cycles can introduce mechanical strains in electrical conductors as well as structural members and so must be accommodated in the appropriate system design.

3. The near-Earth space environment

Far from being a simple void, the space environment is a hostile and dynamic environment within which the satellite must be prepared to operate. Our understanding of near-Earth space has undergone significant revision since the Explorer I first detected the trapped charged particles that became known as the Van Allen belts, and even now this characterization of the environment is being modified as more data on the nature of space and the interactions that occur between the environment and the spacecraft become known.

There are several excellent comprehensive texts on the space environment (Hastings and Garrett, 1996; Tribble, 1995; DeWitt, Duston, and Hyder, 1993) which provide many details not discussed here. In this chapter, we intend only to provide an overview of the near-Earth space environment with emphasis on those elements which may affect the performance of the spacecraft's electrical power system.

The environment consists of many interacting parts, some of which dominate in one region or another. There arises then the somewhat arbitrary decision of how to present the discussion of the environment and its interactions with the spacecraft. For our purposes, we shall consider four categories of the space environment: the neutral atmosphere, radiation, plasmas, and micrometeoroids.

3.1 The neutral environment

The neutral environment includes features which can lead to interactions harmful to the operation of the electrical power system. Virtually every aspect of this part of the environment occurs at low altitudes and generally is no longer a factor above several hundred kilometers. This part of the near-Earth environment is characterized by a very small gravitational acceleration, a tenuous atmosphere whose composition is very different from that found at sea level, and eventually, a hard vacuum. Because of aerodynamic drag, altitudes below about 225 km are too low to permit a satellite to continuously orbit the Earth.

Atmosphere

The density, pressure, temperature, and composition of the Earth's neutral atmosphere all change dramatically with increasing altitude. As seen in Table 2.1, the primary components of the sea-level atmosphere have decreased significantly at 200 km and have disappeared at 800 km. In their place appears atomic oxygen which comprises 80% of the neutral atmosphere at 300 km. Note that while atomic oxygen has appeared, there is no corresponding atomic nitrogen because the dissociation energy for nitrogen is higher than that of oxygen and is above the highest energy of photons from the sun.

Note from Figure 2.8 that the total atmospheric density above sea level decreases approximately exponentially. While atomic oxygen constitutes 80% of the atmosphere

Table 2.1 Percent Composition of the Atmosphere at Various Altitudes

Species	Sea Level	200 km	300 km	800 km
N_2	79	30	18	<0.1
O_2	20	14	<2	nil
O	nil	36	80	24
Ar	1	<0.1	<0.1	nil
He	nil	<0.1	<1	60
H	nil	nil	<0.1	14

at 300 km, it represents significantly fewer atoms per cubic meter than, for example, molecular oxygen at sea level. The actual density of specific neutrals is shown in Figure 2.9 for altitudes between 100 and 700 km. By comparison, the atmospheric density at sea level is about 10^{19} atoms/cm^3. Above 700 km, hydrogen and helium atoms dominate all other species.

The presence of the very reactive atomic oxygen at the LEO altitudes presents a materials problem, especially for those surfaces exposed to the ram direction. As Tennyson (1993) points out, although the atomic oxygen concentration may seem insignificant compared to other neutral atom concentrations at sea-level, because of the large orbital velocity the flux of atoms at the ram surface of the spacecraft is quite large. At orbital speeds of about 8 km/sec, which corresponds to a mean energy of almost 5 eV, the flux of atomic oxygen atoms is of the order 10^{14} atoms/cm^2-sec.

Many materials are susceptible to attack by atomic oxygen, and this erosion can be enhanced even more in the presence of ultraviolet radiation. This is especially worrisome for polymers and composites that may be used in solar arrays and in the thermal management system (Tennyson, 1993).

Table 2.2 summarizes the erosion yield data for a number of materials useful in spacecraft power system design. The units of the yield are 10^{-24} cm^3/atom and are meant to describe the volume of material lost per incident atomic oxygen atom.

ENVIRONMENTAL FACTORS 35

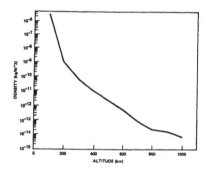

Figure 2.8 Variation of Atmospheric Density (NASA, 1986)

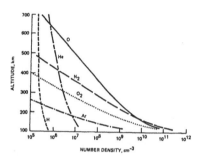

Figure 2.9 Constituents of the Neutral Atmosphere (NASA, 1986)

Table 2.2 Erosion Yields for Selected Materials (Banks, 1990)

		Erosion Yield Range (10^{-24} cm^3/atom)		
0.01 - 0.09	0.1 - 0.9	1.0 - 1.9	2.0 - 4.0	> 4.0
Diamond	Polysiloxane/ Kapton™	Epoxies	Graphite/ Epoxy	Silver
Al$_2$O$_3$		Polystyrene		
Al coated Teflon™ FEP	Siloxane/ Polymide	Most forms of Carbon	Kevlar™/ Epoxy	
Molybdenum	401-C10 flat black		Polyethylene	
	Z-306 flat black		Mylar™	
			Polyester	
			Kapton™ H Polymide	

Tennyson (1993) provides a useful nomogram for estimating the thickness of material lost by exposure to atomic oxygen. The nomogram assumes that the altitude of the satellite and the mission duration are known. Referring to Figure 2.10, the intersection of these two values, altitude and duration, establishes a fluence level, 10^{22} in the ex-

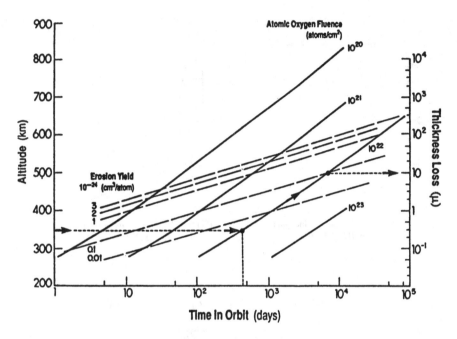

Figure 2.10 Nomogram for Estimating the Material Thickness Lost by Exposure to Atomic Oxygen (Tennyson, 1993, used with permission)

ample indicated in the Figure. This constant fluence curve is then traced until it intersects the erosion yield for the material in question (this yield may be obtained from the preceding table for a number of materials). On the right axis is the estimated material lost to atomic oxygen erosion. This nomogram assumes a zero degree angle of incidence, that is, the material is exposed in the ram direction. The thickness lost decreases to half the value indicated by the nomogram for an angle of incidence of 60 degrees and goes to zero for an incidence angle of about 100 degrees.

As seen in Figure 2.11, the atmospheric pressure from the neutral also decreases rapidly as altitude increases. This decrease is not exactly exponential, but is approxi-

ENVIRONMENTAL FACTORS

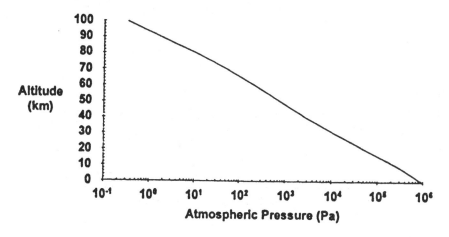

Figure 2.11 Variation of Atmospheric Pressure with Altitude (after Jursa, 1985)

mately so at low altitudes; the deviation from true exponential is due to changing scale factors at different altitudes.

The ambient pressure outside the spacecraft may be quite different from pressures measured inside even if there are a significant number of openings in the exterior of the satellite. One example of this residual pressure is shown in Figure 2.12 which compares the pressure measured at several points in and around the Apollo telescope mount for several hundred hours in orbit.

This may occur for several reasons, including the phenomena of outgassing which has significant implications in the design of power systems. Outgassing, which occurs when many materials are exposed to vacuum conditions, is characterized by a loss of mass due to the escape of volatiles, often water vapor, from the surface of the material. Different materials exhibit different susceptibility to outgassing as shown in Figure 2.13. The outgassing rates and the total volume of volatiles released can be controlled with proper selection of the material and the preparation of the surface prior to flight. The volatiles that escape the surface can deposit on adjacent surfaces and interfere with the thermal management components and electrical and optical systems. The ideal material would not outgas at all or would display a rate that would quickly drop to zero in a very short time.

The outgassing rate is temperature dependent, and the rate can sharply increase if there is a local increase in the temperature. This is a likely occurrence as systems are

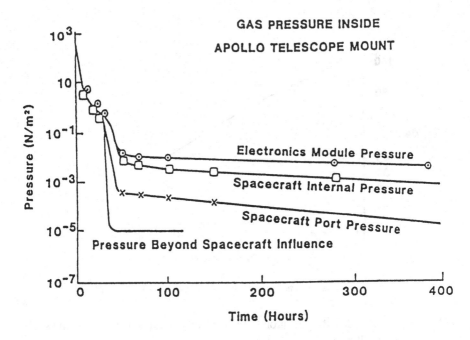

Figure 2.12 Pressure Measurements from the Apollo Telescope Mount

turned on during flight. Within the spacecraft, the generation of these volatiles can increase the local pressure with potentially unfortunate results for high-voltage operations in space. This is related to the Paschen electrical breakdown of gases subjected to high voltages. Figure 2.14 is a plot of the voltage at which hydrogen gas will undergo breakdown (i.e., arc) versus the pressure of the hydrogen gas.

There is a family of curves representing various separation distances of the two electrodes across which the voltage is maintained and between which the pressure is measured. Notice that at very low pressures the gas will tolerate a very high voltage before the onset of arcing, but as the pressure is increased this breakdown voltage decreases, passing through a minimum before rising again. This minimum occurs because at the very low pressures an electron will traverse the entire gap distance without ionizing any hydrogen atom and without initiating an avalanche (breakdown of the gas). At higher pressures, the mean free path between collisions of the electron with a hydrogen atom is so short that the electron cannot gain enough energy between collisions to ionize the atom. At the minimum, less than 300 volts in this example, avalanche ionization occurs and arcing is initiated between the electrodes. While this plot is given for hydrogen, all gases display a similar behavior. As heat is generated onboard an operating

ENVIRONMENTAL FACTORS

spacecraft, the temperature of the various materials will rise and with it the outgassing rate. This outgassing can create uncertain internal pressures within the spacecraft and, if the local pressure in an area of high voltage rises through the Paschen minimum, unanticipated arcing can occur.

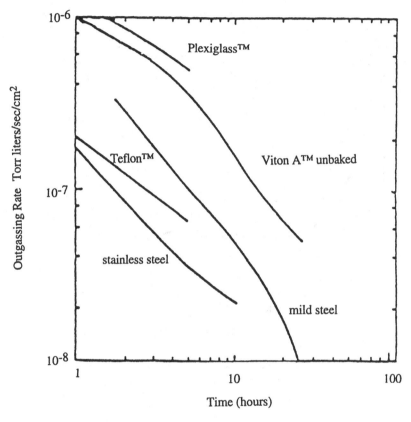

Figure 2.13 Measured Outgassing Rates

A final aspect of the neutral atmosphere, the kinetic temperature, will be mentioned only briefly. The region above approximately 80 km and extending to about 1000 km, known as the thermosphere, displays a sharp increase in temperature with height, rising to well above 1000 K at solar maximum at an altitude greater than 250 km. This increase in temperature is, of course, accompanied by a corresponding increase in the mean particle speed. A more complete discussion of the temperature can be found in Tribble (1995) and Hastings and Garrett (1996).

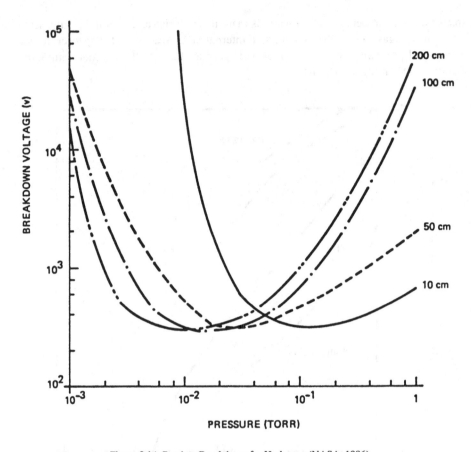

Figure 2.14 Paschen Breakdown for Hydrogen (NASA, 1986)

Gravity

In near-Earth orbit, the gravitational acceleration decreases rapidly from its value at the Earth's surface to a value as small as 10^{-11} g, where g is the gravitational acceleration at the Earth's surface. The actual value will depend on the orbital parameters and perhaps solar activity. Additionally, there will be random accelerating forces due to the operation of attitude controllers and station-keeping thrusters. As Griffin and French (1991) point out, this microgravity environment allows use of very lightweight structural materials, but these materials have very low damping qualities and thus can support large vibrational excitation. Furthermore, these materials generally are not strong enough to support ground-based testing, making design validation much more difficult.

3.2 The plasma environment

Plasmas, often called the fourth state of matter, are the most common form of matter in the universe. More than 99% of all matter across the universe is found in the plasma state, that is, a state in which at least one atomic electron has enough energy to escape the coulomb attraction of the atomic nucleus resulting in the independent motions of the free electrons and the atomic ions. Such a plasma may remain neutral overall if the ion density is equal to the electron density on a macroscopic scale. As a plasma cools, the electron temperature will decrease and with it the average electron energy will likewise decrease until the electron no longer has enough energy to remain free. At that point, the electron may again attach itself to an atomic ion and the plasma will become neutralized returning to a state of matter common to our environment on the surface of the Earth. As before, the term 'temperature' refers to the kinetic temperature which is a measure of the kinetic energy of the plasma particles. The concepts of temperature and particle energy are so closely related in plasma physics that it is customary to express the kinetic temperature in units of energy. An energy of 1 eV corresponds to a kinetic temperature of about 12,000 Kelvin.

While the neutral environment affects only satellites in LEO, the plasma environment can influence the operation of satellites in all orbits. Plasmas are characterized by specifying their density and their temperature, and there are several distinct regions of the upper atmosphere with characteristically different plasma environments. At altitudes above about 60 km, there is sufficient short-wavelength radiation from the Sun to cause significant photoionization of the neutral atoms that are present. The ionization process is also aided by collision of neutrals with energetic charged particles. With the very low densities present, there is no immediate recombination and the plasma can persist for long periods of time. This occurs in the region known as the ionosphere where ions and free electrons are constantly present.

For practical purposes, the ionosphere is said to start at about 60 km in altitude and continue upward to a few thousand kilometers. Below this lower altitude, there is insufficient UV to cause appreciable ionization. Above this rather ill-defined upper boundary are the trapped-radiation zones (the Van Allen belts) starting at about 1.5 Earth radii and extending to about 5.5 Earth radii. Situated even higher are the edges of the magnetosphere where the environment is dominated by the solar wind plasma. Details of this complicated and dynamic structure can be found in Hastings and Garrett (1996) and Cowley (1993). This discussion of the plasma is restricted to those regions of ionized gases in the magnetosphere with relatively low energies, typically less than a few tens of keV. This is to distinguish the plasma environmental effects from those charged-particle interactions that can produce radiation. The radiation environment is discussed in Section 3.3.

The interactions of the plasma environment with spacecraft can pose problems to electrical power systems. These may include electrical effects of spacecraft charging, arcing and electrical breakdown of dielectrics, parasitic currents, and shifts in electrical potential, in addition to a broad range of materials interactions such as sputtering, changes in refractive index, and enhanced contamination collection (Tribble, 1995; Purvis, 1993). The seriousness of these interactions depends on the characteristics of the plasma, the geometry of the spacecraft, and the materials used in its construction. Insulators exposed to the plasma will be the most likely source of problems (Latham, 1993).

The two most interesting plasma regions for our discussion are the low-energy, high-density ionospheric plasmas and the higher-energy, lower-density plasmas associated with geomagnetic substorm activity. In this first case, the ionospheric plasma has a characteristic temperature of about 0.1 eV and a variable density peaking at 10^6 cm^{-3} at an altitude of about 300 km. This variation in plasma density reflects a balance between the increasing amount of UV radiation available versus the decreasing density of the neutral atmosphere available to ionize with increasing altitude (Purvis, 1993). The variation in density of this low-energy plasma with altitudes up to GEO is shown in Figure 2.15.

Because of the role of solar UV in the photoionization of primarily the neutral oxygen and nitrogen, the ionospheric plasma density varies diurnally as well as with

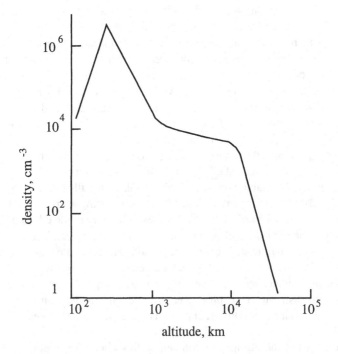

Figure 2.15 The Variation of Plasma Density with Altitude (after Purvis, 1993)

ENVIRONMENTAL FACTORS

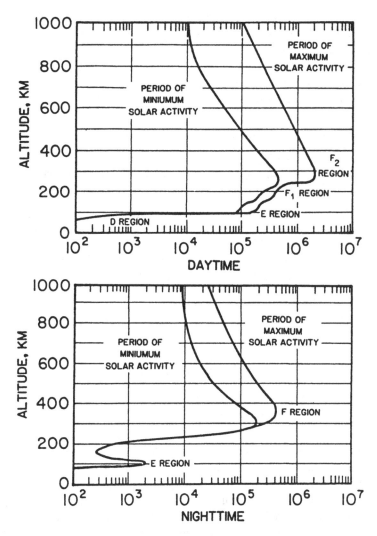

Figure 2.16 Diurnal Variation of Ionospheric Electron Density over a Solar Cycle
(Friedman, 1993, used with permission)

solar activity, as seen in Figure 2.16. Note that the density decreases by more than an order of magnitude at night when the photoionization source term is removed, and that the density can increase by an order of magnitude during periods of high solar activity.

Because of the influence of the Earth's magnetic field on the motion of charged particles, the density will vary with latitude, being greater at the equator than at the poles.

The variation in the thermospheric temperature from daytime to nighttime over a typical solar cycle is presented in Figure 2.17. At the lower latitudes, the kinetic temperature remains relatively low and independent of solar activity, but above about 100 km begins to increase and shows a strong correlation to solar activity.

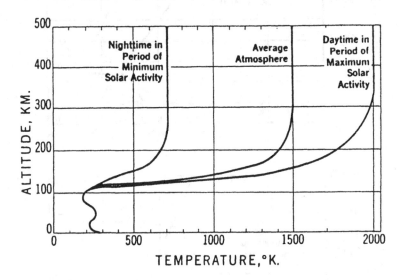

Figure 2.17 Variation of Thermospheric Temperature over a Solar Cycle
(Friedman, 1993, used with permission)

A second plasma environment of significance for spacecraft operations is that occurring on the dark side of the Earth, particularly between about local midnight and dawn in the orbit. The plasmas in this region are of much higher temperature than those found in the ionosphere, typically about 10 keV, but at much lower densities of about 1 cm-3. These plasmas, related to geomagnetic substorms, occur around near-geosynchronous altitudes and are the result of currents injected down the magnetic field lines on the dark side of the Earth (Figure 2.25). The substorm activity will also affect satellites in polar orbits as these low-altitude, high-inclination satellites will encounter the substorm plasmas that follow the geomagnetic field lines as the lines converge at the poles. Satellites will encounter energetic streams of electrons and ions (~ keV) even at low altitudes in these polar zones.

The interaction of spacecraft with these two plasma environments are quite different. As Purvis (1993) points out, in the more energetic substorm plasmas, the Debye length, the scale length over which charge neutrality is established and equilibrium exists, is very large compared to the size of the spacecraft. Thus, space-charge effects can be neglected, system voltages remain small compared to the plasma energies, and the plasma environment determines the electrical interactions. Furthermore, the plasma current density is of the order 10^{-10} A/cm^2 which is comparable to the photoelectron current density for a sunlit surface making the photocurrent an important factor. Shadowed dielectric surfaces will charge negatively due to the excess of environmental electrons impinging while sunlit surfaces will remain at near zero potential due to the balancing effect of the photoelectron current. Photoemission is, in effect, a voltage-limiting mechanism that is interrupted during eclipse. The charging will continue until the shaded side creates an overall negative potential sufficient to suppress the emission of the low-energy photoelectrons from the sunlit surfaces allowing the entire spacecraft to charge negative. Also, during major geomagnetic storms, an increase in the electron flux and energy can lead to deep charging in dielectrics resulting in possible breakdown of, for example, cable insulation (Domingo, 1993).

In the less-energetic, more dense ionospheric plasma environment, the Debye lengths are very small compared to spacecraft dimensions and so space charge plays a critical role in determining the electrical interactions. In this environment, the spacecraft velocity (~ 8 km/sec) is large compared to the ion speeds (~1 km/sec) but small compared to the electron thermal speeds (~200 km/sec). The spacecraft will appear to be at rest to the electrons, but not so for the ions. Only ions in the ram direction, the side facing the velocity direction, will reach the spacecraft, while electrons will continue to strike the spacecraft surfaces from all directions.

Any spacecraft operating in the plasma environment will come to an equilibrium state with the surrounding plasma by acquiring surface charges and establishing surface potentials to reduce the net current flow to zero. The equilibrium process is treated as quasi-static, while in reality it is dynamic, responding to changes in the environment and changes in the operations of systems aboard the spacecraft (Purvis, 1993).

Figure 2.18 depicts the charging process of a dielectric surface exposed to the various currents present in the plasma environment. When sunlit, photoelectrons are emitted from the surface. Ions and electrons from the ambient plasma also lead to the emission of secondary electrons through impact with the surface or through electron backscatter. In addition, there may be leakage currents through the dielectric material. These processes occur point-by-point and may lead to significant potential differences across the dielectric itself. In the case of conducting surfaces, there is, of course, no such differential charging (Purvis, 1993). As Hastings and Garrett (1996) point out, in the case of no photoemission, backscatter, or secondary emission, the high mobility of the electrons in the plasma will result in a negative surface potential (of the order of the electron temperature) to guarantee the condition of no net current flow. If sunlight is present, the surface potential can become slightly positive increasing the plasma electron flow to balance the loss of photoelectrons.

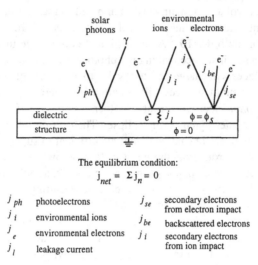

Figure 2.18 Dielectric Charging Due to Interactions with the Plasma Environment
(after Purvis, 1993)

3.3 The radiation environment

There are two aspects of the radiation environment that are important to our discussion of power systems: electromagnetic radiation from the Sun and ionizing radiation created by energetic charged particles impinging on the spacecraft. While all of the electromagnetic radiation originates in the sun, there are two sources of the charged particles. The relatively low energy particles tend to be solar in origin while the more energetic particles are cosmic. By far, the dominant factor in defining the radiation environment is, however, the sun, and so we start this section by discussing its influence on the near-Earth space environment.

The sun

The Earth's orbit around the Sun is an ellipse whose plane, the ecliptic plane, is tilted with respect to the Earth's equatorial plane. The Earth's axis of rotation is inclined to the ecliptic by 23.5 degrees in such a way that the Earth tilts toward the Sun in the northern hemisphere summer and away from the Sun in the northern hemisphere winter. The Earth's orbit has a mean center-to-center distance of about 1.496×10^8 km, being slightly closer to the Sun in the winter (1.47×10^8 km) and farther away in the summer (1.52×10^8 km). In spite of these large distances, the power output of the Sun is so great that solar energy reaching the Earth is the defining factor for our local space environment. A satellite in orbit around the Earth will be affected directly by radiation coming from the Sun and indirectly through the many ways in which the Sun modifies the environment through which the satellite must pass.

The energy source in the Sun is the thermonuclear fusion of hydrogen into helium, a reaction that converts four hydrogen nuclei (protons) into one helium nucleus with the release of about 26.2 MeV of energy. In this process, about 600 metric tonnes of protons are consumed each second resulting in the generation of about 4×10^{26} joules per second. About 4×10^9 kg of mass is converted into energy each second, large but insignificant when compared to the total mass of about 2×10^{30} kg.

Although the energy production within the Sun appears to be constant, the rate at which it is released, both spatially and temporally, is not. As a result of large-scale distortions of the sun's magnetic field, both electromagnetic radiation and the charged particle flux emitted by the Sun and arriving at the Earth can vary significantly. As we shall see, the electromagnetic radiation generally penetrates the atmosphere to various levels while the Earth's magnetic field shields us from the charged particles.

In its active mode, the energy emitted by the Sun is variable on many time scales ranging from solar flares that occur over periods of tens of minutes to the well-known 11-year solar sunspot cycle. This solar activity is characterized in several ways. Solar flare intensity is often monitored by the Doppler shift in the hydrogen-alpha line, the shift being caused by the very high velocity with which the gas particles are ejected from the sun's surface. Traditional sunspot activity is measured by counting the spots (areas of higher magnetic fields and cooler temperatures on the sun's surface) and the radio flux at 10.7 cm wavelength ($F_{10.7}$). This latter measure, the $F_{10.7}$ index, is believed to be related to variations in the extreme ultraviolet radiation. It is used as an indicator of UV flux on the Earth's atmosphere since it can be easily monitored at the Earth's surface, while the actual UV does not penetrate the atmosphere and must be monitored from space itself (Hastings and Garrett, 1996). An index value of about 50 is typical at solar minimum, rising to about 250 at solar maximum. These measurements are important in characterizing solar activity since these changes during the solar cycle can cause several important changes in the characteristics of the plasma environment (Section 3.2) and the temperature and density of the neutral environment. For example, as shown in Figure 2.19, at an altitude of 600 km there can be an increase of almost four orders of magnitude in the atomic oxygen density from solar minimum to solar maximum. As we shall see in the following section, there will also be large changes in the short- and long-wavelength electromagnetic radiation from the sun. As Domingo (1993) points out, the electromagnetic radiation energy bursts at radio frequency, while a predictor of disturbances that may affect the environment in the coming few days, may also interrupt the communications with spacecraft for periods ranging from minutes to hours.

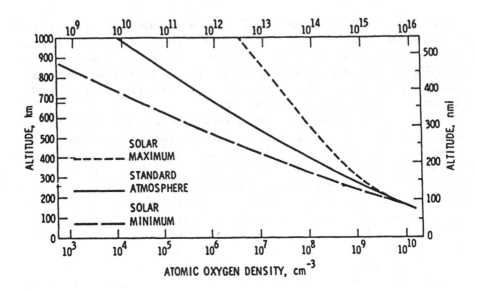

Figure 2.19 The Atomic Oxygen Flux Variation between Solar Maximum and Solar Minimum (Peplinski, 1984)

Electromagnetic radiation

The total amount of electromagnetic energy received just outside the Earth's sensible atmosphere is called the solar constant, or mean irradiance, and is equal to 1367 W/m^2. The solar constant is really not a constant at all, but has been observed to vary by about 0.15% over a period of days and 0.1% during the course of one solar cycle. The eccentricity in the Earth's orbit around the Sun causes about a 6% variation in the solar constant during the course of a year (Domingo, 1993).

The solar spectrum, a plot of the electromagnetic energy as a function of the wavelength, is shown in Figure 2.20. In this Figure, the solar constant is given as 1353 W/m^2, an older accepted value that has now been raised slightly. The Figure also presents several other curves of interest: the solar spectrum measured just above the sensible atmosphere (called the air mass zero spectrum), the air mass one spectrum (the spectrum observed at sea level when the Sun is directly overhead) with and without atmospheric molecular absorption, and finally the radiation curve for a blackbody at a temperature of 5762 K. This blackbody curve will be important in the discussion of thermal management presented in Chapter 9.

As seen in Table 2.3, the variability across the spectrum is quite pronounced, ranging from essentially constant flux in the visible region to variations by factors of 100 or more at the extreme long and short wavelengths. In the radio frequency portion of the

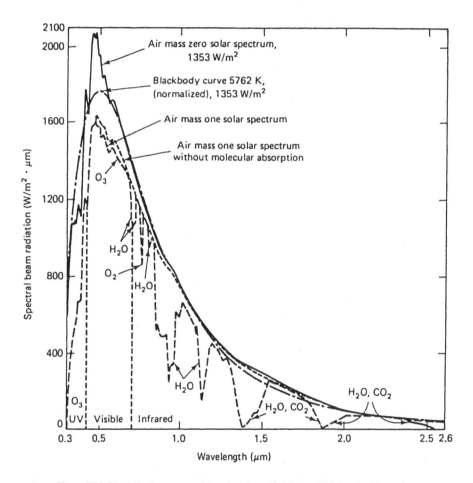

Figure 2.20 The Solar Spectrum, with and without Air Mass and Molecular Absorption
(Thekaekara, 1977)

spectrum there are large rapid bursts of energy during time scales of seconds to minutes, often correlated with solar flares. At UV wavelengths, there are variations of about 25% that correlate to the rotation period of the Sun (27 days) together with variations of a factor of two correlated to the 11-year cycle. Since all radiation with wavelengths shorter

Table 2.3 Flux Levels in the Solar Spectrum

Spectral Region	Wavelength Band	Flux Level across Band J/m^2-s-μm	Variability during Cycle
radio	> 1mm	10^{-11} - 10^{-17}	100 X
far infrared	1mm - 10 μm	10^{-5}	?
infrared	10 μm - 0.75 μm	10^{-3} - 10^2	?
visible	0.75 μm - 0.3 μm	10^3	<0.01
ultraviolet	0.3 μm - 0.12 μm	10^{-1}- 10^2	0.01 - 2 X
extreme UV	0.12 μm - 0.01 μm	10^{-1}	10 X
x ray	< 0.01 μm	10^{-1}	100 X

than 320 μm is absorbed by the atmosphere, these variations in the UV can affect the neutral and ion composition and chemistry of the upper portions of the atmosphere. At the x-ray and EUV wavelengths, there is again variability correlated with the 11-year solar cycle.

On the average, about 10% of the electromagnetic radiation is in the UV and shorter wavelengths, 40% in the visible, and 50% in the infrared and longer wavelengths. Because of the high energy of EUV photons, radiation in this portion of the spectrum will affect spacecraft by directly attacking the satellite materials exposed to it. This is seen, as examples, in the degradation of thermal management surfaces through increased absorptivity (Chapter 9) and in the degradation of optical surfaces through the photo polymerization of hydrocarbon components of organic contaminant materials deposited on the surfaces (Domingo, 1993). In these and in many other cases, the surfaces of the spacecraft may also show increased erosion due to interactions with atomic oxygen when in the presence of EUV radiation.

The extreme ultraviolet portion of the spectrum also affects the operation of spacecraft by modifying the space environment through which the satellite must travel (Domingo, 1993). The solar-cycle effects in the Earth's thermosphere are quite pronounced, as can be seen, for example, in Figure 2.21 which presents a calculation of the altitude profiles of two neutral species number densities for differing solar activity periods. Note that for the heavier component, argon, the number density at 400 km can vary by over 4 orders of magnitude from solar minimum to solar maximum. As Walterscheid (1989) points out, the density at an altitude is a function of both the local and integrated values of the temperature and the composition of the atmosphere, with the integrated effects dominating. This is consistent with the observation that the thermospheric density increases as the temperature increases.

UV is the major source of energy driving the composition, temperature, and chemistry of the stratosphere and upper atmosphere. At lower altitudes, EUV absorption is

ENVIRONMENTAL FACTORS

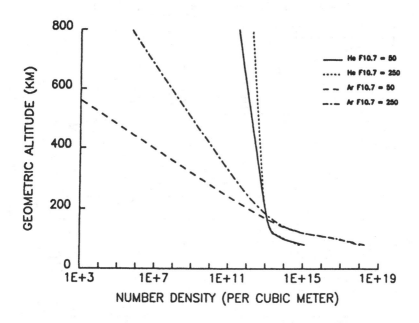

Figure 2.21 Computed Argon and Helium Number Densities at Solar Maximum and Solar Minimum (Walterscheid, 1989, used with permission)

responsible for the high thermospheric temperature not decreasing through normal thermal diffusion. During an average 11-year solar cycle, the EUV and thermospheric temperature can each increase by about a factor of two. As Domingo (1993) points out, because of the different scale heights for different species, changes in the temperature will result in changes in the composition and will take place in a way as to reduce the direct thermal effects. The resulting temperature and composition produce large changes in the density of the upper atmosphere over a solar cycle.

Since the atmospheric drag is proportional to atmospheric density, changes in the density will affect satellite lifetimes in orbit. Figure 2.22 shows satellite lifetimes at various altitudes as a function of the $F_{10.7}$ index. Note, for example, that a satellite in a 500-km circular orbit will have its lifetime reduced from 30 years at solar minimum to 3 years at solar maximum because of the increased density at orbital altitudes. This example is intended to point out the dramatic effects that changes in the solar activity can have on even lower-altitude satellites, recognizing that 30 years extends over more than one solar cycle.

52 SPACECRAFT POWER TECHNOLOGIES

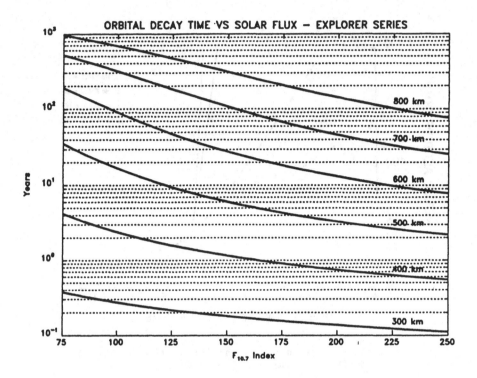

Figure 2.22 Satellite Lifetime versus $F_{10.7}$ Values for Circular Orbits at Several Initial Altitudes (Walterscheid, 1989, used with permission)

Charged particles

In addition to the electromagnetic radiation, charged particles originating in the Sun and from outside the solar system, with energies spanning a very broad range, are present throughout the near-Earth environment. These charged particles are distinct from the ions and electrons constituting the plasma environment (see Section 3.2) since the energies observed in this case are quite large and capable of initiating radiation effects within the spacecraft. These particles, with energies greater than hundreds of keV, include protons and electrons trapped in the Van Allen radiation belts, solar proton flares, and galactic cosmic rays.

ENVIRONMENTAL FACTORS 53

The geomagnetic field

The motion of all charged particles is influenced by magnetic fields, and the charged particles in the near-Earth environment are no different. At distances greater than about 15 Earth radii, the dominant magnetic field is that of the Sun, while at closer distances the Earth's magnetic field becomes important in describing the behavior of the charged-particle environment.

The Earth's magnetic field is approximately that of a uniformly magnetized sphere complicated by the fact that the Earth's magnetic and geographical axes are not exactly aligned. The magnetic dipole axis cuts the Earth's surface at points about 1300 km from the geographic north and south poles because of an 11 degree angle between the magnetic dipole and the axis of rotation and an offset of the dipole from the geographic center (Figure 2.23). This offset gives rise to a weaker magnetic field in the area of the South Atlantic (the South Atlantic Anomaly) as shown in the plot of magnetic field

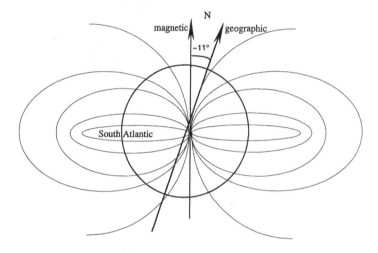

Figure 2.23 The Relative Positions of the Geographic and Magnetic Dipole Centers

intensity given in Figure 2.24. Since the geomagnetic field acts as a shield against the penetration of charged particles, this lower value of the magnetic field is directly related to higher fluxes of charged particles at lower altitudes and thus to higher incidences of radiation damage in the electronics of satellites passing through this area.

The magnetic field pattern would be a relatively simple dipole field (modified somewhat by an interplanetary magnetic field) were it not for the interactions of the solar wind (proton flux) with the dayside of the geomagnetic field. The solar wind is a proton-dominated plasma flow from the Sun toward the Earth with an average density of about

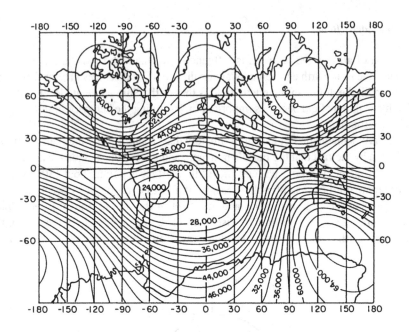

Figure 2.24 Magnetic Field Intensity in Nanoteslas with the South Atlantic Anomaly Clearly Visible (Jursa, 1985)

10 cm^{-3} and a mean particle velocity of almost 500 km s^{-1} and peak velocities in excess of 700 km s^{-1} with somewhat lower densities. At about 10 to 15 Earth radii (R_e) the force of the solar wind is balanced by the outward force of the compressed geomagnetic field and the various plasma currents in the magnetosphere. Most of the solar wind flows around the magnetosphere and so does not reach the Earth. What results is a cavity (Figure 2.25) that is blunt on the sunside and which has a long tail extending to well beyond 50 R_e on the anti-sun side. The solar wind can produce direct effects on the spacecraft from sputtering on mirrors and other optical elements facing the Sun to indirect effects such as spacecraft charging at high altitudes during an orbital eclipse and breakdown of dielectrics due to penetration charging from energetic electrons (Domingo, 1993).

A second source of charged particles affecting the environment is the solar flare which produces very energetic proton fluxes incident on the Earth's magnetosphere. The particles in a solar flare have average energies greater than tens of keV and as large as 1 GeV and the total energy associated with a single solar flare can be as large as 10^{25} joules. The particles can arrive at the Earth in time periods ranging from minutes to days depending on their energy, and the flares can occur over periods as short as tens of minutes. Although solar flares tend to occur infrequently, they can cause serious damage to solar arrays, disrupt microprocessors, and degrade most sensors (Gorney, 1992, and Domingo, 1993). These energetic particles are difficult to shield and since they lose their energy by ionizing the material through which they travel, they are particularly hazardous to semiconductor electronics. The magnetosphere shields most satellites from this intense flux, but satellites in high-inclination orbits can be affected since the charged particles will reach lower altitudes in the auroral regions (Purvis, 1993).

The third major source of high-energy charged particles is the galactic cosmic ray. These are extremely high-energy particles, typically greater than 100 MeV and as great as 10^{20} eV, that originate outside the solar system. Cosmic rays consist primarily of protons (85%) and alpha particles (14%) but some heavy ions can also be present. Inter-

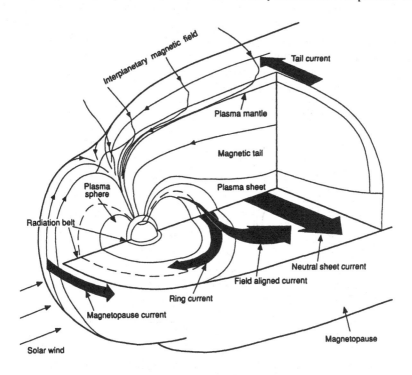

Figure 2.25 The Magnetosphere with the Van Allen Belts Shown
(Mitchell, 1994, used with permission)

estingly, the solar wind affects the galactic cosmic rays by modulating their passage around the sun. Galactic cosmic rays are observed to a lesser degree during solar maximum than at solar minimum (Domingo, 1993).

Within the magnetospheric cavity, the Van Allen belts form toroids of charged particles centered around the equator. These belts contain energetic protons and electrons that are trapped by the Earth's magnetic field into two distinct zones: in the inner zone, located generally between 1.5 R_e and 2 R_e, both protons and electrons are found, while primarily only electrons are trapped in the outer zone at about 5 R_e. These trapped particles spiral around and move along the geomagnetic field lines and are reflected above the north and south poles. This movement along the field lines brings the particles to lower altitudes and higher densities at the poles and thus can create a greater radiation hazard for satellites in polar orbits.

The characteristics of the particles within the two belts differ. The inner belt has a peak proton flux of 10^6 cm^{-2}s^{-1} for particles with energies of greater than 10 MeV. Peak proton energies can reach hundreds of MeV. The inner-belt electron flux is also about 10^6 cm^{-2}s^{-1} for electrons with energies greater than one MeV. The outer belt has an electron flux about three times larger than that found in the inner belt with electron energies generally remaining below about 10 MeV. The proton population is generally stable while the electron populations in both belts are much more variable. The flux contours for protons with energies greater than 10 MeV and for electrons with energies greater than one MeV are shown in Figure 2.26.

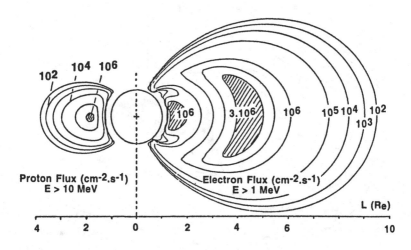

Figure 2.26 Electron and Proton Flux Contours for the Inner and Outer Van Allen Belts
(Bourrieau, 1993, used with permission)

Radiation damage

The concern for highly-energetic charged-particle sources interacting with the spacecraft is the radiation damage they may cause to materials through atomic displacement and ionization. A summary of the effects that radiation can produce in the materials of a spacecraft is shown in Figure 2.27. The chemical and structural changes that lead to

ENVIRONMENTAL FACTORS

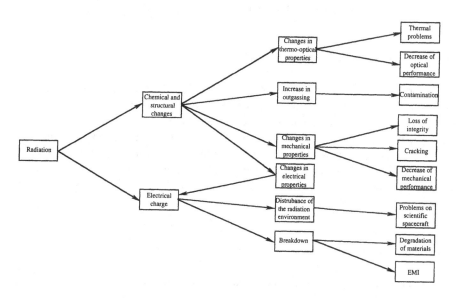

Figure 2.27 Effects of Radiation on Spacecraft Materials
(Paillous, 1993, used with permission)

changes in thermo-optical properties will affect the electrical power management and distribution system as well as the loads presented to the power system, and outgassing can lead to electrical breakdown across vacuum-insulation interface. The most serious problem is the general effect of charge creation at unwanted places such as within dielectrics and semiconductor materials.

While all of these effects are important to the spacecraft design, those involving the electrical components are of greater interest here. For details beyond this brief overview of radiation hazards to electronic materials in space, the reader is referred to a number of excellent texts (DeWitt *et al.*, 1993; Fortescue and Stark, 1995; Hastings and Garrett, 1996; Larson and Wertz, 1992; Pisacane and Moore, 1994; Griffin and French, 1991, among others).

Although much attention is given to the heavier energetic particles such as protons and alpha particles, energetic electrons can also cause damage by penetrating the surface layers of spacecraft and depositing charge inside insulators and on electrically isolated components. As Purvis (1993) points out, the Internal Discharge Monitor on the CRRES satellite recorded radiation-induced discharges on a number of insulating samples, and evidence is mounting that discharges due to energetic electrons are responsible for a variety of spacecraft anomalies. From the perspective of susceptibility of the electrical power system, the major worry remains the damage done to electronic components by the ionizing radiation caused by the heavier particles.

Susceptibility of semiconductor materials

Radiation is responsible for two general types of failure mechanisms of electronic systems aboard spacecraft: a gradual degradation of performance due to total dose and the single-particle effects such as the single-event upset (SEU). The total dose is an integrated value that depends primarily on the orbit and the overall shielding, while the SEU is a random event dependent on the environment and the sensitivity of the electronic component. The naturally occurring quiet environment can generate a low ionization radiation dose rate (< 0.5 rad/s), but proton events related to solar flares can increase that rate by several orders of magnitude.

Degradation of electronics due to total dose

A high-energy particle impacting semiconductor materials such as silicon and silicon dioxide will deposit charges along the particle's path of travel and will locally alter the carefully tailored crystalline structure of the material. After a sufficient number of such events, the accumulation of these trapped charges may cause a malfunction, or as a result of repeated atomic displacements, the semiconductor may cease to be the proper type of material and will no longer be able to function as an electronic device (Maurer, 1994). Stephen (1993) points out that the ionizing radiation may also produce interface traps at the semiconductor/oxide interface, seriously affecting the operation of the MOSFET. In addition to ionization effects, the radiation can also displace the silicon atom from its natural lattice location and move it into an interstitial site. Again from Stephen (1993), these sites are electronically active and will affect the leakage currents of p-n junctions, the current gain in bipolar transistors, and the carrier mobility in MOSFETS.

As an example of the dose to which a satellite might be exposed during its on-orbit life, the total dose versus altitude for a ten-year, circular equatorial orbit is shown in Figure 2.28. The dose is calculated using two thicknesses of aluminum-equivalent shielding, 0.5 and 3.0 g/cm^2 (Al), thicknesses representative of the shielding provided by the skin and structural members of the spacecraft. The peaks in the curves correspond to the locations of the Van Allen belts, and the significant reduction, with even modest shielding, in total dose at an altitude of about $4R_e$ reflects the fact that the outer belt is composed primarily of electrons whose lower energy and smaller mass (relative to the protons found in the inner belt) make them more easily shielded. The total doses in this example, which can exceed 10^6 rad (Si), should be compared to the radiation tolerances

ENVIRONMENTAL FACTORS

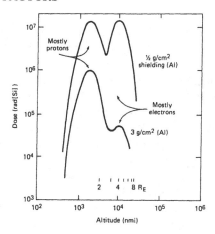

Figure 2.28 Total Dose versus Altitude for Circular Equatorial Orbits
(Griffin and French, 1991, used with permission)

of several semiconductor components given in Figure 2.29. Many of the devices will experience serious difficulties, including failure, at this dosage level. Fortunately for most communications satellites, GEO orbits are beyond the worst radiation zones and even minimal shielding can be quite effective. However, it can be seen from Figure 2.27 that, in a ten-year mission, a lightly shielded component could accumulate a total dose of 10^6 rad.

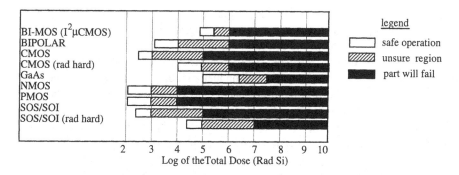

Figure 2.29 Total Dose Tolerances of Selected Semiconductor Devices

The total-dose effects can be aggravated by the radiation intensity. At one extreme, the very-high dose rate that accompanies a solar flare can induce failures well below the total dose levels normally tolerated by a given device. Measurements confirm that failure levels can be as much as an order of magnitude lower (in total dose) when the dose rate is increased by an order of magnitude (Stephen, 1993). This is related to the semiconductor device's inability to anneal at higher dose rates. At the other extreme, recent data reported by Barnes *et al.* (1997) indicate that bipolar integrated circuits (that are an essential part of many space-based electronic circuits) may be more susceptible to ionizing radiation at very low dose rates (~0.003 rad(Si)/sec) than they are to the much higher dose rates often used in laboratory-based validation studies. Their studies of the degradation of the input offset voltage for an LM 324 operational amplifier at various dose rates showed little change from the 50 rad/sec rate to 0.005 rad/sec but a dramatic decrease in the input offset voltage at the 0.002 rad/sec rate. This lowest rate may be a proper equivalent to the rates observed in orbit. The implications that these observations hold for testing protocols are still under study.

The single-particle event

When a charged particle impacts matter, as in a cosmic ray entering semiconductor material, the incoming particle slows down and deposits its energy in the material. The process will lead to the ionization of the material along the path of the incident particle. In silicon, for example, one hole-electron pair is produced for each 3.6 eV of energy that is deposited by the incoming particle. For a relatively low-energy cosmic ray of 5×10^7 eV incident on the silicon, the total charge created by ionization is

$$10^7 \text{ eV}/3.6 \text{ eV/pair} = 2.8 \times 10^6 \text{ electrons} = 0.44 \text{ pC}.$$

This charge, though small, is deposited into the device in picoseconds and can disrupt the microcircuit operation, an event known as the single-event upset. In CMOS circuits the event is called a 'latch up' and can cause the device to pass currents large enough to cause failure of the device (Stephen, 1993). The SEU phenomenon, ironically, is the result of several-decade efforts of the semiconductor industry to increase the speed and to decrease the electrical power requirements of electronic devices by dramatically reducing feature size. The corresponding decrease in the mass and volume made such miniaturization even more attractive. But, as the state-of-the-art in device manufacture enabled integrated circuits with characteristic feature sizes on the order of 1 μm and smaller to be fabricated, the vulnerability of the devices to SEU increased. Small circuits and shrinking transistor junctions imply operation at lower current and charge levels so that device 'critical charge' reached levels in the 0.1 to 1.0 picoCoulomb range,

ENVIRONMENTAL FACTORS 61

as can be seen in Figure 2.30. Since this amount of charge is easily produced by cosmic-ray impact, the vulnerability of microelectronics devices becomes an issue in spacecraft design.

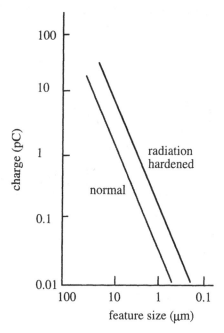

Figure 2.30 Critical Charge versus Device Feature Size
(Griffin and French, 1991, used with permission)

Because the source term, high-energy charged particles, is always present, the probability of an SEU occurring during the extended life of a satellite is almost certain. A number of mitigating design options are available to reduce the effects: use of radiation-hardened devices involving, for example, redundant circuitry, and tailored gate configurations or special materials. These radiation-resistant devices are available and are used in those instances where analyses indicate a particular sensitivity or vulnerability of the system to an SEU (Stephen, 1993). As Griffin and French (1991) point out, even with the error rate as small as 10^{-10}/day, a typical design standard, several upsets could be expected in a one megabit memory unit on orbit for a decade. The galactic cosmic-ray SEU error rate is greater than that for protons in all orbits except ones traversing the central portion of the inner Van Allen belt where the proton flux is greatest.

The SEU error rate increases sharply when the satellite is in the region of the South Atlantic Anomaly. In this region of reduced magnetic field strength, charged particles are able to reach lower altitudes and satellites traversing the SAA are subject to higher dose rates. Specifically, the trapped-proton flux and, hence, the likelihood of a proton-induced SEU event, are greater at lower altitudes than at the same altitude outside the region of the anomaly. Bourrieau (1993) offers a plot (Figure 2.31) of the SEU location (latitude and longitude) from a record of anomalies on the UOSAT2 satellite (inclination 98 degrees, altitude ~700 km) which clearly shows the increased SEU probability around the SAA.

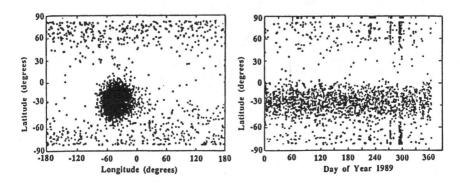

Figure 2.31 SEU Locations of the TMS4416 RAM Anomalies on UOSAT2 Showing the Effect of the SAA (Bourrieau, 1993, used with permission).

Since exposure to some radiation is inevitable at any altitude, shielding becomes critical to the survivability of sensitive electronics. The appropriate amount of shielding for a spacecraft can be determined by computing the dose rate for the desired orbit as a function of the shield thickness using the time-averaged radiation transport for particular orbits needed. Figure 2.32 shows the radiation dose rates for satellites in low-altitude polar orbits for several thicknesses of aluminum shielding. In the case of geosynchronous satellites, protons trapped near GEO have insufficient energy to penetrate 10 mils of aluminum. The protons and heavy ions in this region of space (with energies ~100 keV) will, however, deposit their energy in the spacecraft skin causing a temperature rise sufficient to enhance the infrared background (with heat loads of 0.5 W/m^2 possible). As with the UV, these ions will degrade the effectiveness of paints and protective glasses to which they are exposed (Schultz and Vampola, 1992).

3.4 The particulate environment

The particulate environment is composed of two very different populations: those particles that are naturally occurring, meteoroids, and those that are man-made, debris. The

ENVIRONMENTAL FACTORS

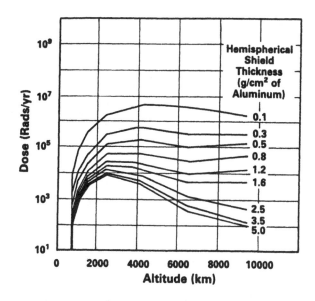

Figure 2.32 Radiation Dose Rates at Various Polar Orbit Altitudes for Several Shielding Thicknesses (Schultz and Vampola, 1992, used with permission)

threat to space systems from either population is an unanticipated hypervelocity impact which, depending on the relative velocity of the particle and the satellite, can pit, spall, or penetrate the satellite's surface. Even minor pitting can be serious to optical systems and may also enhance the effects of exposure of most surfaces to atomic oxygen. Spallation and penetration can be catastrophic to virtually any system. An improved understanding of the particulate environment has come from analyses of damage from hypervelocity impacts on the surfaces of Solar Max, LDEF, the Hubble Telescope, and the European Retrievable Carrier. These have led to estimates of the size and mass distribution of both meteoroids and debris affecting satellites in Earth orbits and to the design of advanced shielding systems.

Naturally occurring meteoroids are part of the interplanetary space environment and their presence in the near-Earth environment has been a satellite design issue since early spaceflight. These meteoroids create a hazard to spacecraft as they pass near the Earth at speeds averaging 20 km/sec. In spite of the great speed, the hazard is manageable because of the small mass and size of the particles which average about 0.01 cm diameter. To emphasize the importance of relative velocity, however, consider that the kinetic energy of a one millimeter diameter aluminum sphere traveling at 20 km/sec is comparable to the speed of a rifle bullet.

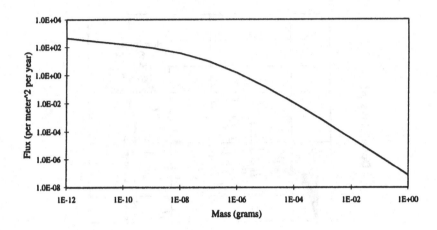

Figure 2.33 Mass Distribution of the Meteoroid Flux Measured from LDEF
(after McDonnell, 1993)

The mass distribution for the micrometeoroid environment is shown in Figure 2.33. This natural environment is not constant, but varies as the Earth revolves about the sun and also undergoes variations due to the focusing effects of the Earth's gravitational field and shielding effects of the Earth itself. There will, therefore, be flux differences depending on the spacecraft orbit.

Man-made space debris differs from the natural meteoroids in two important aspects: it is created and remains in orbit throughout its life rather than just passing through the Earth's environment, and it demonstrates a very different and potentially more dangerous mass distribution than the naturally occurring objects. Although larger masses are present in the debris population, the relative velocity of orbital debris and the spacecraft will not be as great as that seen for meteoroids. The debris originates from previous space operations and includes inactive payloads and items released during satellite operations, from the purposeful or accidental fragmentation of rocket bodies, the deterioration of spacecraft surfaces after extended exposure to the environment, and thousands of kilograms of aluminum oxide released from firing solid rocket motors. Active efforts are underway worldwide to reduce the amount of debris released into Earth orbits although even optimistic projections show increases in flux levels over time.

Figure 2.34 is a projection of the debris flux for two orbital altitudes for particles greater than three millimeter diameter, one set of curves assuming that continued explosions will occur in orbit and one assuming such explosions can be avoided. Notice also the small contribution from the meteoroid environment (Kessler, 1991). A similar projection from the US Department of Defense shows projections about an order of magnitude less for debris sizes one centimeter or greater.

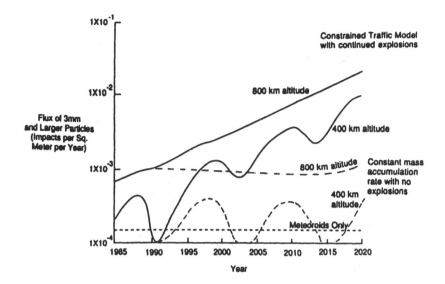

Figure 2.34 Predicted Flux Levels for Two Growth Assumptions
(Kessler, 1991, used with permission)

The orbital variations of the mass/flux spectrum are given in Figure 2.35. Note that the Molniya orbit is exposed to significantly lower flux at all mass levels because of the long time a spacecraft in this orbit spends outside the debris environment. The solid line is an approximation of the meteoroid environment for all orbits. The penetrations predicted as a function of aluminum shield thickness is given in Figure 2.36. Since the relative velocities of the debris and satellite are much smaller at GEO than LEO, damage at geosynchronous altitudes is expected to be less. The threat is further reduced at GEO because of the lower overall debris density.

As Kessler (1991) points out, analysis of retrieved spacecraft indicate that debris flux dominated the population of particulates smaller than 0.01 mm while the meteoroid flux was greater than that from debris by a factor of four for particles between 0.03 and 0.2 mm. Above about 1 mm, the debris flux again dominates.

There are three debris-mass regimes that are important in the design of spacecraft: for the very small particles, less than 0.01 cm in diameter, pitting and erosion are the primary concerns. The long term effects of interactions with these particulates can be particularly destructive to optical and thermal management surfaces. Particles of this size are quite abundant in LEO, and so the threat can be severe. Particles up to about 1 cm in diameter can produce impacts that are capable of penetrating typical shielding found on spacecraft, and often additional spot shielding is mandated for critical components. Electrical insulation is particularly vulnerable, and the impact of particulates on

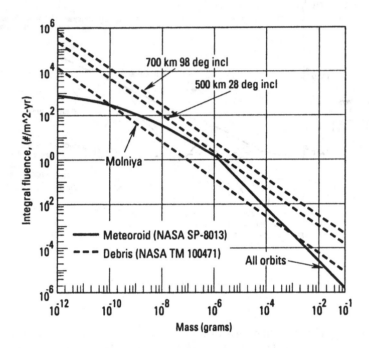

Figure 2.35 The Mass-Flux Spectrum for Orbital Particulates
(Barter, 1992, used with permission)

insulation around conductors operating at 100 volts or higher can produce permanent short circuits (Purvis, 1993). Above particle diameters of about 1 cm, the interaction can be catastrophic and little can be engineered to protect the spacecraft without unacceptable mass penalties.

Structures such as photovoltaic arrays with their large area-time product will naturally have the greater probability of impact, although with such structures some degree of tolerance can be engineered. Other power systems such as nuclear sources may not enjoy that option.

4. References

Banks, B., *Atomic Oxygen*, in *Proceedings of the LDEF Materials Data Analysis Workshop*, (NASA CP 10046, 1990).

ENVIRONMENTAL FACTORS

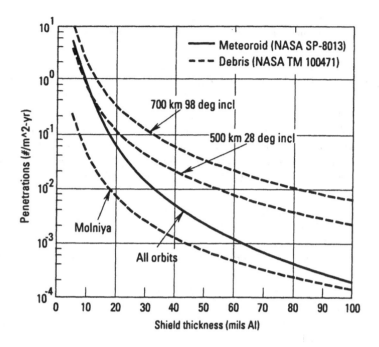

Figure 2.36 Orbital Particulate Penetrations versus Shield Thickness
(Barter, 1992, used with permission)

Barnes, C., Johnston, A., Lee, C., Swift, G., and Rax, B., *Radiation Effects Issues for JPL Interplanetary Missions,* in the *Proceedings of the MEO Platform Workshop,* (Jet Propulsion Laboratory of the California Institute of Technology, internal report, 1997).

Barter, N. J., (ed.), *TRW Space Data,* (TRW Space and Technology Group, Redondo Beach, California, 1992).

Black, H. D. and Pisacane, V. L., *Astrodynamics,* in *Fundamentals of Space Systems,* Pisacane, V. L. and Moore, R. C., (eds.) (Oxford University Press, New York, 1994), 99-171.

Bourrieau, J., *Protection and Shielding,* in *The Behavior of Systems in the Space Environment,* DeWitt, R. N., Duston, D., and Hyder, A. K. (eds.), (Kluwer Academic Publishers, Dordrecht, 1993), 299-351.

Cowley, S. W. H., *The Magnetosphere and Its Interaction with the Solar Wind and with the Ionosphere,* in *The Behavior of Systems in the Space Environment,* DeWitt, R. N., Duston, D., and Hyder, A. K. (eds.), (Kluwer Academic Publishers, Dordrecht, 1993), 147-181.

DeWitt, R. N., Duston, D., and Hyder, A. K. (eds.), *The Behavior of Systems in the Space Environment*, (Kluwer Academic Publishers, Dordrecht, 1993).
Domingo, V., *The Sun - Its Role in the Environment of the Near Earth Space*, in *The Behavior of Systems in the Space Environment*, DeWitt, R. N., Duston, D., and Hyder, A. K. (eds.), (Kluwer Academic Publishers, Dordrecht, 1993), 67-101.
Fortescue, P. and Stark, J., *Spacecraft Systems Engineering*, (Wiley and Sons, Chichester, 1995), 75-102.
Friedman, H., *Near Earth Space, a Historical Perspective*, DeWitt, R. N., Duston, D., and Hyder, A. K. (eds.), (Kluwer Academic Publishers, Dordrecht, 1993), 147-181.
Garrett, H., *Radiation Technologies: The Environment and Its Effects on Space Systems*, (Jet Propulsion Laboratory of the California Institute of Technology, internal document, 1997).
Gorney, D., *Solar Particle Events*, in *Space Mission Analysis and Design*, Larson, W. J. and Wertz, J. R. (eds.), (Microcosm, Inc., Torrance, California and Kluwer Academic Publishers, Dordrecht, 1992), 201-204.
Griffin, Michael D. and French, James R., *Space Vehicle Design*, Prezemieniecki, J. R., (ed.), (AIAA Education Series, 1991).
Hastings, D. and Garrett, H., *Spacecraft-Environment Interactions*, (Cambridge University Press, 1996).
Jursa, A. S., (ed..), *Handbook of Geophysics and the Space Environment*, (Air Force Geophysics Lab, 1985), NTIS Accession AD-A167000.
Kessler, D. J., *Orbital Debris Environment for Spacecraft in Low Earth Orbit*, *Journal of Spacecraft and Rockets*, **28**, No 3, May-June, 1991, 347-351.
Larson, W. J. and Wertz, J. R. (eds.), *Space Mission Analysis and Design*, (Microcosm, Inc., Torrance, California and Klluwer Academic Publishers, Dordrecht, 1992).
Latham, R. V., *Potential Threats to the Performance of Vacuum-Insulated High-Voltage Devices in a Space Environment*, in *The Behavior of Systems in the Space Environment*, DeWitt, R. N., Duston, D., and Hyder, A. K. (eds.), (Kluwer Academic Publishers, Dordrecht, 1993), 467-489.
McDonnell, J. A. M., *The Near-Earth Particulate Environment*, in *The Behavior of Systems in the Space Environment*, DeWitt, R. N., Duston, D., and Hyder, A. K. (eds.), (Kluwer Academic Publishers, Dordrecht, 1993), 117-146.
Maurer, R. H., *Spacecraft Reliability and Quality Assurance*, in *Fundamentals of Space Systems*, Pisacane, V. L. and Moore, R. C. (eds.), (Oxford University Press, New York, 1994), 661-720.
Mitchell, D. G., *The Space Environment*, in *Fundamentals of Space Systems*, Pisacane, V. L. and Moore, R. C. (eds.), (Oxford University Press, New York, 1994), 45-98.
NASA, *High Voltage System Performance in Low Earth Orbit Plasma Environments*, Vol 2., (NASA CR-179641, 1986), also Stevens, N. J., Kirkpatrick M. C., and Stillwell, R. P., (TRW 46870-6006-UT-00, TRW, Inc., Redondo Beach, California, 1986).

Paillous, A., *Radiation Damage to Surface and Structure Materials*, in *The Behavior of Systems in the Space Environment*, DeWitt, R. N., Duston, D., and Hyder, A. K. (eds.), (Kluwer Academic Publishers, Dordrecht, 1993), 383-405.

Peplinski, D. R., *Satellite Exposure to Atomic Oxygen in Low Earth Orbit*, (NASA CP 2340, 1984).

Pisacane, V. L. and Moore, R. C. (eds.), *Fundamentals of Space Systems*, (Oxford University Press, New York, 1994).

Purvis, C. K., *Overview from a Systems Perspective*, in *The Behavior of Systems in the Space Environment*, DeWitt, R. N., Duston, D., and Hyder, A. K. (eds.), (Kluwer Academic Publishers, Dordrecht, 1993), 23-44.

Schultz, M. and Vampola, A., *Trapped Radiation*, in *Space Mission Analysis and Design*, Larson, W. J. and Wertz, J. R. (eds.), (Microcosm, Inc., Torrance, California and Kluwer Academic Publishers, Dordrecht, 1992), 199-201.

Soop, E. M., *Handbook of Geostationary Orbits*, (Kluwer Academic Publishers, Dordrecht and Microcosm, Inc., Torrance, California, 1994), 110-115.

Stephen, J. H., *Hazards to Electronics in Space*, in *The Behavior of Systems in the Space Environment*, DeWitt, R. N., Duston, D., and Hyder, A. K. (eds.), (Kluwer Academic Publishers, Dordrecht, 1993), 407-435.

Tennyson, R. C., *Atomic Oxygen and Its Effect on Materials*, in *The Behavior of Systems in the Space Environment*, DeWitt, R. N., Duston, D., and Hyder, A. K. (eds.), (Kluwer Academic Publishers, Dordrecht, 1993), 233-257.

Thekaekara, M. P., *Solar Irradiance, Total and Spectral*, in *Solar Energy Engineering*, Sayigh, A. A. M. (ed.), (Academic Press, New York, 1977).

Tribble, A., *The Space Environment: Implications for Spacecraft Design*, (Princeton University Press, 1995).

Vampola, A. L., *Analysis of Environmentally Induced Anomalies*, Journal of Spacecraft and Rockets, **31**, No. 2, 1994, 154-159.

Waltersheid, R. L., *Solar Cycle Effects on the Upper Atmosphere: Implications for Satellite Drag*, Journal of Spacecraft and Rockets, **26**, No. 6, 1989, 439-444.

Wertz, J. R. and Larsen, W. J. (eds.), *Space Mission Analysis and Design*, (Kluwer Academic Publishers, Dordrecht, 1991).

Wise, Peter C., *Spacecraft Systems Engineering: Thermal Control Subsystem*, internal lecture notes, Lockheed Martin Corporation (1993).

Wolverton, R. W., *Flight Performance Handbook for Orbital Operations*, (Wiley and Sons, New York, 1961), 2:397-2:421.

CHAPTER 3

SOLAR ENERGY CONVERSION

1. Introduction

Beginning with the launch of the first U.S. solar powered satellite in 1958, solar cells have been the predominant power source in space for over forty years. Although those early solar panels were relatively easy to build and use, their low efficiency was of concern. Almost immediately after that first launch, NASA began to investigate ways to improve solar cells and arrays to meet the already growing demand for space power. Investigations were also begun into other conversion technologies, such as solar thermal systems, with a view toward providing the large amounts of power anticipated for future manned missions. In this chapter we shall discuss only the fundamentals of photovoltaic energy conversion, along with the status of current photovoltaic technology. The reason is straightforward: hundreds of kilowatts of photovoltaic solar power have been placed in orbit on various commercial, civilian, and military satellites, while a solar dynamic system has yet to be launched. Furthermore, the early size limitations on photovoltaic power systems are no longer valid, and multikilowatt systems (with the potential to go up to several hundred kilowatts) are now commonplace. In addition, the design flexibility, reliability, and modularity of space solar arrays are unsurpassed by any other conversion technology. Small arrays have even been transported to, and left behind on, the surfaces of the moon and Mars. In at least one instance a photovoltaic array was in orbit for nearly two decades and still functioned well enough to bring the satellite back to an operational state following several years of dormancy (Mirtich, 1991). Solar cell efficiencies on the earliest arrays were typically around 10% and much had to be learned about the survivability of the devices in the space environment. Enormous progress has been made since 1958, both in understanding the fundamental mechanisms which determine solar cell efficiency and lifetime, and in turning that understanding into tangible cell improvements. In addition, array structural mechanisms have become much more reliable and sophisticated. We will briefly describe the elements of a space photovoltaic power system, review the status of solar cell technologies presently in use, discuss advances in both cell and array technology, and will conclude with a brief discussion of solar thermophotovoltaic energy conversion, an old idea with some new technological developments which give it promise for space applications.

72   SPACECRAFT POWER TECHNOLOGIES

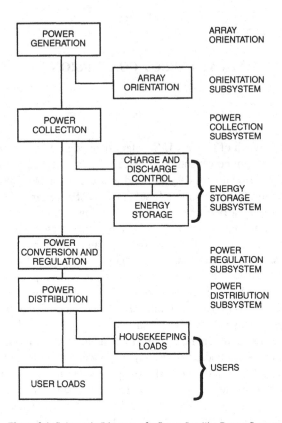

Figure 3.1 Schematic Diagram of a Space Satellite Power System

1.1 Space photovoltaic power systems

A space power system is comprised of a number of subsystem elements, one of which is the solar array. A block diagram of a typical photovoltaic space power system is shown in Figure 3.1. Moreover, the array itself is a set of subsystems, as shown in Figure 3.2. The result is that designing and building even just the solar array requires an interdisciplinary, well-coordinated effort similar to that required to build the complete power system.

There are two figures of merit used to measure the performance of a space solar array, as well as the entire power system: power per unit mass, expressed in watts/kilogram (W/kg), and power per unit area, expressed in watts/square meter (W/m^2). These are referred to simply as specific power and area power density, respectively. The inverses of these quantities are also often used, and are known as specific mass (kg/

SOLAR ENERGY CONVERSION

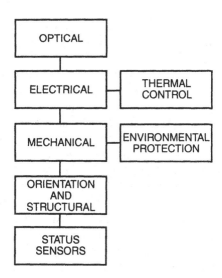

Figure 3.2 Schematic Diagram of a Space Solar Array Subsystem

kW) and specific area (m^2/kW). Typical values for state-of-the-art (SOA) space solar arrays, using silicon solar cells mounted on rigid panels, are 30 to 40 W/kg and 90 to 110 W/m^2 at the start of the mission, or beginning-of-life (BOL). The end-of-life (EOL) values for any given array are dependent on mission time and location. Environmental factors affecting the ratio of EOL to BOL array output include electron and proton radiation induced damage to the cells, along with mechanical and electrical degradation of the cell interconnections and other array components from thermal cycling and interactions with the ambient orbital environment. Elimination, or at least substantial mitigation, of such effects is at the heart of all space photovoltaic device and system research and development efforts and will be described in more detail in later sections of this chapter.

1.2 Space power system applications and requirements

Table 3.1 lists the broad mission categories into which space missions can be roughly divided, qualitative estimates of the power levels required for each, and the primary attributes any sort of power system must have if it is to be considered for use on such missions. The desired attributes are listed in relative priority order for each mission class with the understanding that detailed trade studies are required to establish the actual order of the priorities for any given mission. It is clear, however, that low mass and long lifetime are important power system drivers in virtually all potential space missions. Power system cost and size have greater or lesser importance depending on the mission objectives and operational environment (orbital, planetary surface, interplanetary, deep space, etc.).

For example, low total cross-sectional area is a critical attribute for the space station because of the drag produced by the residual atmosphere in the low altitude orbits in which it will fly (Nored and Bernatowicz, 1986). In this case, the array has an important effect on the total life cycle cost of the mission because it directly affects the amount, and hence the cost, of constantly providing fuel to maintain the space station at its proper altitude. A mission to the lunar surface, on the other hand, would not be subject to such a phenomenon; although there would be a life cycle cost associated with standard array maintenance, it could reasonably be expected to be much less than in the International Space Station case. In the lunar mission, although total area may be important because of other factors such as ease of construction and deployment, it is not a primary driver in selecting a particular technology for the mission. Specific power and resistance to proton radiation damage from solar flares are certainly among the more important factors for selecting a lunar surface solar array technology.

Table 3.1 Qualitative Categorization of Space Missions (Flood, 1989)

Mission Subset	Power Level	System Attributes
Unmanned Near Earth (Leo, Heo, Geo) and Unmanned Planetary	Low to Intermediate	Low Mass, Long Life
Space Station	High	Minimum Area, Low Mass, Low Cost, Long Life
GEO Platform	Intermediate	Long Life, Low Mass
Lunar Base, Manned Planetary	Intermediate to High	Low Mass, Portability, Long Life
Electric Propulsion Orbit Transfer	High	Reusability, Minimum Area, Low Mass

1.3 Space solar cell and array technology drivers

When applied to the system level, it is clear that the desired attribute for low mass translates into high solar cell efficiency. The principle reason is that the solar cells are themselves a relatively lesser fraction of the total mass and cost of a system, while their efficiency and usable lifetime are major determinants of the balance-of-system (BOS) mass and cost. Cell efficiency determines array area, which in turn determines array mass. Array lifetime, however, depends critically on the nature of the mission and the environment in which it occurs and is loosely-defined to be the length of time the array operates before its output power falls to a level below that needed to operate the satellite reliably. In general, assuming that the mechanical aspects of the cell and array have

SOLAR ENERGY CONVERSION 75

been properly engineered to withstand thermal cycling, vibration during launch and orbital maneuvering, and other operational and environmental effects such as interactions with the space plasma (LEO) or spacecraft charging (GEO), array lifetimes are determined by the rate at which solar cell electrical output degrades. The chief cause of electrical degradation is bombardment of the cell by constituents of the naturally occurring charged particle radiation environment. The ability of a solar cell to operate while, or having been, subjected to charged particle bombardment is a measure of the radiation resistance of the cell. It is measured by determining the ratio of the output power (P) remaining after absorbing a given dose of radiation to the initial output power (P_o) determined prior to such exposure (P/P_o). The extent to which a solar cell is radiation resistant depends on many factors: the material from which it is made (i.e., silicon, gallium arsenide, indium phosphide, etc.), its actual device structure, and its ability to anneal (or be annealed) as the damage occurs. In Sections 3, 4, and 5, we shall discuss the mechanisms involved in creating radiation damage in solar cells and the attempts to reduce it or eliminate it altogether as we discuss the various solar cell materials and structures now in use or under development.

Technology drivers for the solar arrays are the mass and stiffness of the deployed structure, its stowed volume on the launch vehicle, and its compatibility with the space environment in which it will fly. With regard to the latter, for very large (multikilowatt output) solar arrays it becomes increasingly more important to consider the use of higher operating voltages instead of the lower voltages commonly used in the past. The reason is straightforward: a considerable contribution to the mass of the array at high power levels will come from the wiring harness. The advanced, flexible blanket, flat-fold arrays originally developed for a solar electric propulsion spacecraft can have a wiring harness that will easily exceed 10% of the entire array mass (Stella and Kurland, 1989), depending on array size and total current. If the power is transmitted at low voltages, the currents will be large and more conductor mass is needed to prevent excessive I^2R losses. To avoid this situation the array must be operated at higher voltages. There are upper limits on the voltages that can be used, however, which are imposed by interactions between the array and its environment. There are two primary concerns: interactions with the space plasma in LEO and spacecraft charging effects in GEO. Array degradation due to both of these environmental interactions will be described in Section 7, along with other related concerns.

2. Solar cell fundamentals

2.1 Introduction

Although Becquerel reported the first recorded experiments with the photoelectric effect in 1839, little progress was made toward achieving today's high efficiency solar cell structure for more than 100 years. It wasn't until after the intensive efforts begun during World War II to improve the properties of Ge and Si semiconductor materials, which resulted in the development of the transistor, that the first working silicon solar cell was patented by employees of Bell Telephone Laboratories in 1954 (Chapin, Fuller and Pearson, 1957). The early cells were p on n devices made by high temperature diffusion of p-type impurities into n-type Czochralski-grown silicon. The basic structure of the silicon solar cell has not changed since that time, although all silicon space solar cells now use the n on p configuration rather than p on n. The reason for the switch has to do with the longer minority carrier lifetimes in p-type compared to n-type silicon; this difference in minority carrier lifetime provides superior performance when the cells are irradiated by the high energy electrons and protons found in the space environment. We shall discuss radiation damage in space solar cells in later sections of this chapter.

Figure 3.3 shows a schematic of a silicon solar cell such as might be found on a modern day satellite. (The books by Hovel and by Sze contain excellent introductions to the material science and physics of solar cells and are highly recommended to those who desire a more rigorous background.) It is typical that the emitter layer is very thin compared to the base layer thickness, so that the bulk of the absorption of the incident light occurs in the latter region of the cell. Figure 3.3 illustrates several features of the modern solar cell, not all of which are used at the same time. Standard aspects are the top surface anti-reflective coating and very precisely defined top contact grid patterns which produce minimal shadowing of the underlying semiconductor material (as little as 3% coverage in a well designed cell intended for non-concentrated sunlight). Graded doping densities at the front and rear surfaces, known as front surface fields and back

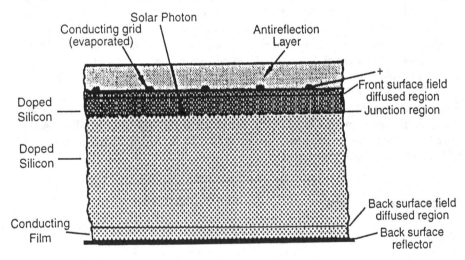

Figure 3.3 Schematic Diagram of a Typical Silicon Space Solar Cell

SOLAR ENERGY CONVERSION

surface fields, can produce enhanced current collection and output voltages while carefully coated back surfaces can produce enhanced light absorption by reflecting uncaptured photons back into the cell. While the nature of these enhancements varies with the material used for the cell, for example the III-V semiconductor compounds GaAs and InP, the basic principles are the same. In the III-V materials, the fields are usually produced by epitaxially depositing a thin layer of a lattice-matched material which has a higher bandgap than the cell material itself. While such minority carrier "mirrors," as they are known, can be very effective, their inclusion adds a certain amount of complexity to cell fabrication, and commercial III-V compound space solar cells often do not have such enhancements included.

The electrical behavior of a solar cell is related to its semiconductor material characteristics in the following ways.

- The polarity of the output voltage of an illuminated solar cell is such that the n-contact becomes negative and the p-contact becomes positive.
- An illuminated solar cell (i.e., diode) connected to an external circuit and delivering power to a load is said to operate in its forward mode.
- Whether illuminated or not, if the positive terminal of an external power supply is connected to the p-contact of the cell and its negative terminal to the n-contact of the cell, the cell is forward biased, and reversed bias if the terminals are switched to the opposite contacts.

There are several important parameters that describe the operation of a solar cell. They are indicated in Figure 3.4 and are the short circuit current I_{sc}, the open circuit voltage V_{oc}, the maximum power P_m, the current and voltage at max power I_m and V_m, and the fill factor. The latter is simply the ratio of the area of the largest rectangle that can be drawn through I_m and V_m to the largest that can be drawn through I_{sc} and V_{oc}, and is a measure of the "squareness" of the I-V curve. In algebraic terms, FF = $V_m I_m / I_{sc} V_{oc}$ = $P_m / I_{sc} V_{oc}$ and is always less than unity. In general, the higher the fill factor, the higher the efficiency of the cell. The efficiency is defined as $\eta = P_m / P_0$, where P_0 is the input solar power; the efficiency can be written as η = FF. $I_{sc} V_{oc} / P_0$. The fill factor is a function of several cell characteristics, such as internal series resistance, shunt resistance, illumination intensity and temperature, among other things.

2.2 Basic theory

A simple pn junction solar cell may be thought of as a diode operating in parallel with a constant current source. Figure 3.5 shows a circuit representation of this model for a solar cell. The internal series and shunt resistances associated with real devices, R_s and R_{sh} respectively, are also shown. The current-voltage relationship is (Sze, 1981)

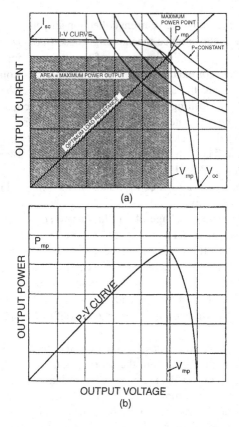

Figure 3.4 a. Output Current versus Voltage for a Typical Solar Cell
b. Output Power versus Voltage for a Typical Solar Cell

$$I = I_s\{\exp[(q/kT)(V - IR_s)] - 1\} - I_L + (V + IR_s)/R_{sh} \tag{3.1}$$

where I_s is the diode saturation current and I_L is the light generated current in the solar cell. The diode saturation current is given by

$$I_s = Aq[D_p p_{no}/L_p) + (D_n n_{po}/L_n)] \tag{3.2}$$

A is the total area of the diode, q is the electric charge, D_p is the diffusion coefficient for holes in the n region, D_n is the diffusion coefficient for electrons in the p region, p_{no} is the equilibrium concentration of holes in the n region, n_{po} is the equilibrium concentration of electrons in the p region, and L_p and L_n are the diffusion lengths (i.e., the average distance a carrier travels before it is lost to an unfilled state or trap) for holes and elec-

SOLAR ENERGY CONVERSION

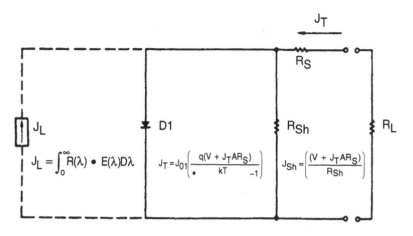

Figure 3.5 Representative Circuit Diagram for a Hypothetical Solar Cell

trons respectively. Minority carrier diffusion lengths are of fundamental importance in considering the suitability of any particular material for use in a space solar cell. As we shall see later, the ability of a solar cell to resist radiation damage degradation while in orbit is critical. The particles of concern are the trapped electrons and protons that circulate in the Earth's magnetosphere and the high energy protons that often accompany a solar flare event. We shall discuss this topic more fully in section 4.2. Returning to our discussion of Eq. (3.2), if the shunt resistance is 100Ω or higher, which is the case for all high performance solar cells, the last term in Eq. (3.1) may be neglected. The power output of the cell then becomes simply

$$P = IV = I\{[kT/q]\ln[(I + I_L)/I_s + 1] - IR_s \tag{3.3}$$

For one sun AM0 conditions, I is essentially equal to I_L in high quality cells with low values of R_s; hence, under short circuit conditions (i.e., $I = I_L = I_{sc}$) when V = 0, Eq. (3.1) becomes

$$\ln[2I_{sc}/I_s + 1] = qI_{sc}R_s/kT \tag{3.4}$$

A typical space silicon solar cell has a thin (≤ 0.5 μm) n-type emitter on a thick p-type base (≥ 200 μm). For all but the shortest wavelengths, the base component of short circuit current dominates the output. For minority electrons in a p-type base,

$$I_s = qAD_n n_{po}/L_n \tag{3.5}$$

In addition, $I_{sc} \gg I_s$ in a good cell, and Eq. (3.4) becomes

$$\ln(1/I_s) = qI_{sc}R_s/kT + \ln 2I_{sc} \qquad (3.6)$$

from which we see that

$$I_{sc} = (kT/qR_s)[\ln(L_n/qD_n n_{po}) + \ln(2I_{sc})] \qquad (3.7)$$

Referring again to Eq. (3.1), the open circuit voltage is simply

$$V_{oc} = (kT/q)\ln[I_{sc}/I_s + 1] \qquad (3.8)$$

where I_L has again been approximated by I_{sc}. If the short circuit current is much, much larger than the saturation current, then Eq. (3.8) becomes simply

$$V_{oc} = kT/q \ln[I_{sc}] \qquad (3.9)$$

Eq. (3.9) will be useful in our discussion of concentrator solar cells later in this chapter. Note that the short circuit current of a solar cell at constant temperature is proportional to the intensity of the incident sunlight for at least moderate concentration levels (up to several hundred suns). Hence, under n-times solar concentration, $I_{sc}(n) = n\, I_{sc}$. As a result, the open circuit voltage under concentrated sunlight increases by a small amount proportional to the log of the concentration ratio.

3. Space solar cell calibration and performance measurements

The ability to determine the size of a solar array for a given mission depends on the accuracy with which individual cell efficiencies are known and on the radiation degradation characteristics of the cells. Since there is a considerable difference in both the intensity and spectral content of solar radiation in space compared to the same quantities at the surface of the Earth, great care must be taken to establish the relationship of the measurements on the ground to the actual conditions in orbit. The usual procedure is to use a Xenon arc lamp and collimating system in a laboratory solar simulator, the output of which is carefully set by a standard cell that has been calibrated in space or in near-space conditions. The space solar spectrum is called the air mass zero (AM0) spectrum, and its intensity at one astronomical unit (AU) from the sun (i.e., at one Earth-sun circularized average orbit diameter) has been established by the World Meteorological Organization to be 1,365 W/m² (Wehrli, 1985). There are several precautions that must be observed, however, to achieve the level of accuracy ($\pm 1\%$ or less) required to make reliable predictions of array performance on orbit.

SOLAR ENERGY CONVERSION 81

3.1 Calibration techniques

Because even the simplest space missions are costly and access to missions which return hardware from orbit is limited at best, alternative techniques have been developed to calibrate solar cells for use in space. The most straightforward approach is to use either a high altitude balloon (Anspaugh and Weiss, 1996) or aircraft (Jenkins *et al.*, 1997). Even so, some manipulation of the data is required to obtain a calibrated value of the short circuit current for each cell. Historically, the two techniques have shown remarkable consistency with their results, not only internally but when compared with each other. Usable results can be obtained by making standard terrestrial measurements and mathematically correcting for the differences in spectral intensity distribution between the observed spectrum and the AM0 spectrum (Bücher, 1997), but the error bars associated with this method can be more than two or three times greater than those with the high altitude measurements, as discussed by Anspaugh and Jenkins.

Figure 3.6 depicts the balloon launch operation as used by NASA. The solar cells and their measuring instrumentation are contained in the top payload, while the transmitting antenna and other flight electronics (GPS position indicator, etc.) are contained in the bottom payload. The top payload also carries a locator beacon to assist in its retrieval when it returns to Earth after the flight since the touchdown can occur between 300 and 400 miles from the launch site. A chase plane is required, as well as a ground pursuit vehicle. While recovery is usually straightforward, bad weather occasionally interferes with the balloon's descent with the result that some damage may occur to the solar cells and panel assembly. Two types of calibration measurements are made: for most cells it is a fixed load (near zero voltage) current measurement, and for a limited number of cells a complete current-voltage (I-V) curve is taken. It is of critical importance to have accurate temperature coefficient data on all of the cells since no attempt is made to control the temperature during the flight. The temperature of the cell panel is carefully monitored, and the data are corrected to the laboratory standard test condition temperature. (At present the test temperature can be either 25°C or 28°C. An ISO 9000 standard is presently being prepared which sets 25°C as the standard test temperature.) The logistical complexity of this technique is such that it is usually carried out only once each year. A second flight is sometimes conducted if circumstances require it, as was the case for 1996 (Anspaugh and Weiss, 1996).

The data in Table 3.2 give evidence of the repeatability of the balloon measurement technique. The data presented are for the immediately preceding 22-year period, although the technique has been in use by NASA for over 33 years. In 1984, NASA flew the Solar Cell Calibration Facility on the Space Shuttle as a one-time check on the accuracy of both its balloon and high altitude aircraft measurement systems. The same set of cells was flown on the 1985 balloon test with essentially identical results (Anspaugh, Downing and Sidwell, 1985), giving a strong measure of confidence in the accuracy of the technique.

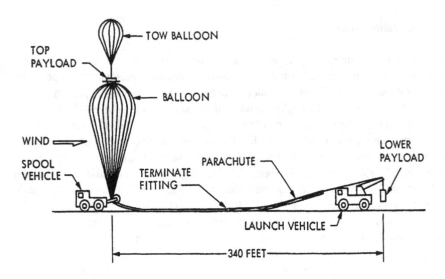

Figure 3.6 Balloon Launch for a Solar Cell Calibration Flight

Comparison of balloon-flown measurements with those obtained using NASA's high altitude aircraft shows the techniques to be equivalent. Although restricted to lower altitudes (approximately 50,000 ft. for the airplane versus 100,000 ft. for the balloon), there are operational differences that make the aircraft technique somewhat more convenient. Among other things, the test cells are not at the same risk for damage as they are with the uncontrolled payload descent following a balloon flight. Additionally, the test platform in the aircraft can be fully temperature controlled so that the measurements can be made at the actual desired test temperature, and both the true short circuit current and a full I-V curve can be obtained for each test cell. No calibrated load resistor is required to be kept with the cell during its subsequent use as a standard to adjust solar simulator intensity for making laboratory test measurements. The disadvantages are that the test area on the current aircraft is about 20% that of the balloon panel, and the measurements must be made in the atmosphere above the tropopause; this limits the opportunities for making the flights in the northern hemisphere to the period from roughly the middle of November to the middle of March each year. The reason for this constraint can be seen from the data presented in Figure 3.7 which show a distinct change in slope and a departure from linearity in the data below the tropopause. With a 50,000 foot altitude limitation imposed by the current aircraft, and the desire to have a 10,000 to 15,000 foot descent for taking data, it is necessary to fly when the tropopause is below about 35,000 feet, which occurs in the winter months in the northern hemisphere. Data are taken as a function of atmospheric pressure. The pressure readings are converted to altitude, which is in turn converted to the air mass remaining overhead at that

SOLAR ENERGY CONVERSION 83

Table 3.2 Repeatability of Balloon Flight Calibration Data for Various Solar Cells

Flight Date	73-182 HEK	80-003 K4 3/4	81-004 K4 3/4 BSR	STS-021 Hi Blue (Millivolts[1])	86-023 Mantech GaAs	86-026 K4 3/4	92-005 GaAs/Ge	95-002 GaAs/Ge
6/6/75	67.88							
6/10/77	67.96							
7/20/78	68.20							
8/8/79	67.83							
7/24/80	68.00	78.69						
7/25/81	67.96		77.55					
7/21/82	68.03		77.52					
7/12/83	68.03							
7/19/84	67.62							
8/84 Shuttle				73.60				
7/12/85				72.85				
7/15/86					58.46	76.25		
8/23/87					59.47			
8/7/88			77.49		58.26			
8/9/89					58.30			
9/6/90			77.43		58.89			
8/1/91				73.08	59.12		62.31	
8/1/92		78.30			58.68	76.29	60.92	
7/29/93	67.71		77.24					
8/6/94	67.77	78.51			58.91	75.82		
8/31/95	67.95				58.69			81.53
6/30/96	67.65				58.51	76.06		79.71
8/8/96	68.26	78.40	77.10	72.85			57.23	79.53
Summary of the Measurements								
Number	15	4	6	4	10	5	3	3
Average	67.65	78.78	77.39	73.10	58.73	76.09	60.15	80.26
Std. Dev.	0.217	0.167	0.179	0.354	0.378	0.189	2.625	1.106
Max. Value	68.37	78.69	77.55	73.60	59.470	76.29	62.31	81.53
Min. Value	67.62	78.30	77.10	72.85	58.260	75.82	57.23	79.53
Max.	0.422	0.215	0.288	0.505	0.741	0.270	2.923	1.273

[1] All measurements are in units of millivolts, although traditionally many such measurements are reported in millliamperes since it is the short circuit current that is used for calibrations. Calibration of the JPL balloon-flown samples is accomplished by measuring the voltage across a fixed resistor when the external load is near short circuit conditions, yielding a calibration in millivolts. The resistor is sized so that the drop across it will never exceed 100 millivolts. The practice is an historical artifact from the early days of making such measurements to assure that the measured signal would not load the input impedance the instruments available then. The cell and fixed resistor must therefore always remain mounted together.

altitude. The data are presented as a plot of the logarithm of short circuit current versus air mass, as shown in Figure 3.7. It is then a simple extrapolation to air mass zero. The standard deviation of measurements taken on a set of control cells for a period of about 20 years is ±0.8% (Jenkins et al., 1997), a result very close to that obtained from the balloon flights.

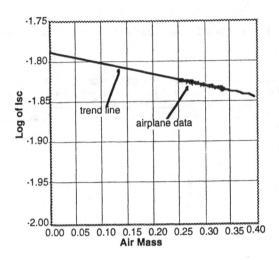

Figure 3.7 Short Circuit Current versus Air Mass for Typical Airplane Calibration Flight

3.2 Laboratory measurement techniques

The solar constant corresponding to the mean Earth-sun distance is 1,367 W/m^2, although because of the eccentricity of the Earth's orbit, this value varies by a factor of 1.0351 at perihelion (approximately January 3rd) and by a factor of 0.9666 at aphelion (approximately July 4th). Figure 3.8 contains a plot of the AM0 spectrum, along with the spectral irradiance for what has become the space solar cell industry standard solar simulator, the Spectrolab X25™. A key issue in measuring the performance of a solar cell with a solar simulator is the amount of mismatch between the simulator spectrum and the actual AM0 spectrum. It is not enough, when making a simulator measurement of space cell performance, to simply insert a calibrated cell of any arbitrary type into the simulator and adjust the beam intensity until the cell produces its AM0 short circuit (corrected by the appropriate seasonal factor.) While doing so does mean that the total simulator irradiance falling on the standard cell is equivalent to 1,367 W/m^2, that is not necessarily true for the test cell. The reason is straightforward: the standard cell provides an integrated response to the simulator spectrum based on its own spectral response which will be different from the integrated response of a test cell with a significantly different spectral response. If, for example, the standard cell has a strong response in the short wavelength (blue/ultraviolet) region of the spectrum, where the simulator output is considerably lower than the AM0 spectrum, the simulator output is set to achieve the standard cell's short circuit current, the long wavelength (red/near infrared) region of the spectrum will actually be more intense than that of the AM0 spectrum. If the cell being tested has a strong spectral response (i.e., output per unit wavelength) in

SOLAR ENERGY CONVERSION

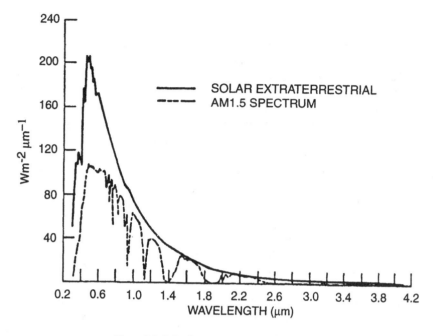

Figure 3.8 Solar Spectral Irradiance Distribution

the red/near infrared region, then its integrated response can easily be too high, and the result will be a measured efficiency that is larger than is actually the case. The simplest way to protect against making such errors is to use a calibrated cell with a spectral response as closely matched with that of the test cell as possible. As a result, it becomes necessary to keep a library of hundreds of calibrated solar cells and their spectral response curves so that a reasonably close match to any test cell can be made. If there is no close match, then it is necessary to fly the test cell on the balloon or aircraft to achieve an accurate result. Once flown, the test cell becomes a standard for use in further testing.

In principle, it is possible to correct for the irradiance differences between AM0 and the output of a solar simulator by mathematically integrating the spectral response of the test cell against the spectrum of interest and not use a balloon or aircraft generated standard cell at all. The accuracy of the test cell performance characterization in this case depends on concurrently determining, with very high precision, the absolute value of the test cell spectral response and the absolute magnitude of the simulator spectral output. The latter requires, in turn, accurate determination of the absolute spectral intensity of a standard lamp, the stability over time of which must also be known with the same very high degree of precision. The time and difficulty involved in using this technique for making measurements that will have the same absolute accuracy as obtained by the balloon and aircraft techniques has precluded its widespread use.

Laboratory cell performance data are taken at a standard temperature of 25°C, despite the fact that array operating temperatures in Earth orbit are always higher. The actual temperature at which a space solar cell will operate is a function of, among other factors, the array thermal and optical design (e.g., planar vs. concentrator arrays), the cell reflectance and absorptance, and its location in orbit at any given time. Predicting array operating temperature is an important issue that will not be dealt with in this chapter, primarily because it is very mission and satellite configuration dependent. In general, accurately known thermal properties of the array materials and components are needed so that complex numerical calculations can be made to predict array temperature. For the cells, that means measuring the various temperature coefficients (i.e., the slopes of plots of power, voltage and current as a function of temperature) of the cell type to be used. Those data, coupled with the 25°C performance data, are then used to predict the array output at the temperatures expected on orbit. In general, the solar cell temperature coefficients are essentially constant over the temperature ranges of interest for all Earth orbiting satellites, so the procedure is reduced to one of measuring I_{sc}, V_{oc}, and P_m, or efficiency at several temperatures, and calculating the slope of the line connecting the data. Table 3.3 gives what little data exist at present for normalized temperature coefficients for some typical silicon, gallium arsenide, and indium phosphide space solar cells from a variety of sources. (The normalized short circuit current temperature coefficient is defined as $1/I_{sc}(dI_{sc}/dT)$, with similar expressions for all the other quantities.)

The calibration and measurement procedures outlined above are quite straightforward when applied to single junction, single crystal cells such as GaAs and Si. It is important to note here, however, that major precautions must be taken with multijunction or multiple bandgap (MBG) cells when determining their efficiency or temperature coefficients. We shall see later that there are two configurations for MBG cells: current matched and voltage matched. At present, all commercial space MBG cells are of the current-matched variety. (See Section 4.2 for an introduction to MBG cells.) That means simply that each subcell in the device must generate the same current in the AM0 solar spectrum, whether at one sun or under concentration. (Concentration introduces an additional complication in that each subcell must be able to carry higher than normal

Table 3.3 Normalized Temperature Coefficients of Selected Space Solar Cells

CELL TYPE	PRODUCER	I_{sc} (ma/C)	V_{oc} (mv/C)	η(1/C)
GaAs/Ge	Spectrolab	+0.714		-2.0
GaAs/Ge	TECSTAR	+0.56		-2.23
InP	N/A	+0.89		-1.59
GaInP/GaAs	Spectrolab	+0.01	-6.4	-0.062
GaInP/GaAs	TECSTAR	+0.007	-5.2	-0.065
Si BSR	Spectrolab			-4.5
Si BSFR	TECSTAR			-4.6

SOLAR ENERGY CONVERSION   87

currents and remain current matched, so series resistance losses must be minimized and balanced throughout the structure as well.) Therefore, it is necessary to measure MBG cell performance in a spectrum that is as close to AM0 as possible to ensure that proper current matching between subcells has been obtained. Any deviation from the AM0 spectral distribution will either overfill or underfill (i.e., generate more or less current in) one or more subcells by differing amounts, and if the cell design is adjusted to achieve current matching under those conditions, the cell will not stay current matched in the AM0 spectrum. Its efficiency and output will fall. Recent progress has been made in the design of multisource simulators and filter systems (Kilmer, 1994; Wilkinson, *et al.*, 1997), so that laboratory measurements of MBG cell efficiency can be made with accuracies comparable to those obtained for single junction cells. It is still critical, however, to have an accurately calibrated MBG standard cell with the same design and spectral response as the cells to be tested to set the intensity of such simulators. Any shift in the spectral response between the standard cell and the test cell can result in misleading data.

Making accurate laboratory measurements of concentrator cell performance, whether the cells are single junction or multijunction, requires that additional precautions be taken. It is not possible, in general, to just place the cell and its associated optics in the beam of a typical laboratory simulator. The optics will simply image the source and not provide uniform illumination over even the small area of most concentrator cells. The non-uniform illumination creates a non-uniform current density distribution, and the regions of higher current density may have higher resistive losses than would otherwise be the case, and the cell's performance will suffer. The only approaches, given that the light source's spectral distribution must also be controlled and must remain similar to that of the AM0 spectrum, are either to fly the cells on a balloon or aircraft, mounted with their optical elements in place, or to increase the intensity of the simulator to the same level provided by the concentrator optics. The latter may be feasible for low concentration levels (perhaps 2X) but not for levels much above that. One technique that gives reasonable laboratory results for single junction concentrator cells is to use an uncollimated pulsed light simulator. The beam from this sort of light source, which is typically a xenon arc lamp, is uncollimated and will have a spread of a few degrees. Hence the intensity will vary with the distance from the arc lamp, and although it is not a point source, reasonable small area uniformity can be maintained over a wide range of intensities. The intensity can easily be varied from 1X to more than 100X by simply changing the distance of the sample from the source. The measurement must always include a standard reference cell with a spectral response matched to that of the test cell. The standard cell should be calibrated at one sun AM0. It can then be used to determine the position in the beam where the desired concentration level is reached, provided its own response to an increasing intensity is linear over the range of interest. If necessary, an absolute intensity cavity radiometer can be used to determine the linearity of the standard cell beforehand.

4. Silicon space solar cells

Until the mid 1980s, silicon space solar cells were the only ones available for use in satellite power systems. Although Si cells are still in heavy demand, they are rapidly being replaced by GaAs and MBG cells for a variety of missions, especially those for which radiation damage is a concern or for satellites with severe array size limitations but high power requirements. As mentioned, the first space solar array was launched on Vanguard I in 1958 and contained eight panels of six cells each. The average efficiency was about 10% at 28°C. During the 1960s, cell design and performance remained fairly static. In the early 1970s, improvements in cell design such as the back surface field and photolithograhically patterned top contact grids had moved efficiencies to about 12%, but it was not until the invention of the so-called Violet Cell (Lindmayer and Allison, 1972) that efficiencies moved to 14% AM0. That development was followed by several others such as the multilayer AR coating (Wang et al., 1973), the drift-field cell (Baraona et al., 1976), and the back surface reflector (BSR) cell (Loferski, 1972). All the above resulted in a cell design that achieved the maximum short circuit current to be expected from a Si solar cell under AM0 illumination. During this same period the 50 micrometer thick cell with a back surface field, or BSF (Godlewski et al., 1973) was also developed (Lindmayer and Wrigley, 1978). A back surface field is formed by creating an abrupt, narrow region of high doping density at the back surface of the solar cell. This design achieved the same performance as the standard 300 micrometer thick cell but with improved radiation resistance. The reason for this will be made clear in the discussion of radiation damage that follows later in this chapter.

The second major thrust was to increase the open circuit voltage of Si cells by using lower resistivity materials. The first generation cells used 10 ohm-cm resistivity material in which open circuit voltages of 550 to 590 mV were obtained. It was known that lower resistivity would increase the open circuit voltage, and cells with voltages in the 610-620 mV range became routine. Standard theory, however, predicted that open circuit voltages of 670-680 mV should be achievable. The problem was discovered to have two parts: reduced minority carrier lifetime in the lower resistivity material and bandgap narrowing caused by the heavy doping in the emitter layers of the cells (Van Overstraeten, 1973; Lindholm, Li, and Sah, 1975). The heavy doping was, of course, the source of the lowered resistivity in the first place. Subsequent improvements in materials and junction formation techniques by commercial vendors have resulted in the commercial availability of cells with greater than 14% efficiency and excellent radiation resistance. The state-of-the-art in silicon space solar cells is embodied in the so-called "Space Station cell" which is 8cm by 8cm in size, 300 micrometers thick, has a BSR (but not a BSF), wrap-through front contacts, and an average efficiency of 14.3% (Baraona, 1990).

SOLAR ENERGY CONVERSION

4.1 Advanced silicon solar cells

While space silicon solar cells can be considered a mature technology by virtue of the length of time they have been available as commercial items, enhancements of their performance which lead to improved mission capability or reduction in mission cost are still being made. Large area cells, such as mentioned above for the Space Station, are an example of the latter. One of the major costs of assembling a space solar array is the cost of mounting the cells on the substrate and interconnecting them. Reducing the number of cells provides a significant reduction in that cost, provided the average cell efficiency has not decreased by going to the larger area cell production. One of the issues that had to be solved in going to the larger space station cell was to prevent an increase in the cell's series resistance which would lower the efficiency of the device. The increased series resistance is caused by the increased length of the contact fingers required to collect the current from the larger cell area. The resistance increase could be avoided by making the contact fingers larger, but the increased shadowing that would result also reduces cell efficiency. Typically, only about 3% of the light absorbing surface in a high quality space solar cell is obscured by the contact metallization. The trade-off between electrical and optical losses in large area cells resulted in the contact configuration shown in Figure 3.9. Rather than bring all the contact fingers to a common busbar located along one edge of the cell, as is customary in smaller devices, they are brought to four circular busbars located in the center of each of the four quadrants of the cell surface. The circular busbars are then wrapped through a hole drilled in their centers and are terminated in a larger contact pad located on the back surface of the cell. The edges of the hole and the area under the back contact pad are covered by an insulating oxide prior to deposition of the contact metals in order to provide electrical isolation from the base region of the cell. This approach has been completely successful and has met the stringent requirements for thermal cycle survivability associated with a long term LEO mission (Smith and Scheimann, 1989).

Yet another important enhancement of space silicon cells has been the previously mentioned development of high efficiency, thin single crystal cells. These devices are 50 to 60 micrometers thick and allow a weight savings at the array level. The impact on array weight depends critically on the basic structure of the array but can be substantial, particularly for the so-called flexible array structures such as NASA's Advanced Photovoltaic Solar Array, or APSA (Stella and Kurland, 1989). (We shall discuss the various array technologies in Section 7 of this chapter.) At 62 micrometers, the cell is too thin to allow full light absorption and additional enhancements are required to maintain cell efficiency. These take the form of light trapping by texturizing the top surface and light reflection at the back surface, which is carefully polished to form a BSR before the back contact is deposited. Semiconductor surface passivation under the front contacts to prevent trapping of the light generated carriers by dangling bonds at the metal-semiconductor interface is also important (Godfrey and Green, 1980).

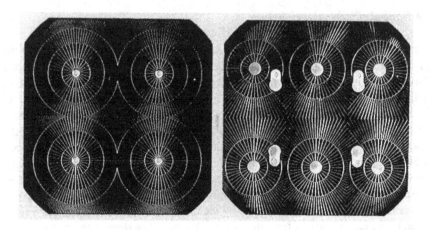

Figure 3.9 International Space Station Solar Cells with Wrap-through Contacts

Finally, it should be mentioned that several of the above features for light trapping and minority carrier confinement, along with surface passivation techniques and microelectronic processing technology, have been combined to produce the highest efficiencies ever achieved in silicon cells (Blakers and Green, 1981; Green *et al.*, 1984). As we shall see in the next subsection, however, many of the enabling features of these high efficiency cell structures are extremely susceptible to radiation damage and for that reason they have not been used in space.

4.2 Radiation damage in silicon solar cells

Resistance to radiation damage is one of the central challenges that must be met in the development of advanced space solar cells. The most important sources of such damage are the trapped electrons and protons that circulate in various regions of the Earth's magnetosphere, as well as the high-energy protons that are often produced by solar flare events. (We shall discuss this and other features of the space environment in Section 7.) With the help of Eqs. (3.7) and (3.8) or (3.9), we are now in a position to investigate the effects of electron and proton radiation on solar cells. Figure 3.10 shows the typical behavior of such cells for the fluence Φ expected in a normal Earth orbiting mission lifetime.

As can be seen, the short circuit current is a slowly varying function of the fluence below about $\Phi = 2 \times 10^{14}$. The logarithm of the short circuit current will vary even more slowly, and it is not an unreasonable assumption to rewrite Eq. (3.7) as

$$I_{sc} = a \ln(L_n) + b \qquad (3.10)$$

SOLAR ENERGY CONVERSION

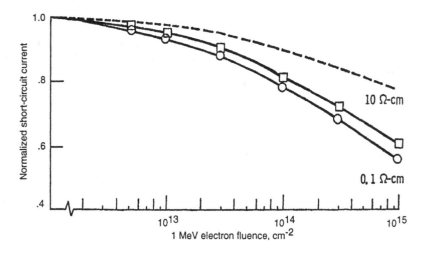

Figure 3.10 Normalized Short Circuit Current for Low (0.1 Ohm-cm) and High (10.0 Ohm-cm) Silicon Solar Cells after One Mev Electron Irradiation

where a and b are constants. An additional assumption is that the diffusion coefficient in Eq. (3.7) is independent of particle fluence, also not unreasonable. If Eq. (3.10) is inserted into Eq. (3.8), we obtain the following expression for the open circuit voltage of a cell under irradiation:

$$V_{oc} = (kT/q)\{[b + a \ln(L_n)]/AqD_n n_{po}/L_n\} \qquad (3.11)$$

Equations (3.10) and (3.11) give the dependence of short circuit current and open circuit voltage on minority carrier diffusion length.

The maximum output power of the cell under illumination is simply

$$P_m = I_m V_m \qquad (3.12)$$

where I_m and V_m are the current and voltage at the maximum power point. The open circuit voltage and short circuit current are related to current and voltage at maximum power by the fill factor, defined simply as the following ratio:

$$FF = I_m V_m / I_{sc} V_{oc} \qquad (3.13)$$

Eq. (3.12) then becomes

$$P_m = FF(I_{sc} V_{oc}) \qquad (3.14)$$

It is well known (Hovel,1975) that cell bombardment by electrons and protons creates an increasing density of recombination centers in the cell material which will reduce the minority carrier lifetime. This inverse dependence is given by

$$1/\tau = 1/\tau_o + c\Phi \qquad (3.15)$$

where Φ is the particle fluence, τ is the minority carrier lifetime, and τ_o is the initial minority carrier lifetime for the particles in question (i.e., electrons or holes). Since the diffusion length is defined by the relationship

$$L = (D\tau)^{1/2} \qquad (3.16)$$

where D is the diffusion coefficient, Eq. (3.15) becomes

$$(1/L)^2 = (1/L_o)^2 + K\Phi \qquad (3.17)$$

where K = c/D is a constant for a particular cell known as the diffusion length damage coefficient. In general, K will be different for different types of particle radiation and will vary as a function of the doping density and dopant type. It will also vary with the energy of the incoming particles. If D is a slowly varying function of accumulated fluence (the usual case for Si cells) and doping concentration (also a good assumption for Si cells), then the damage coefficient K can be considered approximately constant for a given particle energy and type. Under these two conditions it is sufficient to specify only one damage coefficient for electron damage and one for proton damage for the entire range of doping concentrations of interest for Si space solar cells.

Substituting Eq. (3.17) into Eq. (3.10) provides some insight into the dependence of the short circuit current on particle fluence. The result is that I_{sc} decreases with the natural log of the fluence.

$$I_{sc} = b - (a/2)\ln(K\Phi + 1/L_o^2) \qquad (3.18)$$

The expression for the diode saturation current in Eq. (3.6) can be rewritten in the same fashion:

$$I_s = qD_n n_{po} A[K\Phi + (1/L_o)^2]^{1/2} \qquad (3.19)$$

Inserting the results in Eqs. (3.18) and (3.19) into Eq. (3.8) gives the following somewhat complicated expression for the open circuit voltage:

$$V_{oc} = (kT/q) \ln\{[b-(a/2)\ln(K\Phi + (1/L_o)^2)]/qD_n n_{po}[K\Phi + (1/L_o)^2]^{1/2}\} \qquad (3.20)$$

SOLAR ENERGY CONVERSION   93

This is a more complex function of fluence than the expressions for short circuit or saturation current. The equation can be simplified by noting that the numerator varies as the natural logarithm of the fluence and is a much more slowly varying function than the fluence itself which appears in the denominator. Making use of this additional approximation, the open circuit voltage has the following dependence on fluence:

$$V_{oc} = \text{const.} - (kT/2q) \ln\{qD_n n_{po}[K\Phi + (1/L_o)^2]\} \quad (3.21)$$

The expression for the maximum power now becomes

$$P_m = \text{const.} - g \ln[K\Phi + (1/L_o)^2] \quad (3.22)$$

where g may be considered a constant to a reasonable approximation.

The result obtained in Eq. (3.22) is not intended to allow a precise numerical prediction of the degradation of output power with particle fluence, but rather to give a qualitative description of the effects of radiation damage. The output power will degrade approximately as the natural log of the accumulated fluence. Furthermore, the concept of damage equivalence can now be introduced. What is meant by damage equivalence is that the different amounts of damage caused by a given particle type impinging on the cell with different energies can be related to the damage caused by the same particle at a single energy. Rewriting Eq. (3.22), we see that the power remaining after a given amount of fluence is given by

$$P_m = P_{mo} - \text{const} \ln[1 + \Phi/\Phi_c] \quad (3.23)$$

where Φ_c is called the critical fluence and is an arbitrarily chosen constant. It is usually chosen to be the value of electron or proton fluence at a given energy that will degrade the solar cell output to some specified level. It will vary with the energy of the particle and with the doping density of the material. Once the variation of Φ_c with energy is known for a given dopant type and density, however, it becomes possible to determine cell degradation at one energy, and with the use of Φ_c, determine the total fluence required at all other energies to cause the same degree of degradation in the cell. Figure 3.11 shows the variation of the electron critical fluence as a function of electron energy for several base doping densities in n/p silicon solar cells (Tada *et al.*, 1982). The critical fluence has been defined in this case as the fluence for which the output has been reduced to 75% of its initial value. By defining a critical fluence for both electrons and protons for a given cell material it becomes possible to establish an equivalence between the fluences of electrons at one energy and protons at another which cause the same amounts of damage. The energies commonly used in laboratory radiation damage tests are 1 MeV electrons and 10 MeV protons. At these energies the particles travel completely through the cell and damage occurs uniformly, so the approximations made in getting to Eq. (3.22) are reasonably valid. For n/p silicon solar cells the relationship between the critical fluences has been found to be

$$\Phi_c(1\text{MeV e}^-) = 3000\Phi_c(10\text{MeV p}^+) \qquad (3.24)$$

essentially independent of doping density.

A major restriction on the use of the concept of damage equivalence is that the particle energies must be high enough to cause damage uniformly throughout the solar cell. That is generally the case for trapped electrons and for solar flare protons, but it is not the case for low energy trapped protons, which are stopped in the cell. A modified form of the concept may be used which relates the damage at all proton energies to that occurring at the energy for which the relative damage is greatest in actual space envi-

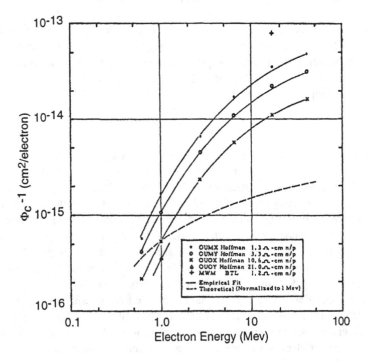

Figure 3.11 Electron Critical Fluences in Various n/p Silicon Space Solar Cells as a Function of Incident Electron Energy

ronments. For silicon cells, the damage caused by ten MeV protons is used for that purpose. Eq. (3.22) or Eq. (3.23) may still be used, but the constants derived from empirically fitting the data are somewhat larger than for one MeV electrons in the same silicon material (i.e., p-type or n-type). For this reason the concept of equivalence between electron and proton radiation damage must be used with care.

The procedure for predicting the output of a silicon solar cell as a function of time for an actual mission is a straightforward but tedious process. It involves making measurements of various cell parameters (for example, I_{sc}, V_{oc}, and P_m) as a function of fluence for several different electron and proton energies, converting the electron data to a one MeV equivalent fluence and the proton data to a ten MeV equivalent fluence, and then consulting the tables in the handbook on radiation damage (Tada *et al.*, 1982) to determine the total fluence, and hence the time in orbit it will take for cell output to fall to a predetermined level. The predetermined level is usually chosen to be the level below which the array output is too low for the mission to continue as originally planned. Such analysis shows that silicon cell output will degrade by about 25% when subjected to a oneMeV electron fluence of about $3 \times 10^{14}/cm^2$, and for a mission in a geosynchronous orbit (GEO) it has typically taken about seven years for that amount of degradation to occur. The occurrence of solar flare events can change the calculated degradation rate dramatically, and there is as yet no reliable prediction technique which takes solar flares into account. Some single flare events can cause power losses on the order of 3-5% in just a few hours, while others will not cause any perceptible degradation. We shall return to the topic of radiation damage when we discuss GaAs and other III-V compound semiconductor solar cells in later sections of this chapter. There we will introduce the concept of displacement damage dose (DDD), and we will see that a simplified procedure for predicting cell degradation will emerge.

5. III-V compound semiconductor solar cells

Although silicon solar cells have been fabricated in the laboratory with one-sun AM0 efficiencies as high as 18%, the fact remains that achieving higher efficiencies requires the use of alternate materials with higher energy bandgaps than the 1.1eV bandgap of silicon. The rapid advances that have been made in epitaxial deposition of a variety of III-V semiconductor materials for microelectronics applications has created the possibility of "bandgap engineering"; a number of binary, ternary, and quaternary compounds can now be grown that allow cell designers to chose the bandgap and lattice constant of the material. Such materials can be tailored to have the optimum efficiency in the AM0 spectrum and can be grown as single junction or multijunction cells. In the sections that follow we will discuss some of the recent solar cells that have been made with these new materials.

5.1 Single junction cells

Efficiency considerations

Figure 3.12 is a plot of the variation of solar cell efficiency at AM0 as a function of the energy bandgap of the cell material (Loferski, 1972). Cell temperature is assumed to be 25°C. The curve is an estimate of the "practical" theoretical maximum efficiency based on an idealized single pn junction device but with realistic cell parameters for series and shunt resistance, contact shadowing, etc. It should be noted that GaAs cell efficiencies exceeding 80% of this theoretical maximum are routinely available from commercial suppliers (Iles and Yeh, 1995). Figure 3.12 shows the bandgap location of some II-VI semiconductors as well. We will consider these materials in the section on thin film solar cells and ultralightweight arrays.

The calculation yielding the curve in Figure 3.12 assumes the cells to be operating under ideal conditions, with no losses from material defects or unpassivated surface states taken into account, and therefore represents a simplified upper limit estimate of efficiency. The presence of a maximum in the curve is easily understood. For very low bandgap cells, most of the incoming solar energy is at energies well above that of the cell bandgap. Hence large numbers of electrons are excited from the valence band into unfilled states well above the conduction band minimum. They quickly interact with the semiconductor lattice, yield their excess energy as heat, and settle into the unfilled states at the bottom of the conduction band from where they can do work in an external

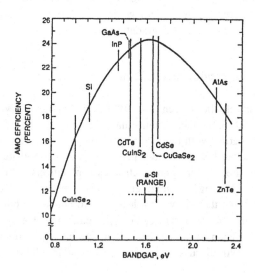

Figure 3.12 Calculated Ideal Photovoltaic Conversion Efficiencies as a Function of Bandgap Energy

circuit. Clearly a cell with a vanishingly small bandgap will lose most of the energy absorbed from the photons to heat and will deliver essentially no net energy. At the other extreme, cells with a bandgap energy higher than that of the incoming photons will simply not absorb them, again resulting in no net energy conversion. The amount of energy lost to heat versus that available to do work in an external circuit may easily exceed 60% of the total incident energy for the single junction cells of interest for space applications. The exact position of the efficiency maximum depends on the shape of the spectral intensity distribution and the cell's operating temperature. For the ideal cell structure (i.e., with no other loss mechanisms) the maximum in the AM0 spectrum occurs for a bandgap of about 1.55 eV.

Laboratory efficiencies of 22.5% AM0, which are very close to the values predicted in Figure 3.12, have been attained in GaAs cells (Ladle-Ristow, 1992). It is important to note that actual cell efficiencies can be higher than the "ideal" values predicted by Figure 3.12 because the calculation does not include any refinements to the basic solar cell structure which can enhance its performance. For example, efficiencies exceeding 20% AM0 have been observed in carefully fabricated silicon cells (Green *et al.*, 1990) which are significantly above the value predicted by the curve in Figure 3.12. Green's Si cells include all the refinements mentioned earlier, such as back surface fields, minority carrier mirrors, textured surfaces, light trapping geometries, etc. Similar results can be expected for GaAs cells with appropriate enhancements, although little development has occurred in this area to date. At least part of the reason for the latter is an issue that arises with the radiation resistance of cells that include efficiency enhancing features. Results obtained in silicon cells to date consistently indicate that the enhancements are strongly affected by radiation damage so that cell output quickly degrades to values typical of "ordinary" cells at higher accumulated fluences (Tada *et al.*, 1982).

The explanation is somewhat complex. The efficiency of a solar cell is strongly dependent on the ratio of the minority carrier diffusion length L to the average distance d the carrier must travel to reach the p-n junction. The ratio L/d must be several times greater than one to achieve high efficiency cell performance (Hovel, 1975). While adding a minority carrier reflector (i.e., back surface field) increases the current collection from the base region of the cell, most of those carriers must actually travel a greater distance to get to the p-n junction than the bulk of the unreflected carriers. Hence, L/d for the reflected carriers starts out lower than for the unreflected carriers. Nonetheless, prior to irradiation the reflected current measurably adds to the cell output and this, according to Eq. (3.9), produces a voltage addition which is proportional to the logarithm of the additional current. As we saw above, however, L is a very strong function of the accumulated fluence incident on the cell and when L/d for the reflected carriers falls below unity, not only does the additional current collection cease, the voltage addition disappears as well. Since the extra current and voltage disappear at a lower fluence than that for which the unenhanced cell output begins to fall (because of the lower L/d), the BSF cell degradation rate is actually higher than it is for the standard cell. However, there are many missions which will not encounter a significant radiation environment (most LEO, low inclination orbits, for example) and in those cases efficiency enhanced cells can be used to good advantage. Whether such enhancements become important for GaAs/Ge cells remains to be seen. As we shall see below, the unenhanced cell performance, with its superior radiation resistance, is already sufficient to make it valuable for large numbers of space missions.

The dependence of cell output on L/d also helps explain the superior radiation resistance of the 62 micrometer Si cell mentioned earlier, at least compared to the standard 300 micrometer thick silicon cell. In this case, the much smaller base region width d means that the minority carrier diffusion length L must fall to much lower values than required in the thicker cells before the cell output is significantly affected. This in turn means that the 62 micrometer cell can absorb a much larger dose of radiation than its thicker counterpart, making it more radiation hard.

Radiation damage

In general, GaAs cells have four major advantages for space application when compared to silicon solar cells. They have a higher efficiency, as explained above, and can operate at a higher temperature by virtue of their higher bandgap. Further, they have the potential to achieve higher cell and array specific power and have demonstrated a higher radiation resistance. All of these things contribute to a potentially lower cost/watt for arrays using GaAs cells instead of Si cells. Figures 3.13 (Woike, 1992) and 3.14 (Flood and Brandhorst, 1987) illustrate this increased radiation reisistance. Figure 3.13 shows the relative radiation damage degradation of silicon and GaAs cells resulting from one MeV electron irradiation under standard laboratory measurement conditions. Figure 3.14 shows similar behavior under proton irradiation for two different temperatures, 22°C and 60°C. The lower temperature is close to the laboratory standard temperature of 25°C, while the higher temperature is closer to actual operating temperatures on orbit. Using the recently introduced concept of displacement damage dose (Summers *et al.*, 1993), we shall show conclusively later in this section that GaAs cells are superior to Si cells. Even though the cost per watt for GaAs is at present about three times that for Si (Curtis, 1998), the total end-of-life (EOL) system cost of a standard, rigid panel, GaAs planar array has been shown to be lower than the EOL cost of the same power level Si array when radiation damage is a factor (Datum and Billets, 1991).

The same is true for EOL array panel mass as well. The power per unit mass of a state-of-the-art rigid panel array using the original "standard" GaAs cell configuration, i.e., a six micrometer active layer on a 200 micrometer thick GaAs substrate, can actually be lower, even at beginning of life, compared to using silicon cells to generate the same power level because the higher efficiency of the GaAs reduces the panel and array

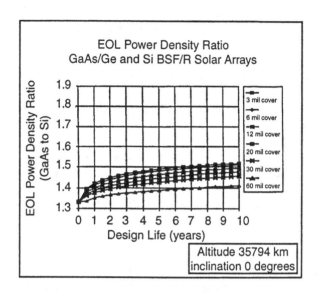

Figure 3.13 End-of-Life Power Ratio ff GaAs to Si Solar Cells for a Geo Mission Environment

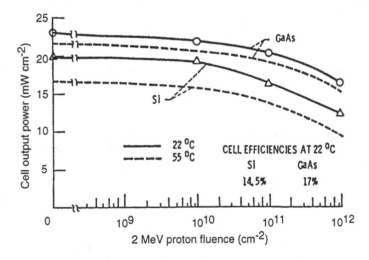

Figure 3.14 Absolute Power Degradation for Si and GaAs Solar Cells after Proton Irradiation
(Note that 55°C is a typical operation temperature on orbit)

size. The potential for lower mass exists at both the array level, as just described, as well as at the cell level. GaAs cells are now commercially available on 75 micrometer thick germanium substrates (Iles and Yeh, 1995) making GaAs all the more advantageous at the array level. Costs are lower than for the "standard" GaAs cell, and the device is substantially lighter and more rugged because of the properties of the Ge substrate.

The last somewhat surprising advantage, given the much higher density of GaAs and Ge compared to Si, derives in part from the fact that nearly 100% optical absorption is achieved in less than five micrometers of GaAs, while at least a 200 micrometer thickness is required in silicon. The higher optical density of GaAs is a key feature of all of the III-V compound materials used for space solar cells: they are all "direct" bandgap materials, while Si and Ge are known as "indirect" bandgap materials. Some III-V materials are also "indirect" bandgap materials, but they are not used as solar cells. The terms "direct" and "indirect" refer to a fundamental electronic property of solids and have to do with the relative positions of the minimum energy state of the conduction band and the maximum energy state of the valence band as a function of the momentum of the charge carriers in each band. (see Sze or Hovel). Basically, in a direct bandgap material the conduction band minimum is located at the same value of inverse electron momentum as the maximum in the valence band, while in an indirect bandgap material the two points are offset by a fixed difference in inverse electron momentum. When a photon is absorbed by a Si solar cell, conservation of momentum at the microscopic level requires that the valence electron that has been excited to the conduction band undergo a momentum transfer with the crystal lattice. This has the effect of lowering the absorption coefficient compared to materials where such a crystal lattice interaction is not required, with the result that a greater thickness of indirect bandgap material is required for complete photon absorption. As a practical manifestation of this effect, high efficiency GaAs cells with less than ten micrometers total thickness have been mounted directly on a standard space solar cell coverglass and achieved over 14% AM0 efficiency (McClelland, Bozler, and Fan, 1980). In a later development, the same sort of GaAs structure was mounted on a 62 micrometer thick Si wafer and demonstrated over 21% AM0 efficiency (Spitzer, 1988).

The expressions derived in Section 3.2 provide the basis for predicting the radiation damage degradation of silicon space solar cells. The procedure for doing so, which has been used for Si cells for decades, is to determine the minority carrier diffusion length damage coefficient for a given material and carrier type, using Eqs. (3.17), (3.19) or (3.20), and (3.21), and the measured dependencies of short circuit density, open circuit voltage, and maximum power as functions of particle fluence at several different energies. Following that, the critical fluence for different energies for the same particle type (electron or proton) are determined using Eq. (3.22). Next, an equivalent critical fluence is established between electrons and protons at two specific energies so that a total equivalent 1MeV electron fluence can be calculated for the orbit and mission duration expected. Finally, the cell degradation is determined. Such a procedure requires an extensive data set, which must include actual space flight data to verify the equiva-

SOLAR ENERGY CONVERSION

lent fluences that have been derived, and requires a significant effort each time a new cell material, such as GaAs or InP, is introduced.

As mentioned earlier, Summers *et al.* (1993) have developed a new approach for predicting radiation damage degradation which is much less empirical and much simpler to implement. They have introduced the concept of displacement damage dose or DDD, which not only enables a correlation to be established between the different amounts of damage caused by the same particle type (either electron or proton) at different energies, but also enables correlating electron and proton damage to each other at all energies as well. This very powerful concept is a direct analog of the concept of ionization dose which is used to describe the energy lost by a high energy particle as it traverses a solid and leaves a trail of ionized or displaced atoms in its wake. The greater share of the energy lost by a charged particle while traversing a solid is, in fact, given over to ionizing the atoms of the solid. The remaining non-ionizing energy loss, or NIEL, is what concerns us here since this is the energy that creates the displacement damage that so strongly affects the minority carrier diffusion length and therefore the damage coefficient. The NIEL values for different energy electrons and protons in various solids are calculated from the following expression (Summers *et al.*, 1993):

$$\text{NIEL} = (N/A) \, \theta_{min}\!\!\int^{180} L[(T(\theta)]T(\theta)[d\sigma\,(\theta)/d\theta\,]d\theta \tag{3.25}$$

where N is Avogadro's number, A is atomic mass, $L[T(\theta)]$ is the Lindhard partition factor (Lindhard *et al.*, 1963) which gives the fraction of transferred energy that is ionizing, $T(\theta)$ is the energy when the incident particle is scattered through the angle θ in the center of mass system, $d\sigma(\theta)/d\theta$ is the differential cross section for elastic scattering of particles into a solid angle increment $d\theta$, and θ_{min} is the scattering angle for which the recoil energy equals the threshold for displacement. Summers *et al.* (1993) have constructed a table of NIEL values for GaAs, InP, and Si for both protons and electrons.

The product of particle fluence and NIEL gives the displacement damage energy loss along the track of the particle in the solid and is the equivalent of absorbed dose. Figure 3.15 shows the experimental curves of the normalized power degradation as a function of fluence for p/n GaAs solar cells for several different proton energies. Applying the definition of displacement damage dose (Summers *et al., 1993)*, the fluence at each data point is multiplied by the NIEL for the energy in question to obtain the absorbed dose D_A at that energy and is given by

$$D_A = \Phi(E) * S(E) \tag{3.26}$$

where $S(E)$ = NIEL and Φ is the particle (in this case proton) fluence. The results are plotted in Figure 3.16 for each of the proton energies displayed in Figure 3.15. As can be seen, the data, when presented this way, fall on a universal curve. This single curve now represents the complete response of p/n GaAs solar cells to protons of all energies. Conversely, and even more significant, it is possible, with the use of displacement damage dose, to measure the degradation of a p/n GaAs solar cell using only a single proton energy.

The degradation of the cell as a function of fluence at any other proton energy can be readily obtained using the following expression:

$$D_A = \Phi_1(E_1) * S(E_1) = \Phi_2(E_2) * S(E_2) \tag{3.27}$$

We have retained the subscripts 1 and 2 for completeness since the relationship in Eq. (3.27) applies to any two particle types as long as they create the same sort of damage in the solar cell. For the rest of the immediate discussion we will drop the subscripts since our attention will be focussed on either proton or electron damage effects and we shall make clear which we are considering. The total absorbed dose from the full spectrum of, in this case, proton energies is determined by integrating the product of the differential absorbed dose dD_A/dE over the spectrum of proton energies. The differential absorbed dose is the product of the proton NIEL as a function of energy, $S(E)$, and the differential fluence spectrum $d\Phi(E)/dE$ with respect to proton energy; hence the total absorbed dose is simply

$$\text{TOTAL } D_A = \int S(E)[d\Phi(E)/dE]dE \tag{3.28}$$

Eq. (3.28) leads naturally to the concept of equivalent fluence, which means that a single energy, such as 10 MeV, can be used to determine the behavior of the cell in the full proton energy spectrum expected to be encountered during the mission. The only

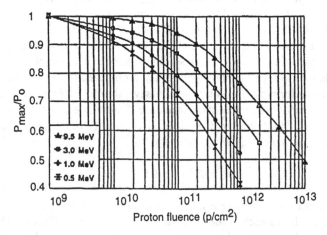

Figure 3.15 Normalized Power Degradation of p/n GaAs Solar Cells for Various Proton Energies

SOLAR ENERGY CONVERSION

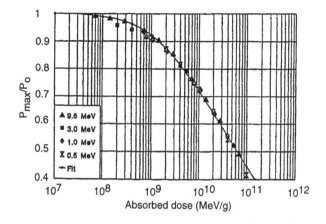

Figure 3.16 Correlated Normalized Power Degradation as a Function of Proton-induced Displacement Damage Dose

caveat is that the proton energies actually encountered must be high enough to reach the active regions of the cell. The 10 MeV equivalent proton fluence can be found by equating the 10 MeV equivalent absorbed dose to the total absorbed dose calculated in Eq. (3.28) and dividing by the NIEL at 10 MeV,

$$\Phi(10) = [1/S(10)] \int S(E)[d\Phi(E)/dE]dE \qquad (3.29)$$

The preceding discussion applies to all high NIEL particles, such as protons and He ions, where the change in photovoltaic parameter has a linear dependence on absorbed dose. According to Summers et al. (1993), a linear dependence is also found for low NIEL particles such as electrons in such materials as n-type Si, GaAs, InP, and perhaps other semiconductors. In all these cases, the preceding development applies completely. In some p-type semiconductors, however, the changes in photovoltaic parameters caused by electron irradiations are observed by Summers and co-workers to have a quadratic dependence on NIEL. This latter case can be handled in much the same way and gives rise quite naturally to a quantity called the effective 1MeV electron equivalent dose, given by

$$D_e(1.0) = [S(E) * \Phi(E)] * S(E)/S(1.0) \qquad (3.30)$$

where $S(1.0)$ is the electron NIEL for one MeV. It should be noted that any normalizing electron energy could be used, but because one MeV electrons are the traditional standard for determining cell performance as a function of fluence it is convenient to use that energy in this case as well. Putting a quadratic dependence of absorbed dose on NIEL in Eq. (3.27) yields

$$\Phi(E_1) * [S(E_1)]^2 = \Phi(E_2) * [S(E_2)]^2 \qquad (3.31)$$

for the absorbed dose resulting from bombardment by electrons at energies E_1 and E_2. From this expression,

$$\Phi(E_1) * S(E_1) = \Phi(E_2) * S(E_2)[S(E_2)/S(E_1)] \qquad (3.32)$$

Eq. (3.32) can be written as

$$D_A(E_1) = D_A(E_2) * [S(E_2)/S(E_1)] \qquad (3.33)$$

and from this expression an effective one MeV electron equivalent dose can be defined as

$$D_e(1.0) = D_A(E_2) * [S(E_2)/S(1.0)] \qquad (3.34)$$

Figure 3.17 (Anspaugh, 1991) shows the dependence of the diffusion length damage coefficients for various electron energies as a function of fluence. The diffusion lengths were derived from experimental measurements of current, voltage, and power degradation versus fluence using expressions for GaAs similar to the ones shown earlier for those quantities in Si. Applying the concept discussed above, when the data are replotted against the one MeV electron equivalent fluence displacement damage dose as in Figure 3.18, all the points fall on a single line. The proton damage coefficients have also been included, and fall on the same line. This line represents the universal response of the minority carrier diffusion length in GaAs p-type material to electrons and protons of all energies. Similar results can be obtained for the current, voltage, and output power for all GaAs solar cells where photon collection is essentially all in the p-region of the cell.

Walters and colleagues have also shown that the preceding discussion pertains to InP solar cells in the same manner (Walters et al., 1995). There is a linear dependence on electron NIEL in n-type material and a quadratic dependence in p-type material. Proton damage is exactly analogous to that in GaAs. This understanding opens the way to much faster determination of electron and proton damage in new III-V compound space solar cells than has been the case prior to now. Once the NIEL values are known as a function of energy for the material in question, cell performance as a function of fluence for the entire electron and proton energy spectrum expected during the mission can be determined by irradiating the cells at one energy and using the methodology described here. Figure 3.19 illustrates the results of such an analysis applied to the

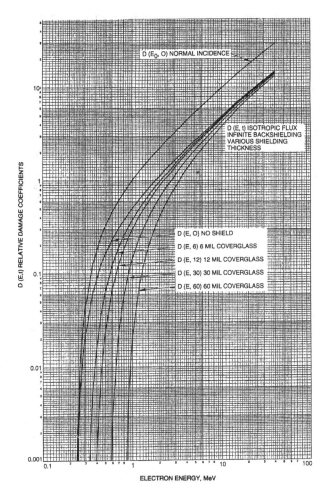

Figure 3.17 Diffusion Length Damage Coefficients for Si Solar Cells as a Function of Electron Energy

degradation of maximum power in InP solar cells fabricated with diffused junctions, while Figure 3.20 shows conclusively the relative superiority of InP cells compared to GaAs cells as a function of displacement damage dose. We saw earlier the relative superiority of GaAs over silicon cells using point comparisons in energy. The same relationship can be shown for all energies using the DDD approach. This result allows the radiation damage degradation of all three cell types to be compared with complete confidence since the cells have been fully correlated by the universal concept of displacement damage (or absorbed) dose.

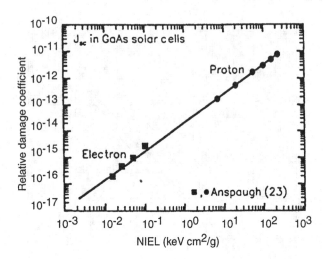

Figure 3.18 Correlated Diffusion Length Damage Coefficients for GaAs Solar Cells

The preceding discussion highlights the principle reason for interest in InP solar cells for space application: superior radiation damage resistance. Although the theoretical solar cell conversion efficiency of InP is only slightly less than that of GaAs (see Figure 3.12), full commercial development of InP for solar cell applications lags well behind that of GaAs and, at the present time, even behind that of the multiple bandgap

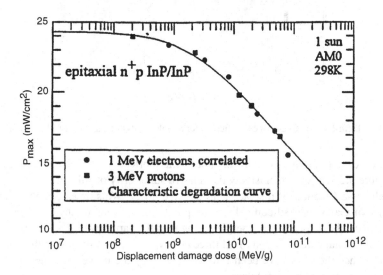

Figure 3.19 Correlated Electron and Proton-induced Power Degradation for InP Solar Cells

SOLAR ENERGY CONVERSION

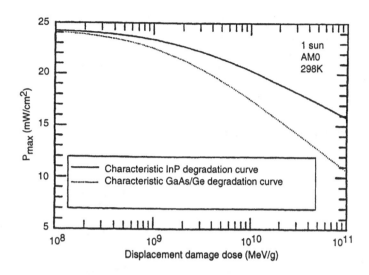

Figure 3.20 Power Degradation of InP and GaAs Solar Cells versus Displacement Damage Dose

cells. The primary reason for this lies with the cost of InP single crystal wafers which have not seen the same price decline as GaAs wafers; the latter is a result of the rapidly growing microelectronic industry demand for that material which has fostered lower cost, higher volume wafer production. For this reason, considerable attention has been given to the problem of growing InP cells on alternate, lower cost substrates, as is done with GaAs on Ge. The problem is that no lattice matched substrates are available for InP; consequently, the heteroepitaxial cells that have been produced with either organometallic chemical vapor phase epitaxy or molecular beam epitaxy have much lower efficiencies than desired, although recent results have been encouraging. InP/Si cells have been fabricated in the laboratory with efficiencies in excess of 13% (Summers, 1995) making them comparable to standard space silicon solar cells, but with much higher radiation resistance than the latter, consistent with the results shown previously in Figure 3.20. It should be noted that the InP cells of Figure 3.20 are diffused-junction n-on-p cells. Walters and co-workers (Walters *et al.*, 1995) have shown that OMVPE-grown, p-on-n, InP/Si cells show the same radiation resistance superiority over GaAs and Si when electron and proton damage are fully correlated with displacement damage dose, just as is the case for homoepitaxial InP cells.

Figure 3.21 presents the results of a comparative analysis of solar cell degradation after a fixed number of years in orbit as a function of orbit altitude, and shows that, even with a lower starting efficiency for InP/Si cells, their EOL performance is definitely far superior for missions where severe exposure to radiation damage is the case. The correctness of this comparison is assured by the results of the DDD analysis. The importance of this result is shown in Figure 3.22 which contains a plot of launch costs as a function of operating orbit altitude for a constellation of communication satellites deployed for continuous global coverage. Total launch costs fall dramatically as a function of orbit altitude and reach a broad minimum which extends through the Van Allen radiation belts and beyond. Clearly, a solar cell with significantly superior radiation resistance is required to take advantage of the lowered costs associated with the reduced size of the constellation at higher altitudes. Whether demand for such a device will arise is yet to be seen, but InP/Si is at least one potential candidate.

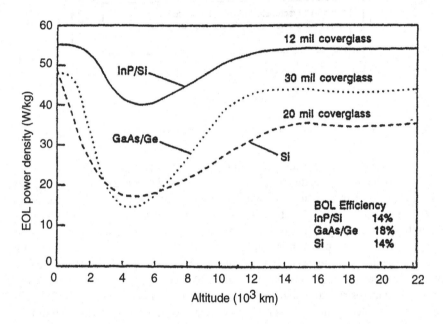

Figure 3.21 EOL Power Densities for InP/Si And GaAs/Ge Solar Cells as a Function of Orbit Altitude

5.2 Multiple junction cells

Although excellent single junction cell efficiencies have been achieved, research continues into making further gains in space solar array performance. Two current approaches are of particular interest: (1) use of concentrated sunlight and (2) multiple bandgap cells. Figure 3.23 shows the effect of concentration on cell efficiency as a function of bandgap at a nominal operating temperature of 80°C (Fan and Palm, 1983). It is readily apparent that a significant efficiency gain can be realized at modest concen-

SOLAR ENERGY CONVERSION

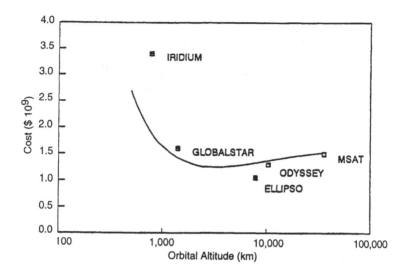

Figure 3.22 Calculated Launch Costs for Various Communication Satellite Constellations as a Function of Altitude, Assuming Complete Global Coverage

tration levels, i.e., at 100X or below. Since cell temperatures must be maintained using radiative cooling, going to higher concentration levels introduces a potentially unacceptable degree of complexity in array design.

Figure 3.24 shows a schematic diagram of a linear Fresnel concentrator/solar cell subassembly of the type to be used on NASA's advanced concentrator array for the Deep Space-1 (DS-1) mission. This mission is the first in a series of specialized missions collectively called the New Millennium Program. The concentrator array consists of two wings, each of which will deliver 1.3 kW at BOL. A concentrator array was chosen because it has the inherent capability to shield the solar cells from the space environment more effectively than a similarly sized planar array. The reason for this is the same reason it can use advanced, high efficiency MBG solar cells cost effectively: the required area of expensive semiconductor material is reduced by the inverse of the concentration ratio. This means the shielded area is also reduced by the same amount so the weight impact of the extra shielding is much less than it would be for the same performance planar array.

Incorporating an advanced solar cell with superior radiation resistance in a concentrator array will enable at least limited use of orbits near, or even in, the Van Allen radiation belts. As shown above, the reason for considering such altitudes, which have until now been a technical impossibility, is the impact it has on total mission cost: the use of fewer satellites with longer lifetimes can substantially reduce total launch and life cycle cost. Referring to Figure 3.22, the minimum in the plot of cost versus altitude arises from the trade-off between global coverage, total satellite costs (i.e., the number

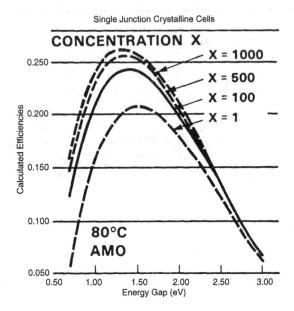

Figure 3.23 Calculated Concentrator Solar Cell Efficiency versus Energy Bandgap, at an Operating Temperature of 80°C

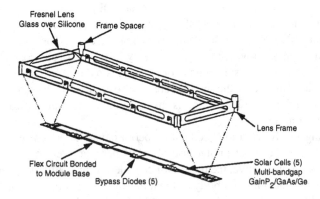

Figure 3.24 Schematic Diagram of a Linear Fresnel Lens, Lens Holder, and Cell Panel Assembly (Scheduled for NASA DS-1, 1998)

SOLAR ENERGY CONVERSION

required to achieve full global coverage), launch costs, and operational costs. The cost advantage of going to the intermediate orbits is evident, provided a solar array is available with superior radiation resistance. The cost-benefit trade-off must be made between cell capabilities at the end of the mission. In almost all cases where significant radiation environments are to be encountered, which includes almost everything except low inclination LEO, cells with superior radiation resistance are often more cost effective even if their initial cost is significantly higher than cells with poorer radiation resistance. In the case of planar GaAs versus planar Si cells, for example, GaAs is the cheaper cell to use even though it is presently about three times more expensive. The same is expected to hold true for different concentrator cells, although the impact may be somewhat less from one cell type to another because of the additional shielding that can be added to a concentrator array at a smaller mass penalty than for a planar array.

Still further gains in array efficiency can be achieved by using multiple bandgap (MBG) solar cells. Figure 3.25 shows the integrated response of a high bandgap space solar cell, such as $GaInP_2$, in the AM0 spectral intensity distribution, along with that of a GaAs cell which has been placed underneath it. Putting the $GaInP_2$ cell on top of the GaAs cell will clearly cut off some of the photons that would otherwise reach the GaAs cell, but the combined output of the two devices will still be higher than it would be from either cell separately. The MBG solar cell has been studied extensively to determine the optimum energy bands to use for a two, three, or four (or an even higher number) junction device, and two junction devices are now commercial realities.

There are two basic types: monolithically grown MBG cells and mechanically stacked MBG cells. The first type is produced as a single crystal structure and requires all layers to be closely lattice-matched throughout the cell to minimize the introduction of crystalline defects which will degrade cell performance. The second type allows each subcell to be produced separately, but then requires careful joining via some external technique such as an optically transparent adhesive. The adhesive may or may not also be conductive, depending on whether the final structure is to be a two terminal or four terminal device. If the current is to pass through the entire structure, as it must in a two terminal cell, the adhesive must be conductive. In a four terminal device each subcell supplies its output in parallel, and the adhesive must be an electrical insulator. Each technique has its advantages and disadvantages. Figure 3.26 shows the isoefficiency contours for a two junction, monolithic cell stack operating at 80°C and 100X concentration. From Figure 3.26 it is apparent that the maximum efficiency is obtained for bandgap pairs approximately centered around 1.7eV and 1.1eV. Figure 3.27 provides the same information as Figure 3.26, but for a mechanically stacked, two junction, four terminal cell configuration. The major difference from the two terminal configuration is the wider range of acceptable bandgaps from which maximum efficiency can be obtained. The central bandgaps are essentially the same as in the monolithically grown, two terminal device, however. The same overall conclusions hold for one sun operation as well.

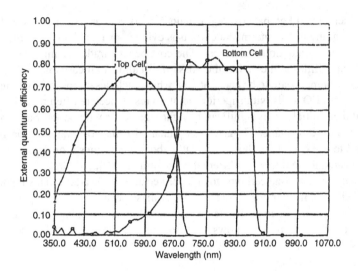

Figure 3.25 Integrated Response of a High Bandgap Solar Cell such as GaInP$_2$ Placed Over a Lower Bandgap Solar Cell such as GaAs

Figure 3.28 is a commonly used plot of the variation of energy bandgap as a function of lattice constant for various compound semiconductors. The positions of Ge and Si are shown as well. The cross-hatched horizontal lines represent the two bandgap ranges for which the efficiency of a multiple bandgap cell is maximized. It can be quickly seen that there is no readily available lattice matched substrate for growing the optimum bandgap pair in a monolithic structure for any of the III-V binary or ternary compounds. The cell now under development is a non-optimum GaInP$_2$/GaAs device grown on a Ge substrate. The two junction device is grown in such a manner that photon collection in the Ge substrate does not contribute to current generation, although attempts are underway to control the formation of a pn junction in the Ge to obtain voltage addition while at the same time meeting the condition for current matching in the three junction structure (Chiang *et al.*, 1995). Doing so will provide an additional 2 or 3 efficiency points over that for the two junction device. Figure 3.29 shows a cross section of the MBG cell developed by Olsen and Kurtz (1994) which forms the basis for the present-day commercially available cells. One sun, AM0 efficiencies in excess of 27% have been measured in laboratory devices (Kurtz, 1994); from the plot in Figure 3.26 it is seen that the calculated efficiency of this bandgap pair is well off the central optimum and can be estimated to be about 28%. In this case, the relative ease with which the lattice matched structure can be grown by OMVPE, which is the standard production technique employed by all current space cell manufacturers, dictates the choice of bandgaps. Production of the GaInP/GaAs/Ge cell is a simple extension of the

SOLAR ENERGY CONVERSION

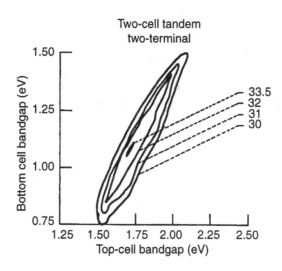

Figure 3.26 Calculated Efficiency Contours of a Monolithically Stacked Two-junction Multibandgap Solar Cell

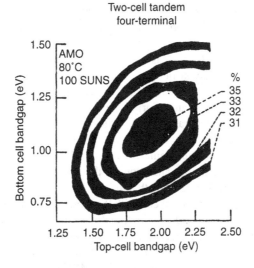

Figure 3.27 Calculated Efficiency Contours of a Two-junction, Mechanically Stacked Multibandgap Solar Cell

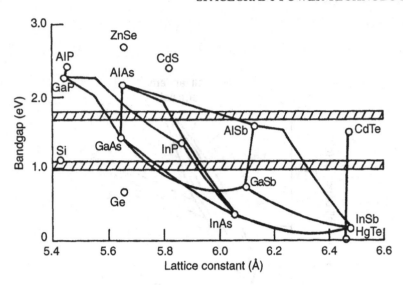

Figure 3.28 Energy Bandgap as a Function of Lattice Constant for Various Semiconductors (The Shaded Bands Represent the Optimum Bandgap Pairs for a Two-junction, Monolithic Device)

production of the GaAs/Ge cell, which is an advantage in controlling the cost of the device. Commercially-available, two-junction MBG cells have AM0 efficiencies in the 21.5% to 22.5% range, and production efficiencies above 24.5% are presently the focus of a joint US Air Force/NASA manufacturing technology (MANTECH) program. Such devices should be commercially available by the end of 1999 (Keener, 1996).

Radiation damage in MBG cells is a much more complex situation than it is for single junction cells and is very difficult to measure accurately, much less predict with any confidence. The difficulty in measuring MBG cell performance is that the spectral response of each subcell will change during the course of the irradiation. Unless the light source is a completely accurate replica of the AM0 spectral intensity distribution, the response of a given subcell might be inaccurately determined by a substantial amount depending on the variation of the simulated spectrum from the actual spectrum. The situation for each subcell is strictly analogous to the single junction case discussed in Section 2.2 of this chapter. The result is that the full MBG cell radiation damage resistance is only as good as that of the subcell with the worst resistance since that cell will fall from its current matched condition first, and the output of the entire device will suffer accordingly. If the radiation resistance of each subcell is to be determined separately, it is still critical to use a light source which accurately reproduces the actual spectrum the various subcells will see as part of the MBG structure. This is most often accomplished by using a filter, or even a finished cell, made of the same material used in

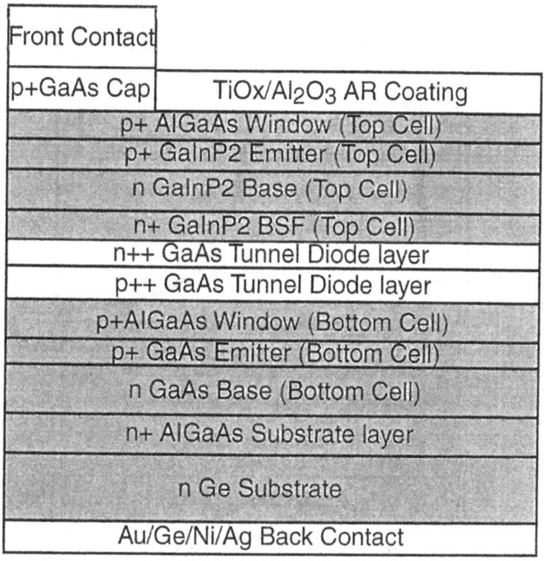

SOLAR ENERGY CONVERSION

Front Contact
p+GaAs Cap   TiOx/Al$_2$O$_3$ AR Coating
p+ AlGaAs Window (Top Cell)
p+ GaInP2 Emitter (Top Cell)
n GaInP2 Base (Top Cell)
n+ GaInP2 BSF (Top Cell)
n++ GaAs Tunnel Diode layer
p++ GaAs Tunnel Diode layer
p+AlGaAs Window (Bottom Cell)
p+ GaAs Emitter (Bottom Cell)
n GaAs Base (Bottom Cell)
n+ AlGaAs Substrate layer
n Ge Substrate
Au/Ge/Ni/Ag Back Contact

Figure 3.29 Schematic Diagram of the GaInP$_2$/GaAs on Ge Substrate MBG Solar Cell

the subcells above the one in question. The so-called filter must also be irradiated to allow for any change in photon absorption that might occur in that material as well. Because of all the variables that must be controlled, simple plots of a cell parameter such as the maximum output power against total fluence are not possible when using the artificial spectrum of a simulator. What remains is to measure the behavior of individual subcells as accurately as possible under the conditions they are expected to encounter and then to calculate the total cell response. The only other option is to fly the cells at high altitude, or in space. Both of the latter are impractical since a flight measurement is required after each irradiation for a single energy, and the time needed becomes measured in years, not hours or at worst days, as is normally the case. Figure 3.30 shows the sort of data that has been gathered on the Olsen/Kurtz MBG cell. Optimizing the cell for radiation resistance becomes just as complex a matter as measuring the degradation. It becomes necessary to adjust the thicknesses and doping densities of the various subcell layers, for example, in order to maintain current matching to as high a fluence (i.e., for as long a time in orbit) as possible.

Using the concept of displacement damage dose would, in principle, reduce the complexity of the problem. The difficulty is the lack of NIEL data for the various subcell materials. The availability of such data would make it possible to actively and accurately compare the radiation resistance of a host of MBG materials and the optimum thicknesses and doping densities for a given mission application. As a further practical consideration, Messenger and co-workers, (Messenger *et al.*, 1994) have shown

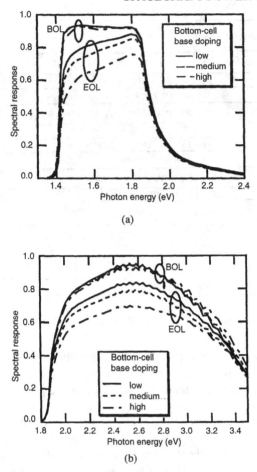

Figure 3.30 a. Spectral Response of the Bottom Cell of Fig. 3.29 after Irradiation
b. Spectral Response of the Top Cell of Fig. 3.29 after Irradiation

that Co^{60} gamma ray irradiations of solar cells can be correlated using displacement damage dose in exactly the same manner as electron and proton irradiations. Figure 3.31 shows the correlation that has been established for n/p InP for both γ rays and 1MeV electrons. It now becomes possible to contemplate carrying the radiation source along with the cell on a balloon, aircraft, or shuttle flight, and making the required measurements while the MBG cell is irradiated in the proper spectrum. Until such time as that is actually done, however, data of the sort shown in Figure 3.30 will be required. No such data are yet available on MBG subcells, but the current results show the promise of the technique.

SOLAR ENERGY CONVERSION

6. Thin film solar cells

Using Figure 3.12, it is possible to conjecture on the efficiency potential of various thin film solar cell materials. The comparison is not altogether valid because the calculation used for the curve shown there is based on so-called "ideal" diode behavior which means that the material is essentially free of any crystalline defects that would enhance the recombination current of the device. That is certainly not the case for the thin film materials which are all polycrystalline in nature; nonetheless, the Figure does illustrate the efficiency potential of the various thin film solar cells that have been investigated and/or developed up to the present, provided a way around the deleterious effects of the polycrystalline nature of the films can be found. Some progress in that regard has been

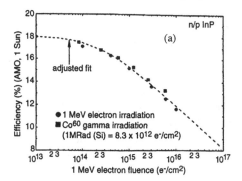

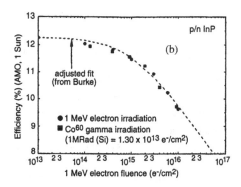

Figure 3.31 Correlated Electron and Gamma-ray Degradation of (a) n/p InP and (b) p/n InP Solar Cells

made in developing thin film cells for terrestrial applications. Various groups have reported very good terrestrial efficiencies for some of the thin film solar cells. Table 3.4 lists the various thin film solar cells that have been considered for space application, their bandgap, the efficiency predicted for that bandgap by Figure 3.12, and the most

Table 3.4 Thin Film Solar Cell Types and Efficiencies

Cell Type	Bandgap	Predicted AM0 Efficiency	Reported Efficiency
$CuInSe_2$	1.0	17%	10%(AM0)
$CuIn_xGa_{1-x}Se_2$	1.0 - 1.2eV	17% - 20%	17% (AM1.5)
α-Si	1.72eV	25%	10% (AM0)
CdTe	1.44	24%	16% (AM1.5)

recent experimental result for either AM0 or AM1.5 global. AM1.5 global is a defined terrestrial standard spectrum and intensity.

There are two primary reasons for interest in thin film solar cells for space application: the potential for low cost and the apparent radiation resistance of the various thin film cell structures. The argument for the first point has been made countless times in the terrestrial photovoltaic development program. It arises from the reduced amount of material required by the thin film devices compared to single crystal solar cells and the fact that the polycrystalline materials can be deposited directly on foreign, potentially low cost substrates such as glass, metal foils, or thin polyimides such as Kapton®, a space qualified material used in a variety of ways on hundreds of spacecraft. Figure 3.32 shows a roll-to-roll manufacturing system for α-Si single junction cells on Kapton®. Although the automated production of such cells is a very attractive way to reduce costs, the efficiencies of the cells made with this process are about 5% at AM0 (Jeffrey *et al.*, 1990). Such solar cells and blankets have not yet been space qualified and it remains to be seen if thin film cells or blankets can truly be low cost and still survive in the space environment. The metal contact stability is a major concern since space solar array in LEO will undergo over 6000 thermal cycles per year, compared to only 365 for a terrestrial array.

The second reason for interest in the thin film solar cells is the possibility that such devices may be inherently radiation hard. While it is true that the already heavily defect-laden active regions of the currently available cell structures can scarcely be degraded further by space radiation bombardment, it is not clear what the case will be as higher efficiency thin film devices evolve; it is of particular concern if such devices rely on cell design enhancements of the sort already discussed for the single crystal cells. That may already be the case for α-Si. Efficiencies approaching values of interest for space applications can be achieved only by using a three-junction MBG cell design

SOLAR ENERGY CONVERSION

Figure 3.32 Roll-to-roll Manufacturing of α-Si Flexible Solar Cells on Kapton™

(Gupta, 1997) whose radiation damage resistance is unknown.

7. Space solar cell arrays

Space solar cell array designs and deployed configurations have undergone a steady evolution from the first array launched on Vanguard I. That array consisted of 6 small panels that were distributed over the outside of the nearly spherical spacecraft body and provided about 1 watt of power for more than six years (Rauschenbach, 1980). Typical early satellites continued to be approximately spherical with small panels distributed evenly over their external surface to assure continuous power generation as the spacecraft slowly spun about its axis. The growth in power demands soon required the entire spacecraft body to be covered with solar array panels, and finally, in order to provide even more power, the satellites were outfitted with small paddles mounted on hinged arms that swung out from the body of the spacecraft. Explorer 6 was the first satellite to use a paddle array system and carried four 51 cm² hinged paddles aloft in August, 1959 (Rauschenbach, 1980). The paddles were oriented to provide continuous power as the spacecraft rotated. Folded and hinged rigid panel arrays quickly became the standard configuration for all spacecraft in the decades that followed. Space solar cell arrays have grown in both size and complexity from the first exploratory launches, and output powers now range from a few kilowatts to tens of kilowatts. The International Space Station (ISS) array, when its assembly is complete, will be the largest such system yet deployed with a planned 150 kilowatts at EOL. The large size of this array, which will consist of separate wings producing 37.5 kilowatts each at EOL, requires use of a stretched membrane panel structure to minimize its launch volume. More will be said about this technology in Section 7.3. Figure 3.33 illustrates the various array options that have historically been available to satellite designers; while each design option has been used at one time or another in the past, the one most commonly used today is the non-concentrator, rigid, one, two, or three axis stabilized tracking array.

7.1 Space solar array evolution

The early spherical spacecraft mentioned above soon gave way to spin-stabilized, cylindrical spacecraft, with the solar cells mounted on the cylindrical surface. The availability of larger and larger launch vehicles through the early years of the space program allowed significant growth in the diameter and length of such satellites, and this configuration was used well into the 1980s. Figure 3.34 shows an artist's rendition of the evolution of a typical telecommunications satellite series during this time.

The rapid expansion in demand for telecommunications services, coupled with the high cost of getting ever larger payloads to geosynchronous (GEO) orbit, eventually brought an end to the widespread use of this sort of satellite and solar array. At the same time, the spin-stabilized satellites and body-mounted arrays were approaching their size limits, the paddle configuration was evolving into a system of large, hinged panels mounted on a rotatable boom which could be extended from the body of the spacecraft. This arrangement not only allowed the full array to be oriented toward the sun for even higher average output, but also allowed the array size and output power to grow within the constraints of the available launch vehicle payload volume and weight capability. Since body-mounted arrays are used on a very limited basis on modern spacecraft, particularly commercial and military Earth orbiting spacecraft, the balance of our discussion will be on the various types of flat panel, deployable array technologies available to current mission planners.

7.2 Rigid panel planar solar arrays

The most commonly used rigid panel construction has been the so-called honeycomb panel which usually consists of two thin-aluminum face sheets glued to a honeycomb-like core. The core consists of a hexagonal cell structure, the cell walls of which are made from thin (approximately 0.02 mm thick) aluminum ribbon. The total thickness of the panel structure can vary from about 6 mm to 25 mm, depending on the mechanical load requirements established for the array. An individual panel is very stiff and

SOLAR ENERGY CONVERSION   121

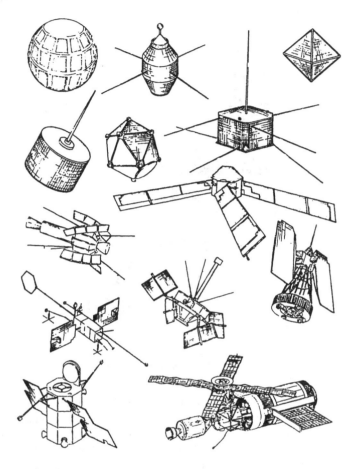

Figure 3.33 Evolution of Solar Array Designs for U.S. Spacecraft

strong, but relatively lightweight. In recent years honeycomb panels have been made from materials other than aluminum, most notably from graphite/epoxy sheets and ribbons. Hybrid panels, which have an aluminum core covered by epoxy/glass facesheets, have also been made and flown. The automatic deployment of rigid panel arrays is accomplished by using springs to actuate motion around a series of hinges between the panels. Once deployed, the panels are locked in place by a set of latches to become a stiffened solar array, and after being locked, such an array cannot be refolded or restowed.

The BOL power density provided by a planar array will depend on the weight and efficiency of the solar cell. In general, BOL power densities range from about 35 to 65 W/kg for silicon cells and from about 45 to 75 W/kg for GaAs/Ge cells. The total array BOL will be about one-third lower than these numbers, mostly because of the additional

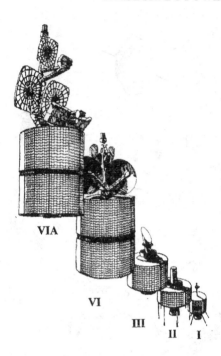

Figure 3.34 Evolution of INTELSAT Solar Arrays and Satellite Configurations

mass of the deployment system. The weight and cost of rigid panel arrays can be reduced by using thinner, lighter weight face sheets and cells and by using large area cells to reduce the labor required to assemble the panels. The initial cells on Vanguard, for example, were 1 cm x 2 cm, and for many years afterward the standard Si cell was still only 2 cm x 4 cm, or sometimes 2 cm x 6 cm in size. Silicon cells now range up to 8cm x 8cm in size, which reduces both the number of cells to be mounted on the panels as well as the number of interconnects that have to be made.

7.3 Flexible, flat panel arrays

The International Space Station (ISS) array represents a major departure from the standard, spring-hinged, fold-out, rigid panel array. The change was necessitated by the need for significantly larger amounts of power than had been the case in previous missions of any sort, whether civil, military, or commercial in nature. There are several important reasons for such a change. The first is the sheer complexity of deploying such large structures with the usual hinge/latching mechanisms used in smaller arrays; the second is the problem of controlling and positioning such large structures once on orbit;

SOLAR ENERGY CONVERSION

and the third is the limited stowage volume available on all launch vehicles other than the space shuttle. If the satellite is to go into other than shuttle-tended low Earth orbits, an expendable launch vehicle must be used, placing severe limitations on total payload volume and weight. The requirement for low mass and stowage volume is especially critical for GEO missions with high power requirements (above approximately 10-15 kilowatts). As a result, NASA initiated the Advanced Photovoltaic Solar Array (APSA) program in the late-1980's (Kurland and Stella, 1989) at about the same time the military initiated the so-called MILSTAR program (Curtis, 1998). Both array designs are based on the same fundamental concept of using polyimide panels stretched between lightweight hinges with the whole structure deployed by an extendible mast. Although there are significant differences in the details of panel construction, the potential end results are the same: array BOL specific powers exceeding 100 W/kg, including the deployment mechanism. The ISS array has the MILSTAR array as its heritage.

Figure 3.35 shows a schematic of the APSA which was originally designed to provide 5.3 kW BOL at GEO. It is important to note that the high specific powers of large, flexible blanket solar arrays do not scale well to lower power levels. The principle reason is that the containment box and deployment mechanism does not scale linearly with array size. As array power levels approach the one kilowatt level from higher values, the total array specific power approaches that of standard rigid panel arrays because the containment box, mast stowage canister, and deployment motors reach their minimum sizes. Other factors can also reduce the high specific power of such array systems, depending on mission reliability requirements, spacecraft orientation and maneuverability capabilities, and any safety requirements if the array is to be used on a manned mission, as on the space station. The result is that the ISS array, for example, achieves a specific power of about 40 W/kg, in large part because the release mechanism on the containment box lid had to be redesigned to allow the astronauts to open it manually if the automatic mechanism failed, and because the mast had to be able to withstand an accidentally hard docking maneuver between the Shuttle and the Station.

The original design of the APSA was for 130 W/kg at 5.3 kW BOL in GEO. The cells baselined for that design are thin (62 micrometer) silicon cells with 50 micrometer covers. The average cell efficiency is 14% AM0. Under those conditions, the structure (mast, release motor, containment box) accounts for about 51% of the array mass. The panel assembly, consisting of the polyimide substrate, cell assembly (cell, coverglass, and interconnect tabs), hinges, and wiring harness make up the rest. This is contrasted with a standard rigid panel array where the panel assembly accounts for 75%-80% of the total mass and the stowage and deployment structure make up the rest. The major difference is the inherently greater mass of the rigid honeycomb panels compared to the polyimide panels of the APSA.

Figure 3.36 shows the impact of various solar cell technologies on APSA specific power at the 5.3 kW design level (Stella and Flood, 1991). Design studies have also shown that extending the length of the array to provide higher output powers will in-

Figure 3.35 Ground Deployment Test of the Prototype of the Advanced Space Solar Array (APSA)

crease the total specific power for a given cell technology and efficiency (Kurland, 1989), primarily because the heavy structural elements such as the containment box, deployment motors, mast housing, etc., do not change appreciably with increasing deployed length. However, such arrays cannot be increased in size indefinitely by simply increasing the length of the array, as might be implied from the above discussion. A very important consideration for the spacecraft designer is overall momentum management of the entire satellite which includes the behavior of the solar array when the attitude control thrusters are activated. The preference is for the satellite to move as a single, rigid body, but as with all mechanical systems there are natural frequencies which must be taken into account. A large, flexible blanket solar array will have relatively low natural frequencies of vibration and torsional oscillation compared to rigid panel arrays. The result is that the attitude control system will be severely compromised when the array aspect ratio (the ratio of length to width) exceeds certain limits. While reducing the aspect ratio by increasing the width of the array can raise the natural frequencies involved, it will also increase the dimensions of the containment assembly, which can have a negative impact on stowage volume and total array mass.

Other flexible array designs have been considered, some of which are presently in use. The Hubble Space Telescope (HST), for example, uses a polyimide blanket in a roll-up stowed configuration. The array is deployed by a bistorable, tubular, extendible member (Bi-STEM) deployment system (Gerlach, 1991). The design is a derivative of an earlier array developed for the U.S. Air Force. The FRUSA (Flexible, Roll Up Solar Array), as the earlier array was called, was flight tested in 1971 for its extension and retraction capabilities. It performed satisfactorily, but was not subjected to any long term testing in the deployed state. The BOL power for the HST array was 5 kW. The first array launched on the Hubble spacecraft exhibited thermally induced vibrations upon coming out of the eclipse each orbit. This problem required an expensive special repair mission to correct and highlights the importance of careful mechanical design to minimize such problems. While flexible arrays are an important mechanical design option for meeting high power system requirements, as the experience with the HST array shows they also present some new concerns with regard to space environmental interactions. We shall return to this topic later in this chapter.

The EOS-AM1 satellite under development by Goddard Space Flight Center incorporates an APSA-type array. Figure 3.37 shows a sketch of the EOS-AM1 spacecraft. The single wing configuration was required to prevent loss of view factor for the instruments at certain times during the orbit, and placed several constraints on the mass of the structure to make it stiff enough to be maneuvered when necessary. The result is a total

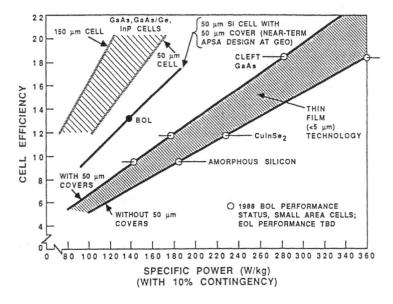

Figure 3.36 APSA Specific Powers Calculated for Various Solar Cell Technologies

array specific power of about 40 W/kg which is no higher than a conventional rigid planar array. The added mass also allows the array to be retracted; however, this was not possible with rigid arrays and was a feature required by the program to allow for on-orbit array replacement so that the same instruments could be used for Earth observations over a 15 year period.

7.4 Concentrator arrays

While the idea of a concentrator space solar array is not new, at this writing none has yet powered a satellite. The first attempt to do so in 1995 ended when the launch vehicle failed to achieve orbit. That first array, known as the Space Concentrator Array with Refractive Linear Element Technology, or SCARLET I, consisted of four concentrator panels mounted on a PUMA™ deployable array structure (Jones and Murphy, 1995) together with two of the planar silicon cell panels that constituted the original array design. Figure 3.38 shows a schematic of the demonstration flight wing. The optical design uses domed linear Fresnel lenses (Piszczor and O'Neill, 1992; Piszczor et al., 1994) that provide about 10x concentration, and the cells are GaAs/Ge cells designed for optimum performance at that concentration level. A second, improved version of this technology, known as SCARLET II, has been built and space qualified for use on the first spacecraft launched in NASA's New Millenium program. Figure 3.39 shows a schematic drawing of the SCARLET II cell/lens module. The 2.6 kW array provides power for an electric propulsion system designed to take the spacecraft out of Earth orbit for an asteroid rendezvous, as well as for normal spacecraft functions.

As we saw earlier, concentrator arrays have two major advantages over planar arrays. The active semiconductor area required for a given power output is reduced by the factor $1/X$, where X is the concentration ratio of the optical element. This has important impacts on array cost and mass. The reduced area allows the use of advanced, high efficiency space solar cells at considerably lower cost than for the same output planar array. To the extent that the optical element cost is also lower than that of the equivalent area solar cell cost, the total array cost is lower as well. The reduced cell area also means that additional shielding against radiation damage can be provided without a major impact on total array mass, giving the concentrator array a potentially significant advantage in EOL performance compared to its planar counterpart. On the other hand, a major concern with the use of concentrator arrays is the requirement for more precise pointing to maintain full output power. Planar array output decreases with the cosine of the angle between the array surface normal and the sun's direction. The SCARLET II array, for example, will provide no output beyond about ±3 degrees from proper alignment along the axis normal to the lens surface. Although the hardware exists to provide the precise control needed (it is routinely used for the on-board antennas, for example), the more important aspect of the issue is the rapid loss of power in the event of sudden

SOLAR ENERGY CONVERSION 127

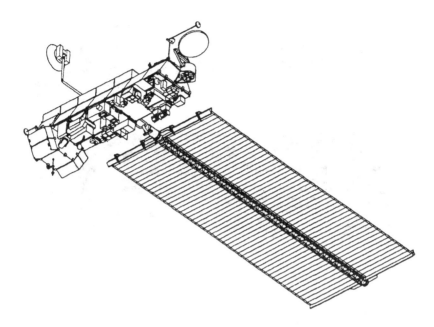

Figure 3.37 Schematic Diagram of the EOS-AM1 Spacecraft

spacecraft misorientation and/or tumbling. In such instances, the satellite systems must rely on a combination of array and battery power until control is reestablished. A planar array can provide some power even at extreme angles of misorientation, while a concentrator array cannot. Nonetheless, the considerable advantages of lower cost and survivability in high radiation environments warrant serious consideration of concentrator arrays when such concerns are important mission drivers.

The domed Fresnel lens design provides high optical efficiency by minimizing internal reflection losses and reducing the effect of shape errors on total transmission. The shape of the lens surface is such that the entrance and exit angles are equal for each ray incident on the lens, which is the condition required to minimize internal reflection losses. The addition of an antireflective coating on the outer surface raises the lens efficiency to over 90% (O'Neill and Piszczor, 1989). Furthermore, the mechanical arrangement of the lenses is such that nearly 94% of the exposed array surface collects light and transmits it to the cells. This is in contrast to a typical planar array packing factor (solar cell area exposed to sunlight as a fraction of total array area) of 85% or so. Part of the reason for the lower planar array packing factor is that room on the front of the array normally must be left for some of the wiring and for differential thermal expansion of the cells, while in the concentrator array nearly 90% of the panel area is uncovered by solar cells in the first place. Another advantage occurs when it comes to

SPACECRAFT POWER TECHNOLOGIES

Figure 3.38 SCARLET I Flight Wing Prior to Installation and Launch

Figure 3.39 SCARLET II Array Under Construction and Test

SOLAR ENERGY CONVERSION

mounting the diodes that are used to protect the solar cell strings from going into reverse bias breakdown during a partial shadowing by some other part of the spacecraft, such as an antenna. Since the diodes need not be in close proximity to the cells they are to protect, they are often mounted on the back of the panel or at the root of the array and connected by bringing the wires through the panel at appropriate locations. Although no additional array front surface is required, the process is relatively complex and requires access to both sides of the panel during assembly. Again, the unused front panel area of the concentrator array provides room to place the diodes near the cells they are to protect. This reduces the complexity and work required and eliminates a significant amount of wire and adhesive as well.

One other concentrator array concept deserves some mention–the trough aperture reflector array–because it is essentially a simple technique to enhance the output of a planar array and, as such, does not require a specially designed concentrator solar cell. A schematic cross section is shown in Figure 3.40. The design consists of two reflectors fastened by hinges to the edges of the panel. In the stowed position, the reflectors are folded against the cell panel and are actuated by either torsion devices or springs to open once on orbit. The concentration ratio is determined by the size of the reflector and the angle at which it is deployed.

Assuming that one constraint is that the reflector width is equal to the panel width, the total additional intensity at the panel is about 1.4 times the one sun illumination. This is in addition to the illumination falling on the panel, so the total panel output can be up to 2.4 times that of the original panel. The actual increase will be less than 1.4 times normal intensity because of losses at each reflector surface and at the cell surfaces as well since the antireflective coating is not optimum for such low angles of incidence. Nonetheless, the relative simplicity of the concept, coupled with a replacement of expensive, active solar cell area with less expensive reflective material, provides a potential cost advantage over the same output planar array.

7.5 Array environmental interactions

The major factor determining the success of a space mission is spacecraft reliability. A major influence on reliability is the interaction of the various spacecraft components and subsystems with the space environment. There are several headings into which such interactions may be grouped: plasma interactions and spacecraft charging; debris and micrometeoroid impacts; chemical reactions with neutral species; radiation degradation of electronic components and thermal control surfaces caused by high energy particles and UV photons, thermal cycling, and solar flare events. Elimination or mitigation of the deleterious effects of such interactions is at the heart of much of the research and development of advanced spacecraft technology. This is certainly the case with the satellite power system and, in particular, the solar array.

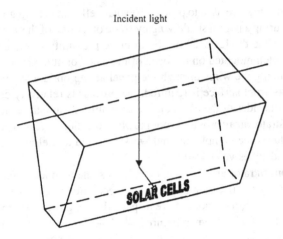

Figure 3.40 Concentrating Trough Solar Array Diagram

Geosynchronous orbit

An interaction of major concern for GEO satellites is the differential charging of different parts of the spacecraft which can lead to high electric fields and subsequent dielectric breakdown and arcing between spacecraft components. The charging can occur by any of several effects, such as secondary electron emission, the photoelectric effect, and electron bombardment from a solar flare or lesser solar substorm. Such differential charging can lead to conditions where potential differences exceeding 10,000 volts can develop between adjacent surfaces and materials, especially on the solar array. When an arc occurs, the resultant current spike can travel through the array and spacecraft wiring and cause instrument latch-up, induce spurious signals in the telecommunication, command and control systems, etc., and in some cases has caused the loss of the satellite. A common design technique is to coat all outside surfaces of the spacecraft with weakly conducting materials to eliminate the possibility of differential charging in the first place (Purvis, Garrett, Whittlesey and Stevens, 1984). This takes the form of a transparent conductive oxide (TCO) coating such as indium tin oxide which is usually applied to all non-conductive surfaces except the solar cell coverglasses.

Prior to the explosive growth of solar array sizes to the multikilowatt level, array output voltage was typically 28 volts dc. As arrays grew in size, array voltages grew as well. The primary reason was to reduce the total current supplied by the array which, in turn, reduces the mass of the array wiring harness needed to carry the current with minimal I^2R power loss. Some recent array failures have been recorded that illustrate the care with which such approaches must be used (Hoeber, Robertson, Katz, Davis and Snyder, 1998). The symptoms of the failures that were observed were development of low impedance shorts between adjacent cell strings (a number of solar cells connected in electrical series to achieve the desired total output is called a cell string) and shorts between the high voltage cells at the end of the strings and array ground. The arrays in question were intended to operate at 100 V. The shorting was observed to develop immediately after a solar substorm occurred which bathed the entire spacecraft with a large negative charge. The charging occurred when the massive electron cloud emitted by a substorm happened to reach the vicinity of the Earth while the satellite was still exposed to the sun. The initial charging took the form of a more or less equipotential surface over the array and the satellite, but subsequent photo and secondary emission soon caused a charge differential to build up between the cell coverglasses and the solar cell edges and panel surfaces. Hoeber *et al.* (1998) performed computer simulation analysis and vacuum tank testing to show that the shorting was initiated by electrostatic discharges between the top surfaces of the solar cell coverglasses and the solar cells themselves. Ordinarily such discharges are very low current and are known to have little effect on array performance when the array output voltage is low. The analysis predicted, and the testing verified, that when the arcing occurs in regions of the array where the voltages between adjacent cells from two different cell strings exceed a threshold value, or where the cell-to-ground voltage exceeds a threshold value, array current is diverted into the arc and sustains it. The much larger current flowing through the arc causes local heating; the heating in turn pyrolizes the polyimide insulating substrate separating the solar cells from the aluminum surface of the array panel and leaves a permanent low impedance path for the array current. The shorting is permanent and progressive.

The situation is actually very complex and dynamic and depends on how long the array has to develop the differential voltage surfaces needed to initiate an arc, the resistivity of the cell coverglass surface, and the electric field (voltage gradient) that exists between a given cell and either spacecraft ground or an adjacent, lower voltage cell. Clearly, the latter is completely controlled by the physical layout of the cell strings. The testing found that for GaAs cell arrays the threshold above which arcing occurred was approximately 60 V, while for Si cell arrays it was about 80 V. No explanation has been developed to explain the difference, but it is suspected that the lower vapor pressure of As in the GaAs may be involved (Ferguson, 1998). It is also presumed that arcing may be more pronounced when the array emerges from the eclipse portion of the orbit. The array temperature can be as low as $-100°C$ when the array emerges from the eclipse, which causes two synergistic effects: the separation between adjacent cell strings (and between cells in a string) is smaller because of thermal contraction of the array panels, and the cell voltages are higher because of the temporarily lower temperatures of the cells as the array emerges from the eclipse. These two factors combine to encourage arc enhancement should one occur while the array has not yet warmed up to its operational temperature. Simple measures taken during array fabrication can prevent such damage from occurring without precluding the use of 100v or higher array output voltages. Arranging cell strings so that the voltage between any two cells on adjacent strings never exceeds the threshold eliminates that failure mode, and effective use of cell bypass diodes can eliminate the cell-to-ground failure mode. Application of an insulating barrier between cell strings (a silicone fillet, for example) can also prevent arc enhancement but can only be used for arrays in GEO where the number of orbital eclipses is typically only a hundred or so during a ten year satellite operational lifetime. Otherwise, bonding the cell strings together can result in physical damage from the relative thermal expansion and contraction that occurs far more often in low Earth orbit. (A typical LEO satellite will experience over 6000 thermal cycles per year.) Although the failure mechanisms mentioned here can be relatively easily controlled, they have been described in some detail because they represent a new manifestation of what was once thought to be a well-understood phenomenon – spacecraft charging. The failures that have occurred to-date have resulted in the loss of several hundred millions of dollars of satellite communication service because of the premature power loss of the solar arrays.

Other major problems in GEO are radiation damage to the solar cells and single event upset phenomena in electronic components in various parts of the spacecraft. The latter are caused by the random impact of extremely high energy, heavy particles from beyond Earth orbit and will not be dealt with here. We have already discussed the issue of radiation damage in solar cells, and predictions of its effects from trapped particle fluxes is reasonably accurate. What cannot be predicted with any certainty is the extent to which solar flare events will occur which also have large fluxes of high energy protons associated with them. Some recent work by Xapsos and colleagues (Xapsos, Messenger, Walters, Summers and Burke, 1998) has applied the principles of extreme value statistics to the problem. The methodology they have developed essentially allows the array designer to estimate the risk that a large solar flare proton event presents to a (in their example, GaAs) solar array in GEO; the risk is discussed in terms of a minimum fluence of protons with energies exceeding some value that could cause significant array degradation. Given that, they are able to calculate the expected degradation that might occur. The accuracy of the prediction depends on the quality of the database for the last few sunspot cycles and can be expected to increase as that database is improved. As a concrete example, Xapsos *et al.*, have shown that there is a 57% chance that the largest event during a solar cycle will have a fluence of 10 MeV protons larger than 10^{10} per cm^2. Such an event can cause a degradation of over 3% of cell output, even with a 75 micrometer (three mils) thick coverglass on the cell.

Figure 3.41 shows the differential energy spectrum of just such a large solar flare proton event and the effects of various solar cell coverglasses on the energy distribution of the protons traversing the body of a typical GaAs n/p solar cell. Figure 3.42 shows the normalized solar cell output power as a function of coverglass thickness. The top curve describes the situation mentioned in the preceding paragraph. The curve labeled Cycle 22 illustrates the total accumulated degradation that would have occurred in a cell exposed to the entire output of Sunspot Cycle 22. Clearly, as shown by the potential magnitude of the degradation from both a single event and a full sunspot cycle, there is great need to be able to characterize and predict the impacts solar flare protons will have on solar arrays. It is worth noting that the statistical estimate of total array degradation caused by solar flare protons from the active part of one solar cycle (~7 years) is nearly the same as that caused by the trapped electron fluence for the same number of years. The difference is, of course, that the electron-induced degradation is predictable and smooth, while the degradation caused by the solar flare protons will be in the form of step function decreases which occur seemingly at random.

Low Earth Orbit

Environmental interactions in LEO have distinctly different manifestations compared to GEO. Differential charging is of primary concern in GEO, while absolute charging is the major concern in LEO. LEO spacecraft surfaces do not normally differentially charge because the thermal plasma current densities are high, and potential differences are bled off by collected plasma currents. However, if there are spacecraft-generated differential voltages of high enough magnitude, the situation can become very complex depending on the amount of high voltage conducting surface that is exposed to the space plasma relative to exposed area of lower voltage surfaces. For example, as array powers continue to grow one of two things must happen. To avoid large Joule heating losses either the array cabling must be massive or the power must be transmitted at higher voltages compared to standard practice for the lower power arrays of the past. (As mentioned above, early spacecraft typically ran at 28 VDC.) All spacecraft surfaces will float at potentials relative to plasma ground that result in no net current collection from the plasma. If conductors at different voltages are not exposed to the space plasma, all of the spacecraft surfaces will float within a few volts of the surrounding plasma potential. If conductive surfaces with different voltages on them are exposed to the plasma, in general the most negative of the surfaces will float negative with respect to the plasma at a potential equal to about 90% of the total voltage difference between the surfaces (Ferguson, 1993). This is of special concern for solar arrays where interconnects between solar cells, and the edges of the solar cells themselves, are all exposed to the plasma with potential differences (relative to spacecraft ground) ranging up to the full output voltage of the array.

If conditions are right, arcing will occur at conductor-insulator junctions (which includes the bases of perforations in the insulation that expose the underlying conductor) or from any dielectric surface where the underlying conductor reaches a negative potential higher than the dielectric breakdown strength of the coating or insulation

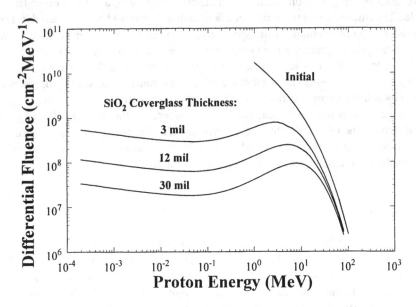

Figure 3.41 Calculated Differential Energy Distribution as a Function of Coverglass Thickness for a Large Solar Flare Proton Event

(Ferguson, Snyder and Carruth, 1990). In such cases, the arc currents may flow out into the space plasma, and the return currents flowing back are distributed over wide areas of other spacecraft surfaces. As with the GEO case described in the preceding section, arcs can also occur between closely spaced conductor surfaces that are at different voltages with respect to one another. Again, such arcs can cause shorting between array solar cell strings, with a resultant loss of power.

Other effects besides arcing and the potential for electrical failure can take place. If a conductor surface is at a high enough negative potential relative to the plasma it will attract high energy ions which might produce some sputtering of the surface. This can lead to the deposition of coatings on nearby surfaces, changing their optical, electrical, and thermal properties. However, if the solar array is attached to the spacecraft so that its potential is forced above plasma ground, an electron current will be collected which can produce localized heating and significant power drain from the array (Stevens, 1978), again leading to reduced spacecraft capability.

The LEO neutral environment is also a matter of concern. The presence of chemically active atomic oxygen can alter and even erode all types of spacecraft surfaces, including conductors, insulators, thermal coatings, etc. Just the additional kinetic energy added by encountering atomic oxygen in the ram direction of the spacecraft's mo-

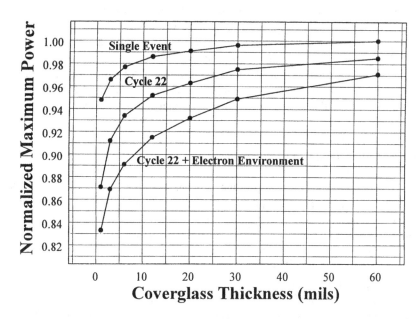

Figure 3.42 Degradation of an n/p GaAs Solar Cell in GEO Caused by a Large Solar Flare Proton Event

tion can enhance its reaction rate with spacecraft surfaces. Charged spacecraft surfaces can also attract chemically active ions, enhancing their reaction rates with the surface materials (Ferguson, 1990.)

The larger the satellite, the more attention must be paid to the preceding concerns. The International Space Station, for example, has been designed to have the structure ground point at the potential of the most negative end of its solar arrays. Since the arrays will produce power at 160 volts, the array and the spacecraft will float at about 140 volts negative relative to its surroundings. Such a potential difference will surely cause major problems. To avoid such problems, a device called a plasma contactor (Ferguson, 1990) has been added to the space station. The plasma contactor essentially uses an electron gun to ionize a neutral particle gas that is then injected as a current into the plasma. The design point is an injected current sufficiently large to keep all points of the structure within about 40 volts of the surrounding plasma.

Care must be taken in the design of space solar arrays to avoid the occurrence of unwanted effects that can occur when operating too close to safe design margins. The period of danger occurs just as the array is emerging from the eclipse, when it is still cold and its dimensions and output voltage have been changed. The increase in voltage that occurs at lower temperatures, coupled with the changes in dimensions, can sometimes result in circumstances more favorable to arcing or high current drain than otherwise occur during the steady-state conditions of the rest of the orbit. NASA has prepared two sets of guidelines for spacecraft design known as NASCAP GEO and NASCAP LEO (Snyder, 1988) which can be used to either avoid or to mitigate the various undesirable effects caused by the many ways a spacecraft can interact with the space environment. NASA has also developed a software design tool for modeling the interaction of a spacecraft with the space environment known as the Environmental Workbench, or EWB (Chock, 1998), that works on a desktop personal computer or workstation.

7.6 Power system design and array sizing

Design considerations will vary depending on whether the satellite is intended for low earth orbit and has low power requirements (Dakermanji *et al.*, 1991) or high power requirements (Winslow *et al.*, 1989; Patil , 1990; Tam *et al.*, 1990). Designs will differ for high Earth orbit and geosynchronous satellites (Lovgren *et al.*, 1989; Malachesky *et al.*, 1991; Winter and Teofilo, 1989). Discussions of generic power systems and solar array considerations may be found in (Moser, 1990; Kenney *et al.*, 1990; Bercaw and Cull, 1991; and Slifer, 1989).

One of the most significant areas where orbit location (LEO, HEO, GEO) affects the power system design is the storage subsystem and its interaction with the array subsystem. Battery charge and discharge cycles differ extensively depending on the sunlight-to-eclipse ratio of the orbit. As a result, much of the attention on power system design is devoted to managing the storage subsystem. The storage subsystem must be designed to provide all the satellite power requirements during eclipse and to be fully recharged during the sunlight portion of the orbit. In low inclination, low Earth orbits, the sunlight-to-eclipse ratio is about two-to-one with approximately one hour in sunlight followed by approximately 30 minutes in eclipse. Battery life is greatly affected by the depth of discharge and the rate of charge while battery usefulness is measured in terms of energy storage capacity in Wh/kg. The extent to which a battery can be discharged, and the rate with which it can be recharged, without causing irreversible damage to the electrodes must also be included. Requiring a smaller number of batteries to provide the eclipse power means that they will have to undergo a deep depth of discharge and then be fully recharged in less than an hour. Increasing the number of batteries will reduce the depth of discharge required but will greatly increase the total mass of the storage subsystem. This trade-off must be made carefully since the storage subsystem mass is often the largest part of the total power system mass. A simple example can illustrate the situation.

Historically, NiCd batteries have been the type most often used for space applications. They are commercially available with energy densities in the range of 10 to 20 watt-hr/kg. For a "typical" LEO eclipse time of 30 minutes, a complete discharge (i.e., 100% depth of discharge, or DOD) over the 30 minutes results in a battery specific

SOLAR ENERGY CONVERSION

power of 20 to 40 W/kg. This value will, of course, be made lower by the mass of the rest of the storage subsystem components (regulators, thermal control elements, wiring, etc.). It will also be made lower by using a lower depth of discharge, a necessary requirement to preserve battery function. At 20% DOD, commonly used for space NiCd batteries to guarantee a five year life in a low Earth orbit (Hord, 1985), the specific power ranges from four to eight W/kg. If the satellite requires two kilowatts of power to function, the battery mass will range from 50 kg to 100 kg at 100% DOD and from 250 kg to 500 kg at 20% DOD.

This should be contrasted with the mass of the solar array. Estimating the array mass must take into account the battery charging requirements along with the load requirements. In our example, the energy consumed from the batteries by the loads during the eclipse is one kW-hr, regardless of the battery DOD, and must be replaced by the array during one hour of sunlight operation. Hence, without allowing for any losses during battery charging, or in the power management and distribution circuits, the array must be able to generate at least three kilowatts of power. (Typically, inefficiencies in the battery charging and load distribution circuits require the array output to be about 2.2 to 2.5 times the power generating capability needed to operate the system directly.) Current rigid panel solar arrays on U.S. satellites have specific powers in the range from 15 W/kg to 30 W/kg (Flood *et al.*, 1989). The array mass in this example ranges from 85 kg to 120 kg which is one-third or less of the battery mass at 20% DOD.

The situation is somewhat different for GEO applications. A LEO satellite will complete nearly 6000 orbital revolutions each year. A satellite in a circular geosynchronous equatorial orbit will complete 365 orbits per year, but will undergo only about 90 eclipses (Rauschenbach, 1980). The maximum eclipse time is about one hour, but since so few battery cycles are required, a much greater DOD may be used. In addition, the charging may take place over a minimum 23 hour period which will help to preserve battery life. The storage subsystem mass may thus actually be lower than that required for LEO, depending on the DOD allowed, and the array need not be oversized by as large a factor as for LEO. It is necessary, however, to oversize the array at the beginning of the mission by an amount at least equal to that expected to be lost because of radiation damage.

Solar array design is a multidisciplinary activity requiring electrical, electronic, thermal, mechanical, and optical engineering. It is also necessary to have as accurate a prediction as possible of the orbital environment in which the array will operate, including a prediction of any shadowing of the array that might occur during its power generating mode. Shadowing can occur because of interference from various other parts of the spacecraft such as antennae or instrument booms, etc. The array operating temperature, the amount of radiation damage expected, and even the potential for thermomechanical degradation caused by repeated thermal cycling or prolonged operation at temperature extremes must be taken into account. This section will conclude by outlining a generic array sizing calculation. No attempt will be made to make an actual orbit-averaged numerical calculation of the total proton and electron fluences the array might encounter.

The procedure is straightforward. A given combination of array components is selected including solar cells, solar cell coverglasses, interconnects (material, size, number per cell, etc.), array substrate (material, size, and shape), array cabling sizes, number and nature of protective diodes, and so forth. The components are then arranged in a candidate configuration, e.g., planar or cylindrical (spinner satellite). The end-of-life degraded maximum output power, P_c, of a solar cell with its coverglass is calculated for the expected operating temperature of the array from the following equation:

$$P_c = P_0 * S' * F_{rad} * F_{Top} * F_M * F_{SH} * F_{BD} * F_{CONF} \quad (3.35)$$

where P_0 is the initial solar cell efficiency at normal incidence AM0 and 25°C; S' is the effective solar intensity, including the effects of coverglass darkening from uv exposure, solar distance, and off-axis pointing; F_{rad} is the solar cell radiation degradation factor, either measured or calculated according to the methodology developed earlier in this chapter; F_{Top} is the operating temperature degradation factor, defined as P_{mpTop}/P_{mp0}, where mp refers to maximum power point values measured in each case, F_M includes all the miscellaneous assembly and other degradation factors (loss of interconnect conductivity, e.g.), typically between 0.95 and 1.00; F_{SH} is the shadowing factor during the non-eclipse portion of the orbit (if there is no shadowing, it equals 1.00); F_{BD} is the blocking diode and wiring loss factor, prorated for a single cell. Typically it can be estimated from the expression $1 - (V_D + V_W)/(V_B + V_D + V_W)$, where V_B is the array bus voltage, V_D is the diode voltage drop, and V_W is the voltage drop in the wiring between the array and the load); and F_{CONF} = the aspect ratio of the array. For flat arrays it is 1.0, and for spinning arrays it is $1/\pi$.

Once all the parameters have been quantified, the calculation proceeds in a straightforward fashion to determine the number of solar cells, N, required to achieve the desired end-of-life solar array output:

$$N = P_A/P_C \quad (3.36)$$

P_A is the required array power and P_C is the single cell output calculated by Eq. (3.35). The array area is then given simply by

$$A_S = (A_C/F_p)N \quad (3.37)$$

where F_p is the packing factor and A_C is the area of a single solar cell. The packing factor is simply the ratio of the area of the array substrate actually covered by the solar cells to the total cross-sectional area of the array substrate itself. Typical values range

SOLAR ENERGY CONVERSION

between 0.85 and 0.92 for planar arrays. Values of 0.95 and higher are very difficult to achieve with flat panel, one sun arrays, in part because it is advisable to allow space between the solar cells for thermal expansion and contraction of both the cells and the substrate. We have also seen the advisability of paying attention to the proximity of solar cell edges and interconnects which are at different potentials to avoid unwanted and harmful interactions with the space environment.

It is important to realize that much more than determining the area of the array goes into its design. Thermal management, total moment of inertia, structural stiffness, and magnetic moment area are also critical considerations and will have a major affect on total array area and mass. With regard to magnetic moment, for example, it is important to avoid a current distribution that makes the array act like a large area magnetic solenoid. It is also important to arrange the physical laydown of the cell strings and wiring to avoid the development of conductive loops which can generate an emf when they pass through the Earth's magnetosphere. Both effects can have serious implications for satellite operation and control.

It will be obvious that we have not included in the immediately preceding discussion anything about determining the mass of a space solar array even though achieving the highest possible power to mass ratio is often one of the most important issues considered when designing a space power system. As we saw at the start of Section 7.6, determining array mass, unlike calculating array size, cannot be done with a relatively straightforward formula. Many of the key factors which affect the mass are mission, or launch vehicle, or even satellite specific. They include such things as the allowable natural frequencies of motion associated with the array structure, i.e., its required stiffness, the speed of the deployment mechanism, and whether or not it must be able to retract the array, the accuracy of the array tracking and pointing subsystem, and so on. Certainly, the total area of solar cells required to achieve the desired array power level has a strong effect, as do both the aspect ratio of the array itself (the ratio of width to length) and the manner in which each wing of the array is supported. Typical power to mass ratios were given earlier when the various classes of space solar arrays were discussed.

The technology of space solar arrays has progressed considerably from the days of Vanguard I. While some of the advances are the result of improved array and satellite designs, the most outstanding areas of progress are the result of advances in the electrical, optical, electronic, and/or mechanical properties of new materials which enable bold new design solutions to be investigated and applied. Yet even these advances are not seen to be enough for the next generation of spacecraft. Masses must be made lower still and costs must be reduced an order of magnitude. The required output power levels are expected to grow by a factor of ten to the multi tens of kilowatt range (and perhaps even to the few hundred kilowatt range), and at the same time to be reduced in size as we strive to explore space with ever smaller and cheaper satellites.

8. Space thermophotovoltaic power systems

Thermophotovoltaic (TPV) energy conversion, although still in an early stage of development, is an attractive possibility for use in space power systems. Figure 3.43 illustrates the basic concepts involved. Thermal energy in space can be provided by concentrated sunlight, a nuclear reactor, or a radioisotope heat source of the sort now used by NASA in its deep space missions. Our interest here is with the conversion technology which is, at least for a fundamental understanding of such systems, independent of the type of thermal energy source. It is, however, very strongly dependent on the source temperature. The conversion technology can be sorted into two distinct subsystems: one for conversion of thermal energy to radiant energy and one for conversion of radiant energy to electric energy. The first conversion occurs when the thermal energy heats an emissive surface and causes it to radiate; the second is accomplished using a spectrally tuned solar cell that converts the emitted radiation into electrical energy with optimum efficiency and/or power density. The first question that naturally arises is to ask why the thermal energy is not simply converted directly into electrical energy in some sort of thermodynamic cycle. The answer is that the TPV system can be simpler than one which uses dynamic thermal conversion techniques, such as Brayton, Rankine, or Stirling engines, with their complex construction. In principle, a space TPV system can be completely passive with few, if any, moving parts in the conversion subsystem. At worst, it could involve a pumped loop cooling subsystem of some sort.

As shown in Figure 3.43, there are two TPV system configurations that are of interest. One uses a blackbody-like radiant emitter, which normally emits over a broad range of wavelengths, in tandem with a bandpass filter to achieve a narrowed spectral output that overlaps the wavelength region where the solar cell has its maximum spectral response. The other uses a selective emitter that naturally emits in a narrow wavelength range that also falls within the cell's wavelength region of maximum spectral response. The corresponding photon energies are normally just above the solar cell's bandgap energy. In the case of the blackbody-like emitter/filter system, enhanced performance requires the unused portion of the radiant energy to be reflected back to the emitter surface. The filter thus makes the broadband emitter behave like a selective emitter, but with one important difference. In the filter TPV system, most of the emitted radiation must be circulated back and forth between the emitter and filter to achieve high efficiency. Therefore, even a small filter absorptivity will result in a major loss of radiant energy for subsequent conversion by the solar cell. In addition, if the thermal emitter does not perform very much like a true blackbody (where the emissivity and the absorptivity both equal unity), the overall efficiency will be reduced. As a result of these losses, it has been shown (Chubb, 1990) that the selective emitter TPV system is more likely to have a higher overall efficiency than the filter TPV system, while the latter may have higher power densities. Since efficiency is of paramount importance for space power systems, we will focus our discussion in this section on the selective emit-

SOLAR ENERGY CONVERSION

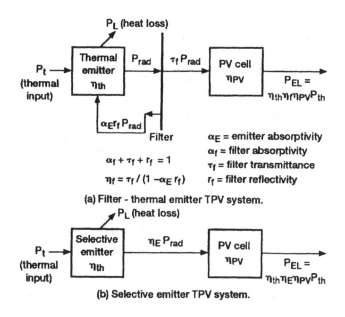

Figure 3.43 Schematic of Thermophotovoltaic System Configurations

ter TPV system. The discussion will be divided into two parts. The first will deal with selective emitters, and the second will deal with spectrally tuned solar cells.

8.1 TPV system efficiency

The value of selective emitters for efficient TPV energy conversion was first recognized by White and Schwartz (1967) who went on to describe a potential solar cell structure for use in a controlled spectrum. Little progress was made in the search for a suitable selective emitter during the two decades that followed, even though it was known that the rare-Earth elements had promise for such an application (Guazzoni, 1972). The physics behind their promise is that the 4f shell electrons lie inside the 5s and 5p electron shells. As a result, when the elements exist as doubly and triply charged ions in a crystal lattice they behave as if they were isolated; the 4f shell does not broaden into a continuum of energies as is the case with most solids but remains as a narrow band capable of sustaining absorption and emission of photons at specific energies.

It was known that ytterbium oxide, or ytterbia, would have strong emission in a narrow band of wavelengths that would be optimum for conversion by silicon solar cells. The spectral emittance work of Guazzoni (1972) did show strong emission bands for several rare-Earths that were suitable for use in TPV energy conversion, but the emittance for photons with energies below the bandgaps of available photovoltaic devices at that time was also high. The resulting emitter efficiency was therefore very low and severely limited the efficiency potential of a TPV system, even if very high photovoltaic device conversion efficiencies were possible. A breakthrough in selective emitter performance was made by Nelson (1986) and by Parent and Nelson (1986). Their emitters were constructed of an open mesh of rare-Earth oxide fibers five to ten micrometers in diameter rather than as solid pillboxes of the sort investigated earlier by Guazzoni. The motivation for Nelson had been to construct a rugged Welsbach mantle for use in gas lighting; the very small characteristic dimension of the fiber emitter resulted in a low emittance for the long wavelength, or out of band, part of the spectrum. This geometrical effect increased the thermal to in-band (i.e., silicon cell usable) radiant energy conversion to nearly 50%.

Chubb (1990) has formulated a three-band model to describe the performance of the rare-Earth emitters developed by Nelson. The emitted radiation is contained in a single band of width ΔE_b centered about energy E_b. Outside the emission band, the emissivity is ε_l for energies below the band, and ε_u for energies above the band. The emitter efficiency is defined as $\eta_E = P_E/P_{RAD}$ where P_E is the emitted radiative power per unit area from the emission band and P_{RAD} is the total emitted radiative power per unit area. Thermal energy lost through conduction or convection, which Chubb labels P_L, is not included in the above expression. A photovoltaic device with energy bandgap $E_G \approx E_b$ (but slightly less than) is capable of converting the radiative energy in the band to electrical energy with high efficiency. The expression derived by Chubb is a complex function of the central band energy, emitter temperature, and ratios of the out-of-band emissivities to the in-band emissivity, $\varepsilon_l/\varepsilon_b$ and $\varepsilon_u/\varepsilon_b$. Figure 3.44 shows a plot of the predicted emitter efficiency as a function of the ratio of the emission band central energy to the emitter temperature. The emissivity ratios have been made equal to each other and varied to generate a family of curves for the case where the bandwidth is equal to 10% of the central energy of the band. Two general observations can be made from Figure 3.43. First, very high radiative efficiency can be obtained if the out-of-band emissivities are kept low. Second, the maximum efficiency occurs at $E_b/kT_E = 4$, from which central band energies can be estimated for a given emitter temperature, T_E. This model predicts that for an emitter temperature of 1500K the central band energy is 0.52 eV, which means that a low bandgap cell will be required for efficient system operation. Table 3.5 gives the characteristics of three rare-Earth oxides that have been observed to have low emission band energies: Nd_2O_3 (Neodymia), Ho_2O_3 (Holmia), and Er_2O_3 (Erbia). Ytterbia has not been included here because the temperature required for reasonable power density and maximum efficiency is considered too high to be of practical use in a space system. It is, however, a primary candidate for various combustion driven systems for terrestrial commercial use.

As shown in Figure 3.43, the total efficiency of a selective emitter TPV system is the product of the thermal efficiency, emitter efficiency, and photovoltaic cell efficiency.

SOLAR ENERGY CONVERSION   143

The thermal efficiency depends on the nature of the thermal source and is outside the scope of our discussion here. As for the photovoltaic cell efficiency, it will be a maximum for photon energies that are near the cell's bandgap energy. Figure 3.45 shows the calculated variation of efficiency with cell bandgap under monochromatic illumination; the wavelength of the incident light corresponds to photons with energies just above the cell bandgap (Olsen *et al.*, 1991). The cell temperature was assumed to be 25°C. Under these conditions, solar cell efficiency will not have an optimum bandgap as was the case for full solar spectrum illumination. Instead, the efficiency will asymptotically approach a limiting value which will be determined by a complex set of cell material, optical, and electronic properties. Lower bandgap devices have lower band edge efficiencies because of the greater values of dark current and lower open circuit voltages associated with the lower bandgaps. As seen, band edge efficiencies for GaAs cells are expected to exceed 60%, but for a 0.5 eV cell the efficiency is about 30%. TPV systems requiring such low bandgaps will, therefore, have limited total thermal-to-electric conversion efficiency even without considering thermal losses and other parasitic effects. Nonetheless, TPV systems represent a potentially significant increase in efficiency for deep space missions compared to what is available with today's SOA radio-isotope thermoelectric generator (RTG) power systems (typically 4% to 6%). Total efficiencies between 15% and 20% have been predicted using a realistic engineering design for a deep space radioisotope TPV (RTPV) system. One obstacle for the RTPV system is the requirement that cell temperatures must be kept near 25°C to achieve high performance, while the RTG system allows the radiator temperature to be as high as 300°C. The lower temperature required for the RTPV systems results in a significantly

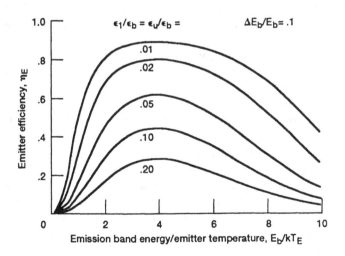

Figure 3.44 Selective Emitter Efficiency

Table 3.5 Emission Band Data for Rare Earth Oxides

Emitter Material	Emission Band Transition	Photon Energy at Center of Emission Band E_b (eV)	Photon Wavelength At Center of Emission Band (μm)	Dimensionless Bandwidth $\Delta E_b/E_b$	Emitter Temperature for Maximum Efficiency K
Erbia (Er_2O_3)	$^4I_{13/2} - {}^4I_{15/2}$	0.827	1.5	0.05	2400
Holmia (Ho_2O_3)	$^5I_7 - {}^5I_8$	0.62	2.0	0.10	1800
Neodymia (Nd_2O_3)	$^4I_{13/2} - {}^4I_{9/2}$	0.52	2.4	0.10	1500

larger radiator area and mass and has a major impact on spacecraft design. Although the problem is not insurmountable, it has deterred aggressive development of the RTPV system thus far.

8.2 Solar thermophotovoltaic space power systems

Solar TPV (STPV) systems have received little attention to date but have some interesting features worth mentioning. A major issue for photovoltaic space power systems is the size and mass of the energy storage subsystem, usually a bank of batteries. Typical energy storage times in Earth orbit are on the order of 30 minutes. An STPV system has the same advantage over conventional space solar arrays with battery storage as the various solar dynamic (SD) systems: it uses an integrated heat receiver/thermal energy storage system. As with SD, this integrated use has the potential to decrease the total weight of the STPV power system and improve the total orbital efficiency compared to an array/battery system. Orbital efficiencies approaching 20% are possible with STPV, compared to the 6% to 10% orbital efficiencies of typical array/battery systems. The difference lies with the higher efficiency inherent with thermal energy storage, at least for the relatively short storage times required in Earth orbit. STPV systems can compete effectively with solar thermodynamic power systems on the basis of total orbital efficiency and are in principle much simpler to build and operate.

The system configuration analyzed here uses an erbium-doped, yttrium aluminum garnet (Er-YAG) selective emitter (Lowe, Chubb and Good, 1994) and a suitable low bandgap solar cell. Thermal energy is stored using the phase change from solid to molten silicon. Figure 3.46 is a schematic drawing of the STPV system. The concentrated solar flux enters the receiver cavity where the thermal energy is transferred to the solid Si (melting point $T_c = 1680K$, heat of fusion $L_s = 1800 J/gm$). The storage medium

SOLAR ENERGY CONVERSION 145

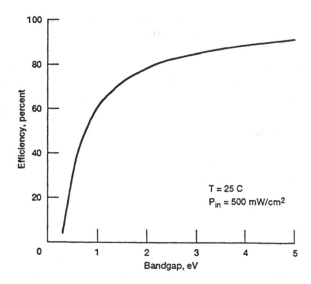

Figure 3.45 Calculated Cell Efficiencies for Band Edge Illumination

conducts the thermal energy to the selective emitter where it is radiated to the photovoltaic module. The Er-YAG emitter and photovoltaic cells are matched so that maximum array efficiency is achieved at the wavelength corresponding to the maximum emissivity of the emitter. When the sun is not available, the receiver door is closed and the system runs on the stored thermal energy.

There are three modes of operation for the STPV system: 1) solar flux is applied as the storage material is melting, 2) solar flux is applied after the storage material is completely melted, and 3) solar flux is not available, the receiver aperture is closed, and the molten storage material is supplying the energy to the emitter. The system performance in each of these modes can be calculated, but for this study we shall examine two limiting cases, namely when the system is producing maximum output and when the system is at minimum output. The former occurs when the emitter temperature T_E is a maximum and the latter when T_E is a minimum. The minimum output will occur when no solar flux is available (receiver aperture closed) and all the storage material has just completed solidification. The value of T_E for the minimum power case was set at 1400K. The radiated power density of the Er-YAG selective emitter starts to fall dramatically below this temperature (Good, Chubb and Lowe, 1995), so the mass of the thermal energy storage systems must be such that the satellite emerges from the eclipse portion of the orbit before this condition is reached.

Performance of the STPV system has been calculated using the model described by Stone *et al.* (1995). The total efficiency, η_T, which does not include power conditioning losses, is given by η_T = (electrical power output)/(solar power input) = P_{EL}/P_{solar} = $\eta_C\eta_R\eta_{th}\eta_{EF}\eta_{PV}$ where η_C is the concentrator efficiency, η_R is the receiver efficiency, η_{th} is the thermal efficiency which accounts for conduction and radiative losses from the high temperature receiver, η_{EF} is the emitter efficiency, and η_{PV} is the photovoltaic (PV) efficiency. The model uses a constant concentrator efficiency, η_C = 0.87, representative of the concentrator used by Stone et al. The receiver is modeled as a black body cavity where radiation loss out the aperture is given by the usual T^4 relationship and is the largest radiation loss possible. Therefore, the calculated receiver efficiency will be conservative. The thermal efficiency was assumed to be η_{th} = .95, and representative values of emittance for rare Earth-doped YAG selective emitters were used. The emitter emission band, centered at photon energy E_b, was coupled to the PV cell bandgap energy E_g to obtain maximum photovoltaic efficiency. In other words, $E_g = E_l + E_b - E_b/2$, where $E_b = E_u - E_l$ is the width of the emission band, E_l is the low energy end of the emission band, and E_u is the high energy end of the emission band. For these calculations, a dimensionless emission bandwidth, $\Delta E_b/E_b = 0.15$, was used. Within the emittance band, the value of emittance used was ε_b = .75. Outside the emission band the emittance values $\varepsilon_l = \varepsilon_b = 0.1$ were used, except for wavelengths greater than 5000 nanometers, where ε = 1.0 was assumed. An "ideal" photovoltaic cell was modeled in which the quantum efficiency at the central wavelength was assumed to be 1.0. The cell reflectance was assumed to be 0.8 for wavelengths beyond the bandgap energy. Finally, the results in Lowe, Chubb and Good (1994) indicate that the optimum bandgap E_g is about

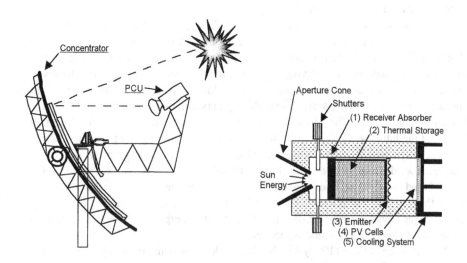

Figure 3.46 Schematic of a Solar Thermophotovoltaic Space Power System

0.7 eV for the STPV system, and this value was used in the calculations. The results are shown in Figure 3.47 as a function of the ratio of the concentrator area to the emitter area, A_C/A_E, for a concentrator area to receiver aperture ratio, A_C/A_R, of 5000. This value is representative of the concentrator used by Stone *et al.* (1995). We also set the emitter area A_E equal to the total cell area A_{PV}. As shown in Figure 3.47, the receiver temperature T_R and the emitter temperature T_E increase rapidly at first and then level off as A_C/A_E increases. Accompanying the emitter temperature increase are modest increases in η_{Ef} and η_{PV}. However, the rapid rise in receiver temperature causes η_R to decrease rapidly as well. The result is that the total efficiency decreases with increasing A_C/A_E. Even though η_T decreases, the rapid increase in emitter temperature means that the output power density, P_{EL}/A_{PV}, increases rapidly with increasing concentrator area to emitter area ratio. These rapid changes with temperature occur because thermal radiation, which has a T^4 dependence, is the controlling mechanism for a TPV energy conversion system.

The high receiver temperature that occurs for large A_C/A_E means that material limitations will determine the maximum ratio that can be used. Silicon carbide (SiC) which has a melting point of 2700°C is a possible receiver material that would allow very high temperature operation for which the output power density is very large. However, as already pointed out, operation at very high temperature means high power density at the expense of high efficiency. Since the PV cells are probably going to be expensive, it may be desirable to operate at high power density to keep cell costs low. However, since η_T will be low at high temperatures, the other component costs may be increased suggesting that the system operating point will not necessarily be at the maximum total efficiency. It may turn out to be desirable to operate in such a way as to minimize the system cost. Similar considerations apply to any solar thermal to electric conversion system. Of importance is the flexibility of STPV system design which, unlike the dynamic systems, need not be fixed on a given modular size of the conversion subsystem. A TPV array is infinitely variable in size and output and can be matched to any size thermal source. Although STPV and RTPV technologies are in their earliest stages of development at present, they have the potential to become significant power system options for future space missions.

9. Conclusion

Silicon cells were the cells of choice in space solar arrays from the time of the first Vanguard array in 1958 until the commercial development of GaAs cells in about 1987, a period of nearly 30 years. During that time, space silicon cell efficiencies rose from about 10% to over 14% and grew in size from 2 cm² to nearly 64 cm². At the same time, the radiation resistance of the cells nearly doubled, thanks in part to the realization that minority carrier diffusion lengths were larger in p-type material than in n-type material,

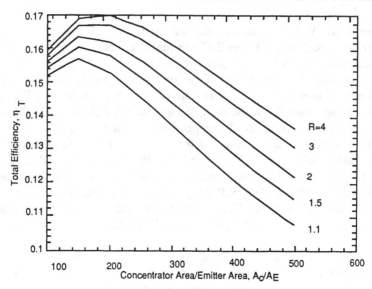

Figure 3.47 Solar TPV System Efficiency

with the advantage that brings in terms of the ratio of diffusion length to cell active region width. Rapid advances in space solar cells using other materials have been made during the past ten or twelve years, however, and now arrays are being launched with 21.5% multiple bandgap solar cells of $GaInP_2/GaAs$ on germanium substrates. The promises of even higher efficiency and stronger radiation resistance have also been realized at the laboratory level with 30% MBG concentrator cells and radiation hard InP cells. Our understanding of the effects of radiation damage in space solar cells has also progressed with displacement damage dose as a unifying concept for correlating radiation damage results from protons and electrons of various energies, thereby simplifying and making more accurate the process of predicting solar cell degradation during the course of an actual space mission. The thin film cells continue to show improvement and, with careful attention paid to their space compatibility, offer the promise of significantly lower cost solar arrays in the future. The potential for a fully encapsulated, monolithically integrated thin film solar array also offers promise for high voltage array operation with array voltages perhaps exceeding 1000 V. Array technology has progressed from the few watts on Vanguard I to the nearly 100 kilowatts planned for the International Space Station. At the same time, solar electric propulsion missions to carry large amounts of cargo to Mars are under study and will require megawatt-class arrays of a sort that will require new and innovative materials, as well as new and innovative deployment and control technology, if such missions are ever to become a reality. In addition, a new class of space power systems, using thermophotovoltaic energy conversion from both solar and non-solar thermal energy sources, will broaden the use of photovoltaic devices and arrays considerably. The future of space solar array research and development holds many challenges and opportunities. It is hoped that this chapter has provided a brief glimpse into some of the fundamental considerations that have been important in achieving the present status of our capability to generate solar power in space.

10. References

Anspaugh, B.E. and Weiss, R.S., *Results of the 1996 JPL Balloon Flight Solar Cell Calibration Program*, JPL Publication 97-6 (1996).
Anspaugh, B.E. and Weiss, R.S., *Results of the 1997 Balloon Flight Solar Cell Calibration Program*, JPL Publication 97-23 (1997).
Anspaugh, B.E., Downing, R.G., and Sidwell, L.B., *Solar Cell Calibration Facility Validation of Balloon Flight Data: A Comparison of Shuttle and Balloon Flight Results*, JPL Publication 85-78 (1985).
Anspaugh, B.E., *Proton And Electron Damage Coefficients For GaAs/Ge Solar Cells*, IEEE 22nd Photovoltaic Specialists Conference Proceedings, New York (1991), 1593-1598.
Arita, T., Suyama, N., Kita, Y., Kitamura, T., Hibino, H., Takada, K., Omura, N., Ueno, and Murozono, M., *Cuinse2 Films Prepared By Screen Printing And Sintering Method*, Photovoltaic 20 (1988), 1650.
Bailey, S.G., Brinker, D.J., Curtis, H., Jenkins, P.J., Scheiman, D., *Solar Cell Calibration and Measurement Techniques*, 15th Space Photovoltaic Research and Technology Conference, NASA CP-97534 (1997).
Baraona, C.R., Brandhorst, H.W. Jr., and Poroder, J.D., *Analysis of Epitaxial Drift Field Non P Silicon Solar Cells*, IEEE 12th Photovoltaic Specialists Conference Proceedings, New York (1976), 9-15.
Baraona, C.R., *Photovoltaic Power For Space Station Freedom*, PVSC 21 (1990), 30-28.
Baraona, C.R., *Photovoltaic Power for Space Station Freedom*, 8th Space Photovoltaic Research and Technology Conference, NASA Conference Publication CP-2475 (1986), 321-330.
Beaumont, B. Garabedian, P., Nataf, J.C., Gibart, P. and Verie, C., *Mechanically Stacked Two Tandem Concentrator Solar Cell Concept*, PVSC 21 (1990), 47.
Bercaw, R.W. and Cull, R.C., Proceedings of the 26th Intersociety Energy Conversion Conference (1991), 332-339.
Blakers, A.W. and Green, M.A., *High Efficiency MINP Silicon Solar Cells*, Applied Physics Letters, 39 (1981), 483-487.
Blakers, A.W., Jigun, S., Kellis, E.M., Wenhem, S.R., Godfrey, R.B., Szpitalak, T.,

Wellison, M.R., and Green, M.A., *Towards a 20% Efficient Silicon Solar Cell*, IEEE 17th Photovoltaic Specialists Conference Proceedings, New York (1984), 386-392.

Brinker, D.J., *High Altitude Aircraft Space Solar Cell Calibration Technique*, NASA, Private Communication (1998).

Bücher, K., *Calibration of Solar Cells for Space Applications, Progress in Photovoltaics*, Vol. 5., John Wiley & Sons, London (1997), 91-107.

Catalano, A., *Advances In A-Si:H Alloys For High Efficiency Devices*, PVSC 21 (1990), 36.

Chapin, D.M., Fuller, C.S. and Pearson, G.L., *Solar Energy Converting Apparatus*, U.S. Patent No. 2,780,765 (1957).

Chiang, P.K. et al., *The Progress of Large Area GaInP2/GaAs/Ge Triple Junction Cell Development at Spectrolab*, 14th Space Photovoltaic Research and Technology Conference, NASA CP-10180 (1995), 47-56.

Chock, R., Private Communication, NASA Lewis Research Center, Cleveland, Ohio 44135, (1998).

Chubb, D.L., *Reappraisal of Solid Selective Emitters for Thermophotovoltaic Energy Conversion*, 21st IEEE Photovoltaic Specialists Conference Proceedings, New York (1990), 1325-1333.

Chubb, D.L., Flood, D.J., and Lowe, R.A., *High Efficiency Thermal to Electric Energy Conversion Using Selective Emitters and Spectrally Tunes Solar Cells*, Nuclear Technologies for Space Exploration, D.M.Woodall, (ed.), American Nuclear Society, Lagrange Park, IL (1992), 281-293.

Curtis, H.B., Private Communication, Lewis Research Center (1998).

Dakermanji, G., *Power System Design for a Small, Inexpensive Satellite*, Proceedings of the 25th Intersociety Energy Conversion Conference, Vol. 1, American Institute of Chemical Engineers, (1990), 47-54.

Datum, G.C. and Billets, S.A., *Gallium Arsenide Solar Arrays-A Mature Technology*, IEEE 22nd Photovoltaic Specialists Conference Proceedings, New York (1991), 1422-1428.

DeMeo, E.A., Goodman, F.R. Jr., Peterson, T.M. and Schaefer, J.C., *SOLAR Photovoltaic Power: A U.S. Electric Utility R&D Perspective*, PVSC 21 (1990), 16-21.

DiNetta, L.C., Negley, G.H., Hannon, M.H., Cummings, J.R., McNeeley, J.B., and Barnett, A.M., *AlGaAs Top Solar Cell For Mechanical Attachment In A Multi Junction Tandem Concentrator Solar Cell Stack*, PVSC 21 (1990), 58.

EOS Reference Handbook, Goddard Spaceflight Center, Greenbelt, Maryland, (1990).

Fan, J.C.C. and Palm, B.J., *Optimal Design Of High-Efficiency Single-Junction And Tandem Concentrator Space Cells at 80 C And 100 Suns*, 6th Space Photovoltaic Research and Technology Conference, NASA CP-2314 (1983), 120-127

Ferguson, D.C., *Interactions Between Spacecraft and Their Environments*, Proceedings of the 31st Aerospace Sciences Meeting & Exhibit, Reno, NV (1993), AIAA 93-0705.

Ferguson, D.C., Snyder, D.B., and Carruth, R., *Findings of the Joint Workshop on Evaluation of Impacts of Space Station Freedom Ground Configurations*, NASA TM-103717 (1990).
Ferguson, D.C., *Atomic Oxygen Effects on Refractory Materials*, in *Materials Degradation in Low Earth Orbit (LEO)*, V. Srinivasian and B. Banks, (eds.), TMS, Warrensdale, PA (1990) 100.
Ferguson, D.C., Private Communication, NASA Lewis Research Center (1998).
Flood, D.J. and Brandhorst, H.W., *Space Solar Cells*, in Current Topics in Photovoltaics, T.J. Coutts and J.D. Meakin (eds.), Academic Press, London (1987), 178.
Flood *et al.*, Proceedings of the European Space Power Conference, ESA SP-294, Madrid, Spain (1989), 643-646.
Gale, R.P., McClelland, R.W., Dingle, B.D., Gormley, J.V., Burgess, R.M., Kim, N.P., Mickelsen, R.A., and Stanbery, B.J., *High Efficiency GaAs/CuInSe$_2$ and AlGaAs/CuInSe$_2$ Thin Film Tandem Solar Cells*, PVSC 21 (1990), 53.
Gerlach L., *Solar Generator for the Hubble Space Telescope-Electrical Performance*, The European Space Power Conference, ESA SP-320 (1991), 725-731.
Godfrey, R.B. and Green, M.A., *High Efficiency Silicon MinMis Solar Cells-Design and Experimental Results*, IEEE Transaction Electron Devices 27 (1980), 737-742.
Godlewski, M.P., Baraona, C.R. and Brandhorst, H.W., Jr., *Low High Junction Theory Applies to Solar Cells*, 10th IEEE Photovoltaic Specialist Conference, New York (1973), 40-46.
Green, M.A., *Coming Of Age*, Photovoltaics:, PVSC 21 (1990), 1.
Good, B.S., Chubb, D.L. and Lowe, R.A., *Theoretical Analysis and Computer Modeling of Thermophotovoltaic Energy Conversion Systems*, 2nd NREL Conference on Thermophotovoltaic Generation of Electricity, American Institute of Physics, New York (1995).
Green, M.A., Coming Of Age, Photovoltaics:, PVSC 21 (1990), 1.
Green, M.A., Lakers, A.W., Jiqun, S., Keller, E.M., Wenham, S.R., Godfrey, R.B., Szpitalak, T., and Willison, M.R., *Towards a 20% Efficient Silicon Solar Cell*, 17th IEEE photovoltaic Specialists Conference, New York, (1984), 386-392.
Green, M. A., Wenham, S.R., Zhao, J, Zolpher, J., and Blakers, A.W., *Recent Improvements in Silicon Solar Cell and Module Efficiency*, IEEE 21st Photovoltaic Specialists Conference Proceedings, New York, 207-210, (1990).
Guazzoni, G.E., *High Temperature Spectral Emittance of Oxides of Erbium, Samarium, Neodymium and Ytterbium*, Applied Spectroscopy, Vol. 26, No. 1 (1972), 60-67.
Gupta, S., *Private Communication*, United Solar Systems Corporation (1997).
Herr, J.L., *A Charging Study of ACTS Using NASCAP*, National Aeronautics and Space Administration Publication CR-187088, (1991).
Hoeber, C.F., Robertson, E.A., Katz, I., Davis, V.A., and Snyder, D.B., *Solar Array Augmented Electrostatic Discharge in GEO*, American Institute of Aeronautics and Astronautics, New York, (1998).

Hord, R.M., *Handbook of Space Technology Status and Projections*, CRC Press, Boca Raton, FL (1985) 56-61.
Hovel, H.J., *Solar Cells*, in *Semiconductors and Semimetals*, Vol. 11, Willardson, R.K. and Beer, A.C. (eds.), Academic Press, New York (1975).
Iles, P.A. and Yeh, Y.C.M., *Silicon, Gallium Arsenide and Indium Phosphide Single Junction, One Sun Space Solar Cells*, in *Solar Cells and Their Applications*, L. Partain (ed.), Wiley-Interscience, New York (1995), 99-124.
Jeffrey, F.R. et al., *Flexible, Lightweight Amorphous Silicon Solar Cells Tuned for the AMO Spectrum*, NASA Contract NAS3-25825, Final Report (1990).
Jenkins, P.J., Brinker, D.J., and Scheiman, D., *Uncertainty Analysis of High Altitude Aircraft Air Mass Zero Solar Cell Calibration*, IEEE 26th Photovoltaic Specialists Conference Proceedings, New York (1997).
Jones, P.A. and Murphy, D.M., *A Linear Refractive Photovoltaic Concentrator Solar Array Flight Experiment*, 14th Space Photovoltaic Research and Technology Conference, NASA CP-10180 (1995), 304-312.
Kenney et al., Proceedings of the 25th Intersociety Energy Conversion Conference (1990), 484-489.
Kilmer, L.C., *Dual Source Solar Simulator*, IEEE Proceedings of the First World Conference on Photovoltaic Energy Conversion, New York (1994), 2165-2168.
Kurland, R.M. and Stella, P.M., *Advanced Photovoltaic Solar Array Program Status*, 24th Intersociety Energy Conversion Engineering Conference, Washington, D.C. (1989), 829-834.
Kurtz, S., *Design of High Efficiency, Radiation Hard, GaInP/GaAs Solar Cells*, 13th Space Photovoltaic Research and Technology Conference, NASA CP-3278 (1994), 181-186.
Ladle-Ristow, M., *High Efficiency GaAs Cells*, Proceedings of the ASME/ JSES/KSES International Solar Energy Conference, Lahaina, Hawaii (1992).
Lee, J.M., and Chiang, C.J., *The Potential Performance Of GaAs Based Mechanically Stacked, Multijunction Solar Concentrator Cells*, PVSC 21 (1990), 41.
Lindhard, J. et al., *Integral Equations Governing Radiation Effects*, Notes on Atomic Collisions, III, Mat. Fys. Medd Dan. Vid. Selsk, 33, No.,10 (1963), 1-10.
Lindholm, F.A., Li, S.S., and Sah, C.T., *Fundamental Limitations Imposed by High Doping on the Performance of PN Junction Silicon Solar Cells*, IEEE 11th Photovoltaic Specialists Conference Proceedings, New York (1975), 3-9.
Lindmayer, J. and Allison, F., *An Improved Silicon Solar Cell-The Violet Cell*, IEEE 9th Photovoltaic Specialists Conference Proceedings, New York (1972), 83-84.
Lindmayer, J. and Wrigley, C.J., *Ultra Thin Silicon Solar Cells*, Proceedings of the 13th IEEE Photovoltaic Specialists Conference, New York, (1978), 450-456.
Loferski, J.J, in *Solar Cells-Outlook for Improved Efficiency*, Report by the Ad Hoc Panel on Solar Cell Efficiency, National Research Council, National Academy of Sciences, Washington, D.C. (1972).

Lovgren et al, Proceedings of the 24th Intersociety Energy Conversion Conference (1989), 13-17.

Lowe, R.A., Chubb, D.L., and Good, B.S., *Radiative Performance of Rare Earth Garnet Thin Film Selective Emitters*, 1st NREL Conference on Thermophotovoltaic Generation of Electricity, American Institute of Physics, CP-321 (1994), 291-297.

Malachesky P.A., Simburger, E.H., and Zwibel, H.S., *Fixed Solar Array Designs for GPS Space Vehicles*, Proceedings of the 21st IEEE Photovoltaic Specialists Conference, IEEE, New York (1991), 1572-1575.

Mayet, L., Gavand, M., Montegu, B., Boyeaux, J.P., and Laugier, A., *Monolithic Tandem Solar Cell Based On GaAlAsGaAs System*, PVSC 21 (1990), 64.

McClelland, R.W., Bozler, C.O., and Fan, J.C.C., *Thin, High Efficiency GaAs Solar Cells by Cleaved, Lateral Epitaxial Film Transfer*, Applied Physics Letters, 37 (1980), 560-563.

Messenger, S.R. et al., *Co^{60} Gamma Ray Irradiation of Solar Cells: A New Way to Predict Space Radiation Damage*, IEEE Proceedings of the First World Conference on Photovoltaic Energy Conversion, New York (1994), 2153-2156.

Mirtich, Michael, Private Communication, Lewis Research Center (1991).

Moser, R.L., *Electrical Power Subsystem Initial Sizing*, Proceedings of the 25th Intersociety Energy Conversion Conference, Vol. 2, American Institute of Chemical Engineers, New York (1990), 66-71.

Moser, R.L., Proceedings of the 25th Intersociety Energy Conversion Conference (1990), 66-71.

Nelson, R.E., *Rare Earth Oxide TPV Emitters*, 32nd International Power Sources Symposium Proceedings, Pennington, N.J. (1986), 95-100.

Nored, D.L. and Bernatowicz, D.T., *Electrical Power System Design For The U.S. Space Station*, Proceedings of the 21st Intersociety Energy Conversion Engineering Conference, Vol. 3, Washington, D.C. (1986).

Olsen, L.C. et al., *GaAs Solar Cells for Laser Power Beaming*, Proceedings of the 11th Space Photovoltaic Research and Technology Conference, NASA CP-3121 (1991), 9-17.

Olson, J. and Kurtz, S., *High Efficiency $GaInP_2/GaAs$ Tandem Solar Cells for Space and Terrestrial Applications*, IEEE Proceedings of the First World Conference on Photovoltaic Energy Conversion, New York (1994), 1671-1678.

Olson, J., Kurtz, S.R., Kibbler, A.E., and Faine, P., *Recent Advances In High Efficiency $GaInP_2/Gaas$ Tandem Solar Cells*, PVSC 21 (1990), 24-29.

O'Neill, M.J. and Piszczor, M., *Mini-Dome Fresnel Lens Photovoltaic Concentrator Development*, Proceedings of the 10th Space Photovoltaic Research and Technology Conference, NASA CP-3107, (1989), 443-459.

Owens, T., Arshed, S., and Hill, R.W., *The Production of Low Resistivity Type CdTe Thin Films Using The Coevaporation Of CdTe With Te*, PVSC 20 (1988), 1662.

Parent, C.R. and Nelson, R.E., *Thermophotovoltaic Energy Conversion with a Novel Rare Earth Oxide Emitter*, Proceedings of the 21st Intersociety Energy Conversion Engineering Conference, Vol. 2, American Chemical Society, Washington, D.C. (1986), 1314-1317.

Patil, A.R., Kim, S.K., Cho, B.H., and Lee, F.C., *Modeling and Simulation of the Space Platform Power System*, Proceedings of the 25th Intersociety Energy Conversion Conference, Vol. 2, American Institute of Chemical Engineers, New York (1990), 96-103.

Piszczor, M.F. and O'Neill, M.J., *Recent Developments in Refractive Concentrators for Space Photovoltaic Power*, 12th Space Photovoltaic Research and Technology Conference, NASA CP-3210 (1992), 206-216.

Purvis, C.K., Garrett, H.B., Whittlesey, A.C., and Stevens, N.J., *Design Guidelines for Assessing and Controlling Spacecraft Charging*, NASA TP 2361 (1984).

Piszczor, M.F. and O'Neill, M.J., *Recent Developments in Refractive Concentrators for Space Photovoltaic Power*, 12th Space Photovoltaic Research and Technology Conference, NASA CP-3210 (1992), 206-216.

Piszczor, M.F. et al., *An Update on the Development of a Line Focus Refractive Concentrator Array*, 13th Space Photovoltaic Research and Technology Conference, NASA CP-3278 (1994), 313-322.

Rauschenbach, H.S., *Solar Cell Array Design Handbook*, Van Nostrand Reinhold Co., New York (1980).

Shock, A. and Kumar, V., *Radioisotope Thermophotovoltaic Energy Conversion System*, First NREL Conference on Thermophotovoltaic Generation of Electricity, American Institute of Physics, CP-321, (1994), 139-152.

Slifer, L.W., Jr., Proceedings of a Conference on Space Photovoltaic Research and Technology, Lewis Research Center, Cleveland, Ohio, NASA Conference Publication 3107 (1989), 204-217.

Smith, B.K. and Scheiman, D., *Thermal Cycling Testing of Space Station Freedom Solar Array Blanket Coupons*, 10th Space Photovoltaic Research and Technology Conference, NASA Conference Publication (1989), 410-420.

Snyder, D.B., Private Communication, NASA Lewis Research Center, Cleveland, OH (1998).

Spitzer, M., Loferski, J.J., Shewchun, J., and Vera, E.S., *Ultra High Efficiency Thin Silicon P-N Junction Solar Cells Using Reflecting Surfaces*, IEEE 14th Photovoltaic Specialists Conference Proceedings, New York (1980), 375-381.

Spitzer, M., Private Communication, Kopin Corporation (1988).

Stella, P. and Kurland, R. M., *The Advanced Photovoltaic Solar Array (APSA) Technology Status and Performance*, 10th Space Photovoltaic Research and Technology Conference, NASA Conference Publication CP-3107 (1989), 421-432.

Stella, P.A. and Kurland, R.M., *The Advanced Photovoltaic Solar Array (APSA) Technology Status and Performance*, NASA Space Photovoltaic Research and Technology Conference, NASA CP 3107, Cleveland, OH (1989), 421-432.

Stella, P.M. and Flood, D.J., *Photovoltaic Option for Solar Electric Propulsion*, IEEE 22nd Photovoltaic Specialists Conference Proceedings, New York (1991).

Stevens, N.J., *Interactions Between Spacecraft and the Charged Particle Environment*, NASA CP-2071 (1978).

Stone, K.W. et al.., *Performance and Analysis of a Solar Thermophotovoltaic System*, Proceedings of the 30th Intersociety Energy conversion Engineering Conference, Vol. 1, American Society of Mechanical Engineers (1995), 713-718.

Summers, G.P., Burke, E.A. et al., *Damage Correlations in Semiconductors Exposed to Gamma, Electron and Proton Radiations*, IEEE Trans. Nucl. Sci., 40, No. 6, New York (1993), 1372-1379.

Summers, G.P. et al., *A General Method for Predicting Radiation Damage to Solar Cells in the Space Environment*, IEEE 23rd Photovoltaic Specialists Conference Proceedings, New York (1993), 1426-1431.

Summers, G.P., *Naval Research Laboratory's Programs in Advanced Indium Phosphide Solar Cell Development*, 14th Space Photovoltaic Research and Technology Conference, NASA CP-10180 (1995), 21-23.

Sze, S.M., *Physics of Semiconductor Devices*, 2nd ed., (Wiley-Interscience, New York (1981).

Tada, H.Y. et al., *Solar Cell Radiation Handbook*, 3rd ed., JPL Publication 82-69 (1982).

Tam, K.S., Yang, L., and Dravid, N., *Modeling of Space Station Electric Power System with EMTP*, Proceedings of the 25th Intersociety Energy Conversion Conference (1990), 212-217.

Van-Overstraeten, R.J., *Heavy Doping*, IEEE Transaction Electron Devices 20 (1973), 290-300.

Walters, R.J. et al., *Correlation of Electron and Proton Irradiation-Induced Damage in InP Solar Cells*, 14th Space Photovoltaic Research and Technology Conference, NASA CP-10180 (1995), 89-99.

Wang, E.Y., Yu, F.T.S., and Simms, V.L., *Optimum Design of Anti-Reflection Coating for Silicon Solar Cells*, IEEE 10th Photovoltaic Specialists Conference Proceedings, New York (1973), 168-173.

Wehrli, C., *Air Mass Zero Reference Solar Spectral Reference Irradiance Distribution*, World Radiation Center, Dorfstrasse 33, Davos Dorf, Switzerland (1985).

Wendt, M. and Huhne, H.M., *Sputtering of Iridium Oxide Films as Protective Layers for Oxygen Evolving Photoanodes*, PVSC 20 (1988), 1656.

White, D.C. and Schwartz, R.J., *PIN Structures for Controlled Spectrum Photovoltaic Converters*, AGARD Colloquium, Gordon Breach Science, New York (1964), 897-922.

Wilkinson, V.A., Goodbody, C., and Williams, W.G., *Measurement of Multijunction Cells Under Close-Match Conditions*, IEEE 26th Photovoltaic Specialists Conference Proceeding, New York (1997), 947-950.

Winslow et al., Proceedings of the 24th Intersociety Energy Conversion Conference (1989), 283-287.

Winter, C.P. and Teofilo, V.L., *Electrical Power System Trade Analysis for Space Surveillance Satellites*, Proceedings of the 24th Intersociety Energy Conversion Conference (1989), 19-24.

Woike, T.W., *EOL Performance of GaAs/Ge and Si BSF Solar Arrays*, 12th Space Photovoltaic Research and Technology Conference, Lewis Research Center, Cleveland OH (1992).

Wolff, G. and Wittmann, A., *The Flight of the FRUSA*, 9th IEEE Photovoltaic Specialists Conference, New York (1972), 240-246.

Xapsos, M.A., Messenger, S.R., Walters, R.J., Summers, G.P., and Burke, E.A., *The Probability of Large Solar Proton Events and their Effect on the Efficiency of GaAs Solar Cells*, Progress in Photovoltaics (to be published).

157

CHAPTER 4

CHEMICAL STORAGE AND GENERATION SYSTEMS

1. Introduction

Spacecraft power can be viewed as the ultimate requirement of portable or remote energy storage devices. Energy storage must be accommodated in the spacecraft power system to provide power for the various mission requirements. The mission requirements determine the various types and levels of energy storage. Both primary (one discharge) and secondary (rechargeable) batteries have been used in space applications. The latter is generally recharged using the photovoltaic array on the spacecraft. Fuel cells provide power for primary Shuttle operations and life support as well as power for other equipment, instruments, and spacecraft in the storage bay. Fuel cells are similar to primary cells in that the stored energy is limited to the fuel and oxidant.

The electrochemical cells in the battery are the basic source of the stored energy. The electrically, mechanically, and thermally connected cells form the battery. Each electrochemical cell is a self-contained device that releases stored chemical energy as electrical energy on demand from an electrical load. When the load is connected across its terminals, one electrode in the cell will spontaneously release electrons while the other spontaneously and simultaneously accepts them. The circuit is closed within the cell by the flow between both electrodes of charged species (ions) in the electrolyte. The number and capacity of the connected cells in the battery determine the energy and power capability.

In a primary cell the reactions are irreversible and therefore the chemical energy can be converted to electrical energy only once. In a rechargeable cell the reactions are reversible, and thus, by reversing the flow of electrons (e.g., from a solar array during the sunlight period), the reactions are reversed, restoring the potential energy difference of the electrodes as chemical energy. The ability to reverse the discharge-charge process thousands of times is a function of the cell chemistry.

The fuel cell system (often referred to simply as the 'fuel cell') has been used for manned missions and is the primary power source for the Shuttle. The fuel cell system includes a number of fuel cells electrically assembled like the cells in a battery to form the fuel cell stack, The remainder of the system includes the external fuel and oxidant tanks, water collection apparatus, and the associated electrical, valving, and plumbing hardware.

The difference between the individual battery cell and the individual fuel cell is that in the former the chemical energy is stored and converted to electrical energy within each cell case. In the fuel cell, the chemical energy is stored in the form of hydrogen gas or more recently methanol (the fuel) and oxygen gas (oxidant) in tanks external to the cells. The energy output of each fuel cell is the result of hydrogen gas or methanol reacting at one electrode releasing electrons on demand from the load and the spontaneous and simultaneous reaction of oxygen gas and the electrons at the other electrode. The circuit is closed within the cell by the flow between both electrodes of charged species (ions) in the electrolyte. The product is pure water which can be used for consumption. Like the battery, the voltage of the fuel cell stack is the sum of the individual fuel cell voltages required for the spacecraft power. However, the fuel cell system energy storage capability is limited in life by the quantity of H_2 and O_2 gasses in the external storage tanks.

2. Inventions

A chronological list of the primary inventors and major events in electrochemical energy storage are given in Table 4.1.

Table 4.1 List of Electrochemical Cell Inventions

Date	Inventor	Type	Couple
1786	Galvani	Galvanic Cell	Cu/Fe
1800	Volta	Voltaic Pile	Ag/Zn
1836	Daniell	Cell in H_2SO_4	Cu/Zn
1839	Grove	Fuel Cell	H_2/O_2
1860	Plante	Rechargeable	PbO_2/Pb
1868	LeClanche	Dry Cell	Zn/MnO_2/C
1900	Edison	Storage Battery	Ni/Fe
1901	Jungner	Sealed battery	Ni/Cd
1945	Ruben	Mercury Cell	$Zn(Hg)/MnO_2$/C
1962	Kummer	Sodium/ Sulfur	Na/S
1961	Miracle	Lithium SO_2	Li/SO_2
1962	Watanabe	Li-Carbon Monofluoride	$Li/(CF)_x$
1965	Dunlop	Nickel-Hydrogen	Ni/H_2
1968	Methlie	Lithium- MnO_2	Li/MnO_2
1973	Will	Nickel/Metal Hydride	Ni/MH
1979	Crotzer	Sodium/Nickel Chloride	$Na/ZnCl_2$
1981	Goodenough	Lithium/Cobalt Oxide	Li/CoO_2
1981	Goodenough	Lithium/Nickel Oxide	Li/NiO_2
1981	Hunter	Lithium/Manganese Oxide	Li/MnO_2
1983	Blomgren	Li-Thionyl Chloride	$Li/SOCl_2$

CHEMICAL STORAGE AND GENERATION SYSTEMS

3. Evolution of batteries in space

Silver-Zinc was the battery of choice in the early days of space missions. Nickel-Cadmium batteries became the major energy storage device over the next 20 years because of their long cycle life. The Nickel-Hydrogen battery started to play a role in the 1980s. Lithium-ion batteries are being baselined for JPL's planetary missions. A chronological history of first uses of batteries in space applications appears in Table 4.2.

The earliest use of a battery in an orbital spacecraft was the primary silver-zinc battery used in the Russian spacecraft, Sputnik I, launched October 4, 1956. The primary battery provided power for communication and spacecraft operations. There were no solar cells available for charging and thus when the energy was depleted, communication was terminated. The Ag-Zn primary battery exhibits reasonably high specific energy, and therefore the battery was intended to provide power to the 84 kg spacecraft for up to three weeks, although the spacecraft actually remained in orbit for three months. The second Sputnik, launched a month later, carried the dog known as Laika. It was six times larger and lasted five months. It also utilized a much larger Ag-Zn battery. In November 1961, the U.S. spacecraft, Ranger 3, containing two 14-cell, 50 Ah batteries for the main power and two 22-cell, 50 Ah batteries for the TV camera power, was placed into solar orbit and took photographs of the Moon. Mariner 2, containing one 18-cell, 40 Ah Ag-Zn battery, launched August 27, 1962, was the first successful interplanetary mission to Venus.

The first U.S. spacecraft to be launched was the Vanguard Test Vehicle 3. It utilized zinc-mercuric oxide primary batteries (in the form of 'D' sized cylindrical cells) and solar cells to provide power. Unfortunately, it failed to orbit. However, Explorer 1, the first of several Explorer spacecraft, was launched February 1, 1958. It was successful in discovering the Van Allen radiation belt. The spacecraft was a cylinder 80 inches in length and 6 inches in diameter containing 5 kg of instruments, batteries (primary mercury type), and a radio.

Explorer 6, launched in August 1959, was the first successful launch of Nickel-Cadmium (Ni-Cd) cells. In August of that year Pioneer, the first stage of the lunar probe, contained Ni-Cd and Ag-Zn cells. The first time Ni-Cd batteries were used for prime power was in February 1960. Transit 1B containing two packs of 28, 5 Ah cylindrical Ni-Cd cells produced by the Sonotone Company, Elmsford, New York, failed to launch. However, in April 1960, the weather satellite TIROS I was successfully launched. It contained three strings of 21 Ni-Cd cylindrical cells. These cylindrical F-5 cells contained glass-to-metal seals to insulate the positive terminal from the metal case. These cells were also fitted with a threaded base which was used to screw into threaded holes in the battery baseplate. The spacecraft operated in a 90 to 110 minute orbit. The spacecraft electrical loads were designed to remove only a conservative 3% of the capacity from the battery during the 30 or so minute eclipse period. The low depth-of-discharge (DOD) is a significant factor in extending life. However, it also means a

Table 4.2 Chronological List of First Use of Batteries in Space

Date (m/d/y)	Satellite	Duration	Type	Comments
10/4/56	SPUTNIK I	3 months	Ag/Zn	1W for 3 weeks
12/6/56	VANGUARD	Failed	Zn/HgO	First U.S. launch
2/1/58	EXPLORER 1	3.8 months	Zn/HgO	Van Allen Radiation Belt
8/6/59	EXPLORER 6	2 years	Cyl Ni/Cd	First earth photos
3/13/61	IMP 1	3.5 years	Ag/Cd	Non-magnetic
1/26/62	RANGER 3	Solar orbit	Ag/Zn	Moon photos
4/26/62	ARIEL I	14 years	Pris Ni/Cd	First LEO mission
8/27/62	MARINER 2	Venus probe	Ag/Zn	Venus mission
6/23/63	SYNCOM-2	Communications	Cyl Ni/Cd	First GEO
5/20/65	APOLLO CM	Short	Ag/Zn	LTD cycle life
6/23/66	NTS-2	5 years	Ni/H$_2$	12 hour polar
9/23/66	USAF	Classified	Ni/H$_2$	LEO
2/14/80	SOLAR MAX	8 years	Ni/Cd	Standard battery
4/4/83	STS-3	Days	Li-BCX	Astronaut use
5/19/83	INTELSAT V	14 years	Ni/H$_2$	GEO
4/6/84	LDEF	6 years	LITHIUM	Exposure to space
10/18/89	GALILEO	Hours	Li-SO$_2$	Jupiter probe
4/25/90	HST	In orbit	Ni/H$_2$	NASA LEO
6/10/90	LEASAT	Orbiting	Super Ni/Cd	GEO
1/25/94	CLEMENTINE	5 months	SPV Ni/H$_2$	Lunar mapping
1/25/94	TUBSAT-B	4 years	2 Cell CPV	Store messages
5/19/95	CENTAUR	1st mission	Li-SOCl$_2$	28V, 250AH battery
5/5/96	IRIDIUM-1	Commercial	50Ah SPV	34 to date-LEO
12/4/96	Mars Lander	JPL mission	Ag/Zn	40AH rechargeable
12/4/96	Mars Rover	JPL mission	Li-SOCl$_2$	3 'D' cell batteries
11/19/97	FLIGHT EXP	USAF experiment	Na/S	Wakeshield platform

CHEMICAL STORAGE AND GENERATION SYSTEMS 161

much larger battery is required for the low DOD requirement. This is a high price to pay in inefficiency and cost.

In November 1964, the first prismatic Ni-Cd cells produced by Gulton Industries were flown on Explorer 23. Each cell was a box-like configuration which provided a means for producing an efficient battery design in which the cells were lined up in a close knit battery pack held together with end-plates and metallic rods. The cells also were designed with two insulated terminals such that the cells were electrically floating. General Electric (G.E.) of Gainesville, Florida, and Eagle-Picher (E-P) Company in Joplin, Missouri, succeeded in developing and flying Ni-Cds in space.

In 1966, the OAO-series of spacecraft developed by the prime contractor, Grumman Aerospace, utilized three batteries of 20 Ah prismatic Gulton cells uniquely assembled into two battery frames. Pairs of cells were interspersed between the two assemblies to minimize temperature variation. This power system utilized a V_T (temperature-compensated voltage) charge control system that applied a constant voltage to batteries during charge. The preset V_T limited the charge, resulting in a safe and reliable charge condition. This is a condition in which the batteries return to full charge but are not excessively overcharged. The selected voltage limit was based on a parallel set of temperature-compensated voltage curves (V_T curves). The V_T curve selection used to limit the charge voltage provided flexibility to account for unexpected high depths of discharge and/or imbalance between cells and/or batteries.

In the mid 1970s, NASA undertook a program to develop 'standard' flight hardware including 'standard' cells and batteries, a Modular Power System (MPS), an Attitude Control (ACS), Computer and Data Handling (C&DH), tape recorders, etc. To this end, Goddard Space Flight Center (GSFC) was responsible for developing the 'standard' Ni-Cd cell and then the 'standard' battery. Four companies were given the opportunity to develop prismatic 20 Ah 'standard' Ni-Cd cells. Cells from the four manufacturers would then be capable of being assembled into the 'standard battery' structure produced by McDonnell Douglas with the end use in the MPS. The battery was designed to meet all NASA mission and launch requirements including vibration and shock. The competing manufacturers included General Electric, Eagle-Picher, Yardney Technical Products (YTP), Pawcatuck, Connecticut, and Saft Corp., Valdosta, Georgia. Each company submitted cells for evaluation; however, only the G.E. cells were selected as 'NASA standard cells.' The cells were accompanied by a 'Manufacturing Control Document' (MCD) which was intended to provide a consistent process and reliable product. The first lot of 'standard' 20 Ah cells in 'standard' 20 AH batteries were flown successfully on the Solar Max Mission (SMM) for more than eight years. Subsequently, the technology was extended to 50 Ah cells which were used on several NASA spacecraft, e.g., Landsat, TOPEX, UARS, and GRO.

Also in the 1960s, a technology was discovered that made use of the NiOOH electrode from the Ni-Cd cell and the H_2 electrode from the fuel cell. The Individual Pressure Vessel (IPV) Ni/H_2 cell was contained in a pressure cylinder configuration due to the buildup of hydrogen during charge; as much or greater than 400-800 psi. The replacement of the cadmium electrode with a hydrogen electrode reduced weight and increased energy, thus almost doubling the specific energy over the Ni-Cd cell. However, because of the cylindrical configuration and the wider spacing of the cells on the baseplate, the energy density (Wh/liter) of the battery was similar to that of the Ni/Cd battery. However, the Ni-H_2 system offered the capability of extended life at higher DOD. Comsat was the first to develop this battery and use it in the Intelsat spacecraft in a GEO mission. In 1983, Eagle-Picher was successful in combining two cells in the same cylinder. This Common Pressure Vessel (CPV) cell was first used in JPL's Mars Global Surveyor mission in 1994. The next step resulted in the development of a rechargeable Single Pressure Vessel (SPV) Ni-H_2 battery in which 22 cells were mounted in the same structure. It was used for the first time in 1994 in Clementine, a Navy satellite which circled the Moon. Ni-MH cells using chemically bonded hydrogen in the form of a hydride have been used in a few rocket experiments, but have not been used in any major flight program. However, because of the low pressure these cell cases do not require high pressure cylinders.

Lithium primary cells were used in space during the 1980s. Li-$(CF)_x$ was one of the first used for range safety on launch vehicles. Li-BCX (Li-$SOCl_2$ cells with bromine chloride additive), produced by Wilson Greatbatch Limited, Clarence, NY, was selected by NASA/Johnson Space Center for use in astronaut equipment, specifically the helmet lights and TV camera. Later, Li-SO_2 batteries from Alliant TechSystems were selected for the probe on the Galileo mission to Pluto that required nine years of storage before use. Seven kWh Li-$SOCl_2$ (250 Ah, 28 V, batteries were developed by Saft and Yardney for the Air Force Centaur launch vehicles to replace the Ag-Zn batteries in order to extend operating time in placing payloads in orbit. Smaller 'D' size cells based on this technology were used by JPL in the Mars Rover in 1997.

In the 1980s, JPL initiated development of a rechargeable Lithium cell in an in-house program. The Lithium-Titanium Disulfide (Li-TiS_2) used pure lithium as the anode. The specific energy achieved 100 Wh/kg or 2 times the NiH_2 or Ni-MH system and was cycled more than 1000 times at 50% DOD. However, the use of the metallic lithium foil concerned the users. In the follow-on lithium-ion cell development effort in the 1990s, coke or graphite replaced the Lithium anode foil and several cathode materials, e.g., cobalt, nickel or manganese oxides, replaced the undesirable titanium disulfide. There was no lithium metal in the cell. This new system made use of the difference in concentration of Li ion between anode and cathode. The potential of each cell is 4.0 volts and the specific energy greater than 125 Wh/kG. This new technology is being considered for a number of NASA and AF programs. A diagram of the Ni-Cd, Ni-H_2, and Li-Ion batteries stages is shown in Figure 4.1.

CHEMICAL STORAGE AND GENERATION SYSTEMS 163

	1958-69	1970-79	1980-89	1990-97
NI-CD	'59 EXPLORER-6 (CYLINDRICAL) '62 ARIEL-1 LEO (PRISMATIC) '63 SYNCOM2 GEO		'80 SOLAR MAX NASA 20 AH STD '82 LANDSAT -D STD 50 Ah	'90 LEASAT GEO SUPER NI-CD
NI-H$_2$		'77 NTS-2 & AF FIRST IPV USE	INTELSAT 5 IPV GEO	'90 HUBBLE IPV '94 CLEMENTINE SPV '94 TUBSAT CPV
LITHIUM			'83 STS- LI-BCX '89 GALILEO Li-SO$_2$	'95 CENTAUR '96 PATHFINDER

Figure 4.1 Major Steps in the Development of Ni-Cd, Ni-H$_2$ and Li-Ion Batteries

4. Fundamentals of electrochemistry

Electrochemistry, for the most part, is based on thermodynamics. As given in Eq. (4.1), electrical work (W) equals the quantity of electricity (Q) multiplied by the electric potential or electromotive force (E).

$$W = Q \times E = I \times t \times E \quad (4.1)$$

In the operation of an electrochemical cell, Q is equal to the current (I) integrated over time (t). This value of I x t is generally defined as the product of ampere-seconds (coulombs) or ampere-hours. It is related to n, the number of moles of electrons produced by one mole of reactant, E, the electrochemical potential, and F, the Faraday constant (units of ampere-seconds or coulombs):

$$W = n F E \quad (4.2)$$

If the components are at unit activity, then $E = E°$, where $E°$ is the Standard Electrochemical Potential, and furthermore if the cell operates reversibly, then $w = w_{max}$. Because electrical work is non PV work (useful work), $w_{elec,max} = - \Delta G$ (Gibbs Free Energy). Thus, Eq. (4.2) becomes:

$$\Delta G° = - n F \Delta E° \quad (4.3)$$

4.1 Standard electrode potential and free energy

E^o is the standard electrochemical reaction at unit activity and standard temperature and pressure conditions (25°C and 1 atm) given in Table 4.3. Thus, ΔG^o, a thermodynamic property, can be determined from the difference (ΔE^o) between half-cell reactions. By the adopted conventions, if ΔG^o is a negative value the reaction is spontaneous and exothermic (heat released). If ΔG^o is positive, the reaction will be non-spontaneous and endothermic (heat required). Standard half-cell (single electrode) electrode potentials are generally shown as reduction reactions, i.e., reactions use electrons to produce a product having a lower valence state. Reactions written in the reverse will have reverse polarity. A spontaneous reaction will occur within a cell when the reaction is written in such a way that the ΔE^o is positive. To obtain a positive ΔE^o, the reactions have to be

Table 4.3 Representative Standard Electrode Potentials
(In Acid Solutions)

Element	Oxidized Form		Reduced Form	Reduction Potential E^o (V)
Potassium	$K^+ + e^-$	=	K	- 2.925
Calcium	$Ca^{2+} + 2e^-$	=	Ca	- 2.86
Sodium	$Na^+ + e^-$	=	Na	- 2.614
Magnesium	$Mg^{2+} + 2e^-$	=	Mg	- 2.36
Aluminum	$Al^{3+} + 3e^-$	=	Al	- 1.66
Manganese	$Mn^{2+} + 2e^-$	=	Mn	- 1.18
Zinc	$Zn^{2+} + 2e^-$	=	Zn	- 0.663
Iron	$Fe^{2+} + 2e^-$	=	Fe	- 0.440
Cadmium	$Cd^{2+} + 2e^-$	=	Cd	- 0.040
Cobalt	$Co^{2+} + 2e^-$	=	Co	- 0.266
Nickel	$Ni^{2+} + 2e^-$	=	Ni	- 0.250
Lead	$Pb^{2+} + 2e^-$	=	Pb	- 0.126
Hydrogen	$2H^+ + 2e^-$	=	H_2	0.000
Copper	$Cu^{2+} + 2e^-$	=	Cu	+ 0.336
Iodine	$I_2 + 2e^-$	=	$2I^-$	+ 0.536
Silver	$Ag^{2+} e^-$	=	Ag	+ 0.699
Mercury	$Hg^{2+} + 2e^-$	=	Hg	+ 0.854
Oxygen	$O_2 + 4H^+ + 4e^-$	=	$2H_2O$	+ 1.229
Chlorine	$Cl_2 + 2e^-$	=	$2Cl^-$	+ 1.356
Gold	$Au^+ + e^-$	=	Au	+ 1.68
Fluorine	$F_2 + 2e^-$	=	$2F^-$	+ 2.86

(In Basic Solutions)				
Hydrogen	$2H_2O + 2e^-$	=	$H_2 + 2OH^-$	-0.828
β NiOOH	$\beta NiOOH + H_2O + e^-$	=	$Ni(OH)_2 + OH^-$	+0.490
Oxygen	$O_2 + 2H_2O + e^-$	=	$4OH^-$	+0.401
Silver	$Ag_2O + H_2O + 2e^-$	=	$2Ag + 2OH^-$	+0.345
Cadmium	$Cd(OH)_2 + 2e^-$	=	$Cd + 2OH^-$	-0.809
Zinc	$ZnO_2^{2-} + H_2O + 2e^-$	=	$Zn + 4OH^-$	-1.216

CHEMICAL STORAGE AND GENERATION SYSTEMS 165

written in such a manner that the sum is positive. This requires that reactions be written such that the electrons used are the same number as that released. As an example, using Table 4.3, the following oxidation and reduction reactions can be arranged to show a positive $E°$.

Zn	$\rightarrow$	$Zn^{++} + 2e^-$	0.663 V	(4.4)
$Cu^{++} + 2e^-$	$\rightarrow$	Cu	0.336 V	(4.5)
$Cu^{++} + Zn$	$\rightarrow$	$Cu + Zn^{++}$	1.100 V	(4.6)

The reaction is balanced and both electrons released in the Zn oxidation reaction (4.4) are used in the Cu reduction reaction (4.5). The +1.100 V for $\Delta E°$ of the net reaction (4.6) will result in a negative ΔG and therefore the reaction as shown is spontaneous to the right. This is also expected in practice wherein copper will plate out in the presence of an active metal such as zinc (provided the electrolyte contains Cu^{++} ions). $\Delta G°$ can be calculated for this reaction as follows:

$$\Delta G° = -n F \Delta E° \qquad (4.7)$$

using 23.06 kcalories/volt for the Faraday constant,

$$\Delta G° = 50.73 \text{ Kcal} \qquad (4.8)$$

4.2 The Nernst equation

The Nernst Equation for a half cell reaction is given in Eq. (4.9). The terms $a_{products}$ and $a_{reactants}$ refer to the activity of the constituents. The approximation shown in Eq. (4.10) can also be used for the oxidation/reduction reactions for a full cell. It takes into account the variation from standard conditions due to temperature and concentration of reactants and products shown as moles per liter.

$$E = E° - RT/nF \ln a_{products} / a_{reactants} \qquad (4.9)$$

or

$$\Delta E = \Delta E° - 0.059 \log [\text{products}] / [\text{reactants}] \qquad (4.10)$$

where ΔE is the calculated voltage, $\Delta E°$ is the reversible equilibrium voltage (difference in half cell reactions) for the net reaction, R is the gas constant (1.986 Cal/deg C/mole), T is the absolute temperature (25°C = 298 K), n is the number of electrons per mole, and F is the Faraday constant (23,06 Cals/Volt).

$$\frac{RT}{nF} \ln\left(\frac{Zn^{++}}{Cu^{++}}\right) = 0.59 \log\left(\frac{Zn^{++}}{Cu^{++}}\right) \quad \text{at } 25°C, \ n=1 \qquad (4.11a)$$

In the above example of the Cu and Zn reactions:

$$\Delta E = 1.100 - 0.059 \log [Zn^{++}] / [Cu^{++}] \qquad (4.11b)$$

Representative values for theoretical voltages, specific energy, and energy density are shown in Table 4.4, along with the two types of fuel cells for comparison.

4.3 Capacity and the Faraday relationship

The capacity available from stored active materials is usually expressed in coulombs (ampere seconds) or ampere hours (Ah). It is based on the Faraday expression which relates the number of coulombs (96,494) required to plate 1 gram equivalent (112.41g) of silver (Ag) from a solution at unit activity and concentration of silver ion (Ag^+) according to Eq. (4.12).

$$Ag^+ + e^- \rightarrow Ag \qquad (4.12)$$

Cell and battery capacity is generally described in terms of ampere hours. The Faraday constant is 26.8 Ah per equivalent. For reactions where more than one electron is involved, the atomic or molecular weight is divided by the number of electrons to determine the equivalent weight. In the example of the copper reaction, Eq. (4.5), the reaction requires two electrons to plate 63.5 g of copper. The atomic weight of copper is 63.5. Therefore, 26.8 Ah would result in 31.8 grams of plated copper.

5. Cell and battery mechanical design

The discussion below provides the basis for understanding cell and battery design concepts.

5.1 Cell design

An electrochemical cell consists of two electrodes (anode and cathode), electrolyte, separator, insulator, insulated seal(s), terminals, and case. The anode and cathode electrodes, each comprised of one or more plates, contain active material of opposing potential energy and polarities. Plates of different polarities are alternated and separated from each other with a separator which provides electrical insulation and in some cases serves as a container for the electrolyte. Each plate contains electrochemically producing ac-

CHEMICAL STORAGE AND GENERATION SYSTEMS

Table 4.4 Theoretical Voltage and Capacity of Major Battery Systems (from Linden, 1984)

Battery	Reaction Mechanism	V	Theoretical Capacity g/Ah	Ah/kg	Wh/kg
PRIMARY CELLS					
LeClanche	$Zn + 2MnO_2 \rightarrow ZnO + Mn_2O_3$	1.6	4.46	224	358
Alkaline MnO_2	$Zn + 2MnO_2 \rightarrow ZnO + Mn_2O_3$	1.5	4.46	224	336
Mercury	$Zn + HgO \rightarrow ZnO + Hg$	1.34	5.26	190	254
Silver Oxide	$Zn + Ag_2O + H_2O \rightarrow Zn(OH)_2 + 2Ag$	1.6	5.55	180	288
Zinc/Air	$Zn + 2O_2 \rightarrow ZnO$	1.65	1.55	658	1320
Lithium/Sulfur Dioxide	$2Li + 2SO_2 \rightarrow Li_2S_2O_4$	3.1	2.64	369	1175
Lithium/Thionyl Chloride	$4Li + 2SOCl_2 \rightarrow SO_2 + 4LiCl + S$	3.6	2.48	403	1471
Li/Manganese Dioxide	$Li + Mn^{4+}O_2 \rightarrow Mn^{4+}O_2(Li^+)$	3.5	3.50	286	1001
SECONDARY CELLS					
Lead-Acid	$Pb + PbO_2 + 2H_2SO_4 \rightarrow 2PbSO_4 + 2H_2O$	2.1	8.32	120	252
Nickel/Cadmium	$Cd + 2NiOOH + 2H_2O \rightarrow 2Ni(OH)_2 + Cd(OH)_2$	1.35	5.52	181	244
Silver/Zinc	$Zn + AgO + H_2O \rightarrow Zn(OH)_2 + Ag$	1.85	3.53	283	523
Nickel/Hydrogen	$H_2 + 2NiOOH \rightarrow 2Ni(OH)_2$	1.5	3.46	289	433
Nickel/Metal Hydride	$MH + NiOOH \rightarrow M + Ni(OH)_2$	1.35	6.50	206	278
Silver/Cadmium	$Cd + AgO + H_2O \rightarrow Cd(OH)_2 + Ag$	1.4	4.41	226	318
Li / Cobalt Oxide	$LiC_6 + CoO_2 \rightarrow 6C + LiCoO_2$	4.00	6.08	164	656
Li /Nickel Oxide	$LiC_6 + NiO_2 \rightarrow 6C + LiNiO_2$	3.90	6.08	164	640
Li/Manganese Oxide	$LiC_6 + MnO_2 \rightarrow 6C + LiMnO_2$	4.50	9.18	109	491
Sodium/Sulfur	$2Na + 3S \rightarrow Na_2S_3$	2.10	2.65	377	792
Sodium/Nickel Chloride	$2Na + NiCl_2 \rightarrow 2NaCl + Ni$	2.58	3.28	305	787
FUEL CELLS					
H_2/O_2	$H_2 + O_2 \rightarrow H_2O$	1.23	0.336	2965	3662
$MeOH/O_2$	$CH_3OH + 3/2 O_2 \rightarrow CO_2 + 2H_2O$	1.21	0.199	5025	6080

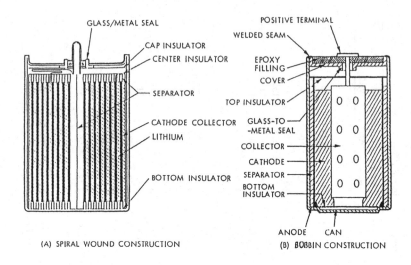

LITHIUM PRIMARY CYLINDRICAL CELL CONFIGURATIONS

Figure 4.2 Diagrams of Cylindrical Cell Construction

tive material in contact with a current collector from which the electrons either enter or exit from the cell. A porous conductive substrate is used in some cases to house the active material. Cylindrical cells are either spirally wound to provide higher rate capability or of bobbin construction which utilizes a center anode or cathode with the alternate electrode surrounding the center rod. A diagram of the two types of cylindrical cells is given in Figure 4.2.

A cell pack consists of the electrodes and separator in close contact to minimize ohmic losses. Exiting from the plates is an electrically conducting metallic tab for making electrical contact to the terminal inside the cell. The tabs of the same polarity are connected together directly or via a busbar and can easily be welded to their designated terminal within the cell during the cell assembly process. An example of a Ni-Cd cell pack is shown in Figure 4.3.

The non-woven nylon or polypropylene separator, which is chemically inactive in the cell environment, is used to maintain separation of plates thus avoiding shorting. It also has the function of providing a path for ions in the electrolyte to diffuse between plates and thus maintain optimum conductivity. In the case of a 'starved' (semi-dry) cell design wherein the electrolyte volume is limited in the cell, the separator can also serve as a sponge to maintain contact between the plates. The separator material must be chemically and thermally stable as well as have a minimal effect on the resistance to flow of ions within the cell.

CHEMICAL STORAGE AND GENERATION SYSTEMS

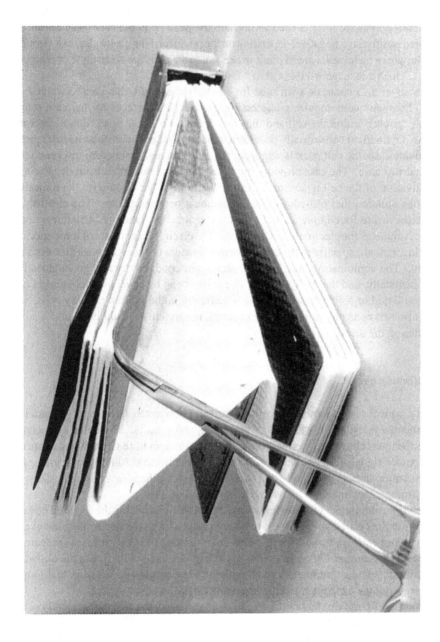

Figure 4.3 Prismatic Cell Constituents

The cell pack is inserted into a polymeric or metallic type case. Hermetically sealed cells desired for space applications use low carbon 304 stainless steel for the case and cover and a polymeric liner is used to insulate the pack from the case The terminals designated positive and negative, depending on the polarity of the electrodes, exit through the cover where metal seals are in place to prevent leakage of the electrolyte or reaction products. Such a leakage would lead to a failure of the cell.

The cover also contains a fill tube for electrolyte addition that can be welded or sealed. Prismatic cells require compression on their flat surfaces to maintain interelectrode spacing within the cell and thus avoid expansion of the case due to internal pressure. Cylindrical cells usually provide a structure that can withstand internal pressure. In this case, the cell pack is wrapped taut around a mandrel before insertion into the cylindrical case. The case provides the strength to maintain interelectrode spacing.

Activation of the cell is accomplished by addition of the electrolyte, an ionically conducting solution that provides the ion movement between plates. The electrolyte provides the means for completion of the electrochemical oxidation/reduction reactions and thus completes the electrochemical circuit. The electrolyte consists of ionic species (salt, acid, or alkali) specific to the electrochemical nature of the electrodes dissolved in a solvent. The combination must provide maximum conductivity and be chemically, electrochemically, and thermally stable. The solvent can be aqueous, inorganic or organic, and liquid or solid. In some cases, specifically in the lithium primary cells, the solvent also serves as the cathode active material, referred to as 'catholyte,' and is consumed during the discharge.

5.2 Battery design

A battery consists of a number of cells electrically connected in a series or parallel arrangement, a housing and baseplate, connectors, and sensors. The battery structure contains the intercell wiring, thermal fins, voltage and temperature sensors, and connectors for power, signals from the sensors, and voltage monitoring. Although surprising to some, the battery is only a structure that contains the electrochemical cells in the desired voltage arrangement and does not store energy in and of itself.

The prime requirement for a battery is to be capable of providing the required power and energy at the desired voltage and over the required period of time. Among the overriding requirements are those of minimum size, volume, and cost to meet spacecraft requirements. *Caution:* On occasion the spacecraft designers will dictate the allowed volume, mass, and cost prematurely and without due consideration for the real long-term energy storage capability and the usual growth of the power budget.

From a configuration standpoint, for space use a battery must be mechanically configured to withstand a wide range of shock and vibration, be capable of dissipating heat and maintaining a uniform temperature between cells and across the battery, utilize wire

CHEMICAL STORAGE AND GENERATION SYSTEMS

size and intercell connectors to minimize voltage drop, be equipped with sensors to provide the power system with the data with which to make decisions, and include connectors to interface power, sensing, and controls to the spacecraft.

The size of a battery will depend on the voltage and capacity required. These in turn determine the number and size of cells. The design of a battery can be initiated after selecting (described later) the cell capacity and configuration (prismatic, cylindrical) required to meet mass, volume, voltage life, and energy storage requirements with adequate margin. It is essential that the cells in a battery and between batteries in parallel be of the same type and capacity to avoid discontinuous performance leading to early battery failure. In addition, it is most important that the cells be from the same manufacturing lot so as to assure consistent performance across and between batteries and to assure mission reliability.

The battery base-plate size will be determined by the cell configuration and spacecraft requirements. Prismatic cells offer the best packaging factor because the cells can be placed in a close-packed arrangement. Thermal dissipation and control occur through the wide flat face of the cell through thermal fins between the cells to the baseplate. Endplates on both ends of the cell stack maintain compression on the cells and connection to the baseplate. The battery-to-cell weight ratio for a prismatic cell battery can be expected to be as low as one to 1.15. However, for a cylindrical cell battery the ratio is more like one to 1.5. Although each cylindrical cell is structurally sound, mounting the cells into a battery configuration onto a baseplate leads to inefficient packaging. The cylinders also have to be arranged so that adequate heat is removed through the baseplate which adds to the inefficiency and complexity of heat removal.

6. Performance metrics

Considerations for the selection of a cell and battery to meet mission goals depend on performance characteristics and other practical and safety factors. These are listed in Table 4.5. The electrical parameters used to characterize battery performance are described below.

Table 4.5 Battery Selection Factors

Mass	Voltage	Capacity	Specific Energy
Volume	Rate Capability	Cycle Life	Energy Density
Design	Maintenance	Availability	Depth of Discharge
Cost	Efficiency	Environmental	Temperature Range
Safety	Storage	Voltage Profile	Voltage as f(Temp)

6.1 Voltage

The number and type of cells in a series or parallel arrangement determine the battery voltage. Therefore the discussion that follows relates to cell voltage. The voltage of a cell depends on the potential energy difference of the active materials stored in the anode and cathode as described above.

Open circuit

The open circuit voltage of a cell is the voltage measured without a current drain. It is near the equilibrium voltage of the anode and cathode reactions in contact with the electrolyte. However, even at open circuit, the voltage of a cell will depend to a small extent on the relative quantities of charged and discharged active materials in those electrodes. For a fresh or fully charged cell at open circuit, the potential energy difference between both electrodes is at its highest, and this results in a high open circuit voltage. The nearer to the end of discharge, the closer the potential energy of the electrode active materials and therefore the lower the equilibrium open circuit voltage. The measured equilibrium voltage is also dependent on temperature. The Nernst Equations (4.9, 4.10) describe the effect of temperature on voltage.

Voltage during discharge and charge

Cell voltage during discharge and charge is affected by several factors. Once discharge or charge is initiated, voltage moves from the open circuit value. During discharge, cell voltage will fall continuously toward zero volts where there is little potential energy difference between the electrodes. During charge, electrons are being forced into the cell to return the active materials in the anode and cathode to the state of highest potential energy difference. Therefore, the cell voltage will increase above its equilibrium voltage. If the charging continues without a charge control mechanism, additional reactions can occur within the cell which result in undesirable products that can exceed the safe operating conditions.

The term 'polarization' is used to describe the voltage deviation from equilibrium caused by current flowing through the cell and its constituents. The voltage offset is due to electrical, ionic, and/or kinetic 'impedance' within the cell or its chemical reactions. The three types of polarization experienced in electrochemical reactions are: ohmic (η_{IR}), activation (η_{Act}), and concentration (η_{Conc}) polarization. All three are also influenced by temperature, the higher the temperature the less the effect of the polarization and the less the offset from equilibrium voltage. Figure 4.4 describes how the three factors affect polarization.

CHEMICAL STORAGE AND GENERATION SYSTEMS 173

The voltage change introduced by ohmic polarization (η_{IR}) is caused by current flowing through the inert parts of the cell from one terminal to the other. This change occurs the instant the current is drawn or applied and is sometimes used to measure the resistance of a cell by subjecting it to a current pulse ($\Delta E/\Delta I = R$). The η_{IR} remains relatively constant during charge and discharge.

Activation polarization (η_{Act}) is a kinetic phenomenon that is related to the charge transfer step of the electrochemical reaction. In viewing this phenomenon on an oscilloscope, a curve similar to the capacitor discharge curve is seen. This mimics the ca-

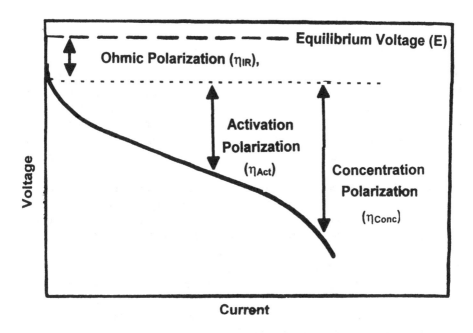

Figure 4.4 The Types of Polarization

pacitor action in that the ions and electrons are aligned at the electrode surface and undergo conversion to the desired products at the active electrode site.

Concentration polarization (η_{Conc}) refers to the effect on conductivity of the ionic species in the electrolyte and active material. It is most predominant at low concentrations and at the end of discharge where the quantities of the charged active materials are low. At higher concentrations the electrolyte tends to be more conductive than at lower concentrations. Also, aqueous electrolytes tend to be more conductive than organic or other inorganic electrolytes. The higher the concentration the less the affect on voltage. As the active material in either electrode is depleted, the voltage during discharge will fall slowly until one of the active ingredients has been exhausted. At that time the voltage will fall precipitously toward zero volts.

The greater the capacity removed during discharge, the greater the effect of the polarization factors. Because there is less active material to convert, the electrolyte may become depleted, and/or some active materials may become passive, thus increasing the resistance to current flow. This is another way of bringing up the subject of depth of discharge. The greater the quantity of total cell capacity removed on a percentage basis (DOD), the closer to the depletion of active material, the lower the cell voltage, and the closer to cell failure. The reverse is true during charge.

6.2 Capacity and energy

The capacity refers to the number of hours the required load current can be sustained during the discharge (for cells and batteries the units are ampere hours). It is derived from the Faraday relationship. (See Section 4.3 for a description of the basis for capacity determination.) Except for laboratory testing, rarely is the current constant during a mission. With equipment and heaters turning on and off and pulsing from the controls, the battery cells have to be capable of providing a wide range of current within the time or orbit requirement.

Electrical engineers can view the battery capability to store and deliver energy as related to a capacitor. Capacitors store energy like a battery; however, the energy is released in fractions of a second. The energy storage materials are on the surface of the electrode and are controlled by the dielectric within the case. They are rated in terms of microfarads and exhibit characteristics relative to known discharge/time relationships. On the other hand, the comparable metric for energy storage is on the order of Farads. It equates to the quantity of stored energy in a cell or battery released over a time period of minutes to hours. As long as there are chemically active stored materials within both electrodes, the cell voltage will be maintained and the battery will release or store energy over a period of time depending on rate.

A list of the theoretical capacities and specific energy of electrochemical cells used in space was given in Table 4.4. The theoretical specific energy (Wh/kg) is determined by the product of the voltage and the value for Ah/kg. The practical values for these is shown in Table 4.6.

The list includes a number of primary cells that have been used in Shuttle instruments and astronaut equipment. The secondary cells include types that have either been used for space or have been evaluated for use in space. The Lead-Acid characteristics are listed for comparison. In addition, the capacity numbers are provided in the Table.

CHEMICAL STORAGE AND GENERATION SYSTEMS

The use of H_2 or methanol offers a ten-fold increase of energy over that of primary and rechargeable batteries.

Capacity is affected by temperature, discharge rate, charge rate, and charge control methodology. It is also dependent on the relative quantities of active material in the two electrodes. The electrode with the limiting capacity is the one that determines which electrode will have its active material depleted first.

Energy is simply the integral of the voltage and current over time. In general, an estimate of the energy of a battery can be made by taking the product of the average voltage, average current, and time.

6.3 Specific energy and energy density

Important metrics to the spacecraft engineers are specific energy (SE), a function of the total energy of a battery or cell per unit weight, and the energy density (ED), the total energy of a battery or cell per unit of volume. These parameters enables the engineer to

Table 4.6 State of Art Operational Characteristics of Major Cell Types

Battery System	Anode	Cathode	Nominal Cell Voltage (V)	Specific Energy (Wh/kg)	Energy Density (Wh/l)
PRIMARY CELLS					
LeClanche	Zn	MnO_2	1.5	85	165
Alkaline-MnO_2	Zn	MnO_2	1.5	125	330
Mercury	Zn	HgO	1.3	100	460
Silver Oxide	Zn	Ag_2O	1.6	120	500
Zinc/Air	Zn	O_2(air)	1.5	340	1050
Li/SO_2	Li	SO_2	3.0	260	415
Li/$SOCl_2$	Li	$SOCl_2$	3.6	320	600
Li/MnO_2	Li	MnO_2	3.0	230	550
Li$(CF)_n$	Li	$(CF)_n$	3.0	220	410
SECONDARY (RECHARGEABLE) CELLS					
Lead-acid	Pb	PbO_2	2.0	35	60
Nickel-Cadmium	Cd	NiO_2	1.2	35	80
Nickel-Metal hydride	(MH)	NiO_2	1.2	50	65
Nickel-hydrogen	H2	NiO_2	1.2	55	60
Silver-Zinc	Zn	AgO	1.5	90	180
Silver-cadmium	Cd	AgO	1.1	55	100
Zinc-Air	Zn	O_2 (air)	1.5	150	160
Lithium-Ion	C	$LiCoO_2$	4.0	90	125
Lithium-Organic	Li	Mn_2O_4	3.0	120	230
Lithium-Polymer	Li	V_6O_{13}	3.0	200	150
Sodium- Sulfur	Na	S	2.0	160	110
Zebra	Na	$NiCl_2$	2.3	120	110

make an estimate of what can be expected of a particular battery chemistry. Generally, primary batteries are higher in both parameters than rechargeable. For example, a Lithium-Thionyl Chloride primary cell has a theoretical specific energy of 1471 Wh/kg compared with 656 Wh/kg for Lithium-Cobalt Oxide. The actual specific energy and energy density can be expected to be 20-30% of the theoretical value. Lithium-Thionyl Chloride has demonstrated capability for 300-600 Wh/kg (depending on rate) while the Lithium - Cobalt Oxide cell has demonstrated 100-120 Wh/kg.

The theoretical values in Table 4.4 are based on the active materials only, and therefore do not include the electrolyte, terminals and mechanical hardware. Specific energy and energy density can be determined by multiplying the capacity per unit weight or volume by the voltage. Considering that cells comprise a case, electrode structures with current collectors, busbars, and other components, these values are not actually achievable. A rule-of-thumb for a well-developed electrochemical cell is 25% of the theoretical energy. Practical operating voltages and demonstrated values of specific energy and energy density are given in Table 4.6.

6.4 Life and performance limitations

The metric that best describes the life of a cell or battery in an aerospace application is the cycle or point in time when the energy storage requirement of the spacecraft can no longer be met. That implies that the voltage and current (power) during discharge will be depleted and unable to meet planned spacecraft operations.

The process for avoiding this condition or assuring that the mission can be met is to incorporate enough margin to the selection of the cell capacity and voltage so that the expected cell degradation will be offset by excess capacity and voltage. Selection of a larger capacity cell and battery impacts efficiency of mass and volume. An optimum in balance between the efficiency and life requirements needs to be considered when making a selection for a mission. This implies that for a GEO mission, where only 100 cycles per year are required, the battery DOD can be high (as much as 75%). For a LEO mission requiring 5000 cycles per year, the DOD usually is below 30%. Figure 4.7 provides a generalized relationship between cycle life and DOD.

The factors that play an important role in cell life are rate, temperature, depth of discharge, charge control, and voltage limits.

Effect of rate

Rate plays an important role on capacity and battery life because it influences the conversion of products to reactants and vice versa (for rechargeable cells). The effect of

CHEMICAL STORAGE AND GENERATION SYSTEMS 177

cell charge and discharge rate on life is related not only to the current flowing through the cell but more specifically to the current density (ma/cm^2) at the surface of the plates.

Some electrochemical cell types, particularly primary cells, generally operate at low current densities (<1 ma/cm^2). Rechargeable Ni-Cd and Ni-H$_2$ cells operate reversibly and relatively efficiently at 10 ma/cm^2. At high rates, some of the active material in either or both electrodes may not be accessible because of plate construction or polarization losses which prevent the desired reaction from taking place. The result is that some of the material is disconnected from the active material and becomes 'inactive'. Under this condition the cell will appear to have an unusually low voltage and will exhaust one of the electrodes sooner resulting in cell failure. Thus, the general trend is that higher rate results in lower capacity as shown in Figure 4.5. However, most electrochemical cells are relatively forgiving when operated within the rates for which they were designed.

The higher the discharge rate, the greater the voltage drop due to ohmic losses that exist in the cell and electrical circuitry including wire and internal and external cell connections. During charge, the measured battery voltage is higher at higher rates. With regard to 'state-of-charge' as the cell or battery is depleted, the voltage on discharge will steadily drop (less in cells with a flat discharge voltage). However, there will be a sharp drop in voltage when the capacity is close to depletion (approaching 'failure') indicating the potential differences are approaching equality (0.0 volts). If the load on the battery continues, the battery voltage will drop below the bus voltage resulting in a significant spacecraft problem.

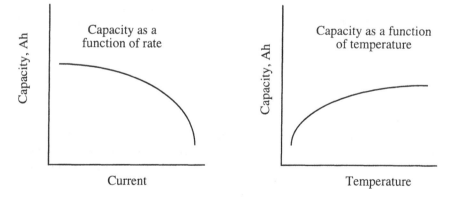

Figure 4.5 General Trends for the Effect of Rate and Temperature on Capacity

Low rate generally results in high capacity. However, at very low rate, products may be formed which are not reversible. In the case of rechargeable cells the long term effect of operating at low rate can result in shorter operating life.

Effect of temperature

Temperature plays a role similar to rate. The two are actually related in that high rate tends to result not only in higher IR polarization but I^2R heating as well. Increased heating can reduce activation polarization losses because it improves the kinetics and reduces concentration polarization losses by enhancing the conduction of the ions. The result is a lowering of the cell voltage charge (i.e., closer to the equilibrium voltage). However, excessive heating can also result in permanent loss of reactants or products and in fact can lead to a condition in a rechargeable cell known as 'thermal runaway'.

Thermal runaway can occur if the V_T charging method (see Section 6.5) is used. The charge voltage limits are set by the V_T curves, shown above, which are in the power system. Heating results in a lower cell voltage which will result in higher current from the solar array to raise the battery voltage to the preset V_T charge voltage. This results in more heat which lowers the battery voltage even further, causing the power system to impose higher current to reach the voltage set point, and so on. This produces 'thermal runaway' which most often results in venting, at times with violence. This can also happen if one cell in a battery experiences a short. Unless this is recognized and the V_T level is lowered, the power system will continue to view the full battery as a 28 V. system and overcharging of the remaining cells will occur.

The effect of temperature on a Ni-Cd cell is given in Figure 4.6. Note that lower temperature (0-5° C) extends Ni-Cd life significantly. Low temperatures generally result in increased polarization resulting in lowering of reaction rate and ion flow within the electrolyte; therefore, at temperatures below ambient, the voltage of a battery will be lower on discharge. On charge, because the current (electron flow) is reversed, it takes a higher voltage to return the same current to the cell or battery.

Depth of discharge

The depth of discharge is the percent or fraction of the cell or battery capacity removed during a discharge. Primary cells are usually discharged completely; thus, this parameter is used for rechargeable cells. It is well known that there is a strong relationship between DOD and life. The greater the depth of discharge on a regular basis, the sooner the cell will fail to deliver the required voltage for the time period required. The relationship can be described as asymptotic. An example of this is the ability of a Ni-Cd cell to undergo 40,000 cycles at 13% DOD in the Solar Max Spacecraft that orbited the earth

CHEMICAL STORAGE AND GENERATION SYSTEMS

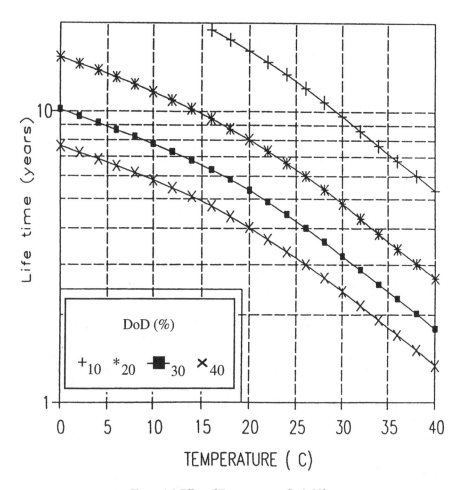

Figure 4.6 Effect of Temperature on Cycle Life

for eight years and 1000 cycles at 50% DOD. It should be recognized that some batteries have a stronger relationship than others. This metric also depends on cell construction, e.g., thickness and number of plates and quantity of active material which affects the current density. An accepted relationship between depth of discharge and cycle life is shown for a Ni-Cd Cell in Figure 4.7.

SPACECRAFT POWER TECHNOLOGIES

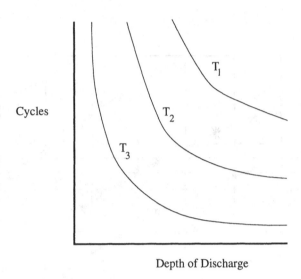

Depth of Discharge

Figure 4.7 Generalized Relationship Between Depth of Discharge and Cycle Life

6.5 Charge control

To recharge a cell is to return the active materials to the condition of highest potential energy difference between anode and cathode. This requires forcing electrons from the solar array, or other source of electrons, into the battery. The voltage of the solar array and the voltage at the input of the battery has to be large enough to overcome polarization, however, not so large that the input voltage forces unwanted reactions to take place. From this it is easily seen that charge control methodology is an essential contributor to the battery life.

A number of methods have been used for charge control. All types make use of a voltage limit to avoid undesirable reactions that can be caused by excessive voltage. To a great extent the mission orbital schedule determines the size of the battery, and the size of the solar array determines the rate of charge of the battery. In a LEO orbit, the energy removed during the 35 minutes of eclipse has to be replaced within the 65 minutes in sunlight without exceeding the safe voltage limit. Charge control is a major consideration for this orbit that occurs 5000 times per earth year. In the GEO orbit, the spacecraft is in the Sun for three full months; then the batteries provide energy in increasing amounts each Earth day to a maximum of 45 minutes, then in decreasing amounts until the spacecraft returns to full Sun. This occurs twice each Earth year resulting in only 100 charge/discharge cycles per year.

Methods used for charge control include constant current, maximum current to a temperature compensated voltage limit followed by current taper (V_T control), and mul-

CHEMICAL STORAGE AND GENERATION SYSTEMS 181

tiple step charge. The former has been used in GEO orbits because the sun time is in excess of 23 hours during each shadow period. During the sun period trickle charge current is used to offset the self-discharge experienced. This also minimizes the chance for excessive voltage.

The V_T method is used most often for Ni-Cd and other types of battery charging using the maximum solar array power. A large fraction of the solar array power is used for the loads. The remainder is used to charge the battery. When the battery voltage reaches a preset level determined by laboratory tests, the voltage is held constant to minimize the overcharge. Any of several voltage levels can be selected for operation depending on the condition of the batteries and the spacecraft operational conditions. Also, they can be used for operation when unusual battery characteristics are present, i.e., high or low DOD or a cell short condition. The V_T levels used in the Modular Power Subsystem (MPS) flown in TOPEX, Landsat, UARS, Solar Max, and other space conditions are shown in Figure 4.8.

During the constant voltage period, the current is allowed to taper to a low value depending on the ability for the battery to accept current at that voltage. This provides a means of reducing the current when the cell reaches the fully charged condition where the inefficiency results in gassing, heating, and/or undesirable reactions. The V_T level is selected so as to minimize or eliminate overcharge. For a Ni-Cd battery the recommended charge to discharge ratio as a function of temperature is shown in Figure 4.9. The ratios take into consideration the charging inefficiency due to the O_2 competing reaction that occurs when charging at temperatures above $0°C$.

Figure 4.10 shows the use of a V_T operation of a Ni-Cd battery at Level 5 and $10°C$. This was the selected condition for operating a healthy Ni-Cd battery in space for 25% DOD. If the voltage is set too high, the maximum current available from the solar array would continuously charge the battery until the higher voltage level was reached.

As described above, if the voltage is too high and the current is high, heating occurs and thermal runaway results. The multi-step method provides a mechanism for reducing the current in steps as the charge voltage reaches preset limits. It takes the place of an automated V_T operation except that the current is decreased in a stepwise manner when preset voltage limits are reached.

6.6 Efficiency and thermal properties

Coulombic (Electrochemical) and energy efficiencies are the key factors in determining cell and battery performance. The former is used to describe the reversibility of the electrochemical reactions as a function of temperature, the effect of competing reactions, and the effect of self-discharge factor. Primary cells contain reactants that are mostly irreversible. Rechargeable cells are by their very nature highly reversible. How-

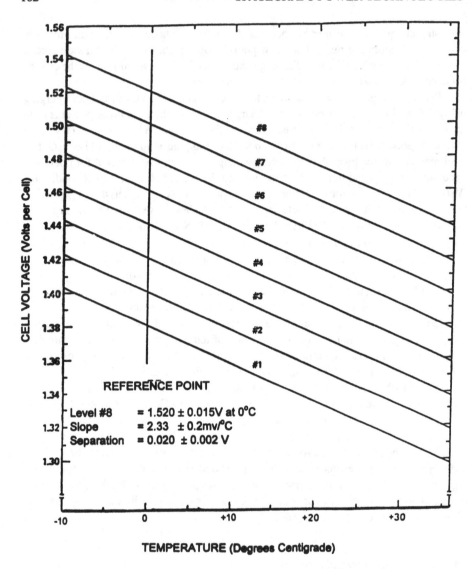

Figure 4.8 Temperature-Compensated Voltage Curves Used for V_T Charging

ever, increases or decreases in temperature can change the reversibility and result in competing reactions.

For example, the reactions of the electrodes in the Ni-Cd cell are quite reversible. However, near the end of the charge, the oxidation of $Ni(OH)_2$ to $NiOOH$ competes with the oxidation of OH^- to O_2. The competition results in the inefficient use of elec-

CHEMICAL STORAGE AND GENERATION SYSTEMS 183

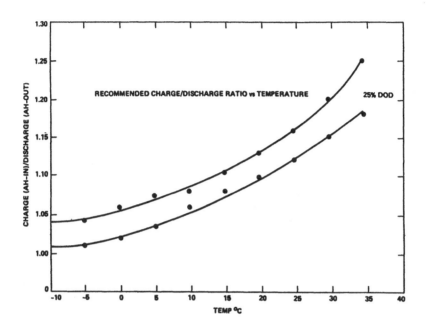

Figure 4.9 Recommended Charge to Discharge Ratio for a Ni-Cd Cell

trons for the $Ni(OH)_2$ reaction. To account for this, the charge process is therefore continued beyond that of 100% reversibility. The extent of overcharge is strongly affected by temperature. As the temperature rises, the overpotential of the OH⁻ reaction increases faster than the $Ni(OH)_2$ reaction resulting in larger quantities of O_2 production. Thus, the higher the temperature the greater is the electrochemical inefficiency. For this system the coulombic efficiency decreases from 0.98 at 5°C to 0.85 at 20°C. The Li-Ion cell is practically free of competing reactions to the point at which the cell is fully charged over the same temperature range. Thus, its coulombic efficiency is relatively constant at 0.98.

Energy efficiency depends on the same coulombic factor and the voltage of the cell. Cell voltage, affected by polarization, increases during charge and decreases during discharge. The ratio of the integral of the voltage and current during discharge to that on charge provides the energy efficiency. For the Ni-Cd cell, the energy efficiency at 5°C is 0.84, while that of the Li-Ion cell is 0.95. The result of inefficiency is heat generation that is taken into consideration in the power system thermal design.

184 SPACECRAFT POWER TECHNOLOGIES

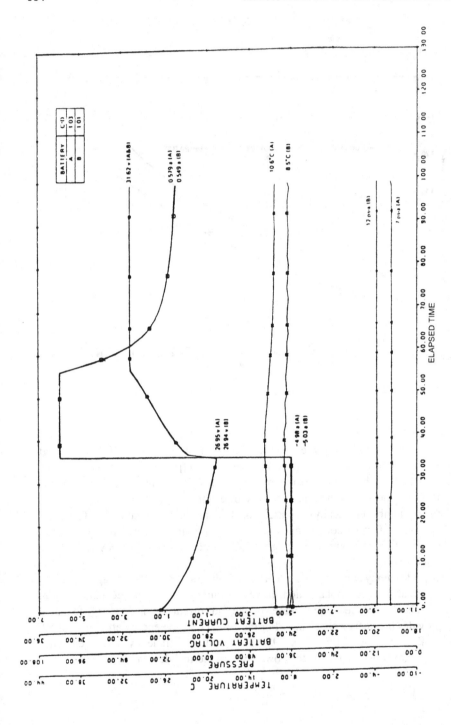

Figure 4.10 The Performance Characteristics of a Ni-Cd Battery During V_T Operation

CHEMICAL STORAGE AND GENERATION SYSTEMS

7. Electrochemical cell types

Except for a limited number of cases, primary cells can only be discharged once. These include the lithium primary battery systems. However, Ag-Zn cells on the Shuttle backpack can be recharged a few times. This helps to provide the engineers with the information that the batteries can meet the capacity requirements and are acceptable for flight. However, most primary systems do not have this capability. They are considered a primary cell because the reactions during discharge cannot be reversed.

7.1 Primary cells

Initially, the short demonstration missions used primary cells for prime power to minimize cost and complexity. Several space applications still require the use of primary

Table 4.7 Primary Cells Used in Shuttle Missions

Zinc Anode Primary Cells/ Batteries
Zinc - Manganese Dioxide-LeClanche (Zn - MnO_2)
Zinc - Manganese Dioxide Alkaline (Zn - MnO_2)
Zinc - Mercuric Oxide (Zn - HgO)
Zinc - Oxygen (Zn - O_2)
Zinc - Silver Oxide ('Silver-Zinc') (Zn - Ag_2O)

Lithium Anode-Soluble Cathode Primary Cells/ Batteries
Lithium - Sulfur Dioxide (Li - SO_2)
Lithium -Thionyl Chloride (Li - $SOCl_2$)

Lithium Anode-Solid Cathode Primary Cells/ Batteries
Lithium - Carbon Monofluoride (Li - $(CF)_x$)
Lithium - Manganese Dioxide (Li- MnO_2)

batteries. Some of these are used in Shuttle applications to support astronaut equipment, for instrument power, and for experimental packages such as the Get-Away Special (GAS). Small-size rechargeable batteries have also been used for Shuttle applications in OEM and instrument power. In addition, they have been used in landers, penetrators, and probes in planetary missions. A list of the primary cells used in Shuttle missions is given in Table 4.7.

The broad spectrum of power and energy storage requirements for payload power provisioning requires the use of many types of batteries. All batteries require approval by the Johnson Space Center Safety Office before use. Many will require various degrees of design modification and safety tests to make them acceptable for aerospace payload applications from both a performance and a safety viewpoint. Proper consideration must be given to required battery enclosures, seals, vents, type and amount of electrolyte, type and amount of gas generated on stand and during operation, operating temperature capability, pressure environment, and many other factors involved in a battery selection activity. Careful consideration should be given to the applicable Department of Transportation shipping requirements. Listed below are primary types that have been used in payload applications:

A comparison of specific energy of primary batteries used in space is shown in Table 4.4 for theoretical energy and in Table 4.6 for actual use. From the initial use of these cells, new technologies have resulted in an increase in voltage and energy capability. Specific characteristics of the zinc anode cells are shown in Table 4.8.

Zinc anode primary cells/batteries

Many of the zinc anode cells have been and are available to the consumer for use in toys, flashlights, and other well-known applications. Generally, they are low in cost, used in low rate (<C/100) applications, and are relatively safe. These include the familiar Zn-MnO$_2$-LeClanche cell, the Zn-MnO$_2$-alkaline cell, and the Zn-HgO oxide cell. All the zinc anode cells, except the LeClanche, type use alkaline electrolyte (aqueous potassium hydroxide) which enhances the conductivity and improves rate capability and voltage during discharge.

Zn-O$_2$ (air) and Zn-Ag$_2$O cells are higher in energy density, and the Zn-Ag$_2$O is capable of operating at higher rates than typical primary cells. Zn-O$_2$ batteries have been used in space in the Shuttle cabin. The zinc cells are attractive because of their high volume energy density. Primary Zn-Ag$_2$O batteries are available in low rate (button cell) designs and also have been used in space in both flooded electrolyte, starved, and reserve configurations. These offer the highest discharge rate capability of the zinc anode cells. In zinc anode cells, the zinc anode is always the negative electrode because it provides the electrons to the load during discharge. A more detailed description of each of the types follows.

Zinc-Manganese Dioxide-LeClanche (Zn-MnO$_2$)

This cell, sometimes called the zinc-carbon cell, utilizes an aqueous solution of ammonium chloride (NH$_4$Cl) and/or zinc chloride (ZnCl$_2$) as the electrolyte. Its reaction is given as:

$$Zn + 2MnO_2 + 2NH_4Cl \rightarrow 2\ MnOOH + Zn(NH_3)_2Cl_2 \qquad (4.13)$$

CHEMICAL STORAGE AND GENERATION SYSTEMS

Table 4.8 Zinc Anode Primary Cell Characteristics

Name	Cell Designation	Nom. Volt. (V)	OCV (V)	Energy Density (Wh/kg)	Energy Density (Wh/l)	Temp. °C	Comments
LeClanche	Zn // NH_4Cl / $ZnCl_2$ // MnO_2 / C	1.6	(1.3)	65	100	-5 to 45	Low cost, sloping voltage
Alkaline	Zn //ZnO /KOH // MnO_2 / C	1.6	(1.3)	95	220	-20 to 55	Greater drain rate, sloping voltage
Mercury	Zn // ZnO / KOH //HgO	1.35	(1.3)	105	325	0 to 55	Level voltage
Zinc-Air	Zn // KOH/NaOH // O_2 (Air)	1.5	(1.4)	290	905	0 to 50	Highest available energy of Zn anode cells
Silver-Zinc	Zn // KOH / NaOH // Ag_2O/AgO	1.8	(1.6)	200	515	-20 to 55	High drain rate, high cost

However, the reaction products differ depending on the rate of discharge and electrolyte composition. The metallic zinc amalgamated electrode forms the inner wall of the cell case. Carbon is mixed with the MnO_2 powder to form the cathode. The separator is a gelled paste or gel-coated paper.

It is generally found in the cylindrical (bobbin) and wafer type constructions and therefore is of limited electrode area, restricting it to low rate (<C/100) applications. This cell has wide consumer applications in toys, flashlights, radios, flashers, and instruments. One can remove 50 percent of the capacity at the C/50 rate and 100 percent at the C/400 rate. This type of cell has been used in some Shuttle applications for instruments, flashlights, and radios.

The advantages include: 1) low cell cost, 2) low cost per watt-hour, 3) large variety of shapes, sizes, battery voltages, and capacities, 4) wide distribution and availability, and 5) reliability.

The disadvantages include: 1) low energy density, 2) poor low-temperature service, 3) leakage under abusive conditions, 4) low efficiency under high current drains, 5) limited shelf life, and 6) sloping voltage.

The potential hazards include: 1) H_2 gas accumulation during discharge needs relief (do not encapsulate), and 2) cells can leak salt-solution electrolyte (avoid contact to metal).

Zinc-Manganese Dioxide-Alkaline ($Zn-MnO_2$)

This cell is similar to the Leclanche cell except that it utilizes a strong solution of potassium hydroxide (KOH) for electrolyte. Its reaction is given as:

$$Zn + 2MnO_2 \rightarrow ZnO + Mn_2O_3 \qquad (4.14)$$

In this cell, the zinc anode is a pressed powder amalgamated with small amounts of mercury and a binder to form a gel or pressed as a dry powder. The mercury suppresses the hydrogen gassing. The cathode is similar to the LeClanche cathode mixed with acetylene black. The separator is a microporous woven or felted material. The 25 to 50 percent KOH electrolyte is immobilized in a gel.

Available in a wide variety of sizes in the consumer market, it is used for higher drain rate application than the LeClanche cell. Applications in space are primarily for OEM equipment including radios and recorders. It is produced in cylindrical and button cell configurations. It has a factor of four times the rate capability of the LeClanche cell and therefore has a broader range of applications.

The advantages include: 1) improvement of low-temperature service over LeClanche, 2) higher efficiency under high discharge loads, 3) good shelf life (4-year storage with

80 percent of capacity), 4) wide availability, 5) moderate cost (higher than LeClanche), and 6) sloping discharge curve (better than LeClanche).

The disadvantages include: 1) hydrogen gassing rate higher than LeClanche, and 2) shorted cells (high rate) can produce high temperatures (100°C).

The potential hazards include: 1) H_2 gas accumulation during discharge needs relief (do not encapsulate), and 2) leakage of corrosive alkaline electrolyte.

Zinc-Mercuric Oxide (Zn-HgO)

This cell utilizes either 30 to 45 percent potassium hydroxide (KOH) or sodium hydroxide (NaOH) saturated with zinc oxide (zincate) as the electrolyte. The reaction is given as:

$$Zn + HgO \rightarrow ZnO + Hg \qquad (4.15)$$

The cell is available in different forms that determine the structure of the zinc electrode, e.g., corrugated strips or pressed powder amalgamated with Hg. The cathode consists of mercuric oxide mixed with graphite. Layers of cellulose are used for the separator. The zinc-mercuric oxide cell is manufactured in three basic structures: a) wound anode, b) flat pressed powdered electrode, and c) cylindrical pressed powder type.

The applications: The zinc-mercuric oxide cell is available in a button cell or cylindrical cell configuration. Because of a higher volume-energy density than the previous two systems, it has applications where rate is low and volume is minimized. It also exhibits a level voltage during discharge which makes it ideal for a memory storage or time standard. It is also used in pacemakers, hearing aids, detectors, and sensors.

The advantages include: 1) higher volume-energy density, 2) long shelf life, 3) flat discharge curve over a wide range of current drains, and 4) high resistance to impact, acceleration, and vibration.

The disadvantages include: 1) somewhat higher cost than $Zn-MnO_2$, 2) disposal of Hg produced on discharge, and 3) electrolyte creep may result in leaks.

Potential hazards include: 1) because of zinc oxidation, a tendency for H_2 to accumulate during discharge (do not encapsulate), 2) the toxic nature of Hg vapor (avoid cell damage during handling, shipment, and storage to prevent Hg loss), 3) short circuit or any excess heating of the cell can result in Hg vapor release (do not solder leads directly to the cell terminals without proper precautions), 4) the need to open and/or dispose of the cell under controlled conditions to prevent the escape of Hg vapor (if there is a need to open, one should place the entire contents under water in a polyethylene or polypropylene container), and 5) possible electrolyte (strong alkali) leak from cell (neutralize with a saturated solution of boric acid or large quantities of water).

Zinc-Oxygen (Zn-O_2 or Zn-Air)

This cell also utilizes a potassium hydroxide (alkaline) electrolyte and an amalgamated zinc powder sometimes with a gelling agent as the anode. The cathode referred to as an air cathode remains intact (does not undergo reaction) throughout the discharge reaction. Oxygen from the air enters the cell through openings in the case, flows through diffusion and hydrophobic membranes, and is reduced at the carbon or wintered nickel structure impregnated with a catalyst, sometimes platinum. The reaction is given simply as:

$$Zn + \Omega O_2 \rightarrow ZnO \qquad (4.16)$$

Large zinc-air batteries with the appearance of an automobile battery having sheet zinc anodes (low rate) have been used for railroad signal switching, telecommunications, and beacons. Recently, they have been utilized by NASA in the Shuttle cabin, in a button cell configuration. The small cells have been used in portable communications gear and hearing aids.

The advantages include: 1) high volume-energy density, and 2) flat discharge voltage at low rates, while the disadvantages include: 1) capacity highly dependent on load, 2) cell drying out due to accessibility to air, 3) anode area limits power output, 4) dependent on environmental conditions, and 5) pulse capability limited.

There are two potential hazards: 1) electrolyte leakage (use saturated solution of boric acid or dilute vinegar to neutralize), and 2) H_2 evolution can occur because of the electrochemical reduction of zinc (do not hermetically seal).

Zinc-Silver Oxide (Silver-Zinc) (Zn-Ag_2O)

This cell utilizes a porous electro-formed amalgamated zinc electrode, a porous silver oxide electrode, and a 35 to 45 percent solution of KOH. To prevent silver migration in the cell, the separator system is constructed of multiple layers of cellophane and nylon fabric. Sometimes radiation-grafted polyethylene is used to extend wet life. The cell reaction is given as:

$$Zn + Ag_2O \rightarrow 2Ag + ZnO \qquad (4.17)$$

Because of its high energy density it is widely used in electronic equipment, hearing aids, watches, calculators, and other instruments requiring extended life. In space, it has been used in several important applications: as a primary battery for launch vehicle controls and communications, to power the tools to repair the Solar Max Mission (SMM) spacecraft, and for the Manned Maneuvering Unit (MMU) required to propel the astronaut during extravehicular activity (EVA). This battery, if assembled with adequate separator and design, is considered a limited use rechargeable cell. A cell/battery can be recharged a limited number of times to allow checking of performance or associated equipment. As a reserve battery it has been used in some long term missions where the electrolyte is maintained in a bellows leaving the cell dry until activation. In space, most of the primary batteries on Apollo and Skylab were zinc-silver oxide, a limited cycle life rechargeable battery. A 40 Ah, 28 V. battery was recently used to provide prime power for the Mars Lander to meet a 30 day mission. It actually lasted 100 days. A photo of the battery appears in Figure 4.11. The Zn-Ag_2O system is available in button, prismatic, and reserve configurations.

The advantages: 1) high energy density, 2) good low-temperature operation, 3) good shock and vibration resistance, and 4) can be charged and discharged to determine capacity. The disadvantages: 1) relatively high cost, 2) active cell shelf life limited to 1 year or as little as 30 days (except for reserve types), and 3) two voltage plateaus associated with Ag_2O and AgO discharge product.

The potential hazards include: 1) strong alkali may leak through pressure relief valve (use boric acid solution to neutralize), 2) need to vent H_2 accumulated on open circuit stand and at low voltages (< 0.3V, Ag has low H_2 overvoltage), and 3) zinc dendrites can form (on charge) creating internal short (internal shorts can produce heat which increases pressure within the cell).

Lithium anode primary cells/batteries

Various cells with lithium anodes and non-aqueous electrolytes have gained importance for consumer and space related applications. The reason is that they offer the highest specific energies (Wh/kg), highest volume energy densities (Wh/L), and longest storage life of any electrochemical cell thus far developed. Like the zinc anode cells described above, most of the lithium anode cell technologies in use today are for primary cell applications. Rechargeable lithium cells are presently being developed for space and commercial use.

The main reasons for the continued growth in interest in lithium cell technology over the past several years is large energy storage capability and lengthy shelf life of lithium itself. Ironically, the same reasons are also the cause of the safety hazards associated with this technology. The bases for lithium's energy storage capabilities are the magnitude of its oxidation potential (3.01V) compared to other metals such as zinc (0.66V) and its large capacity per unit weight (3.86 ampere hours/gram), again compared with zinc (0.82 ampere hours/gram). Together they provide the largest watt hour/ gram material known. The basis of the lengthy storage life is the protective film formed on the lithium surface when it is in a suitable non-aqueous media. The film protects it against corrosion and loss of energy experienced in all aqueous electrochemical systems.

Figure 4.11 The 40Ah, 28V Ag-Zn Battery Used in the Mars Pathfinder Lander

CHEMICAL STORAGE AND GENERATION SYSTEMS

The lithium cell groups can be divided into three categories:
(1) Soluble cathode cells.
(2) Solid cathode cells.
(3) Solid electrolyte cells.

Lithium cells use either solutions of organic or inorganic electrolyte. In the soluble cathode type, the solvent containing a conductive salt, 'catholyte', also serves as the active energy producing material for the cathode reaction. A porous inert carbon electrode provides the reaction site and stores the product deposit. Lithium metal foil is the source of the anode reaction. Compatible salts are dissolved in the solvent to enhance conductivity. The basis of the soluble cathode technology is that the cathode material is used up during the discharge reaction. These cells have the capability of being discharged at rates as high as C/10 and higher. The majority of safety issues are related to this group of lithium anode cells. In these cells the active material of the cathode is also in contact with the anode, thus providing capability for relatively high discharge rate for primary cells but also concern for safety. These cells are used by the DOD and NASA but are not available for commercial use.

The soluble cathode types used in space applications include lithium-sulfur dioxide (Li-SO_2) and lithium-thionyl chloride (Li-$SOCl_2$). Lithium-sulfuryl chloride (Li-SO_2Cl_2) has been considered, but presently the Li-SO_2 and Li-$SOCl_2$ have been used for various applications. A variation of the Li-$SOCl_2$ cell (Li-BCX) containing bromine chloride has been utilized in astronaut Shuttle applications.

The inherently large energy storage capability is quite desirable when the electrochemical cell system is properly designed for the application and used within the limits for which it was intended. However, if subjected to abnormal operations or conditions, the large quantity of energy can be released quickly, sometimes with violence, resulting in venting or, on rare occasions, explosion. For this reason, the soluble cathode lithium anode cell should be used only in applications where the zinc anode cell does not meet performance requirements.

Solid cathode cells ranging in voltage from 1.5 V to 3.3 V have been developed over the years. Although a number of types have been identified, only a few have become common. These are gradually replacing the zinc anode cells because of the higher energy storage capability. They utilize a cathode material that remains in the solid condition during operation, and they are used in applications where rate requirements are low (<C/100).

One of the few solid electrolyte cells in use is the lithium-iodine cell. It has the lowest rate capability of the lithium systems because of the conductive path through solid electrolyte. Thus, it is the safest of lithium cells, except at temperatures above 186°C (lithium melting point) where it, like the other lithium cells, can be hazardous.

Because lithium is an active metal, it reacts with moisture. Therefore, the assembly of the lithium cells must be accomplished under moisture-free, dry room conditions to optimize performance and safety. Thus, the quality control of materials and processes is critical to success.

Finally, as in the zinc anode cell designation, the lithium anode is the negative electrode relative to the opposing electrode because it provides the electrons to the load during discharge. The three types of lithium electrochemical cells are described in Table 4.9.

Lithium Anode Soluble Cathode Cells/Batteries

The electrochemical cell types comprising this group are lithium-sulfur dioxide (Li-SO_2), lithium-thionyl chloride (Li-$SOCl_2$), and lithium-thionyl chloride with bromine chloride (Li-BCX). They are similar in that they comprise cathode active materials that are used up during the discharge process sometimes forming reactive discharge products. They are capable of rates up to C/10 and higher for short periods because the active materials are in direct contact with the carbon current collector. The carbon electrode is porous and serves as the storage site container for some of the discharge products.

In addition to the improved energy and storage performance capability, these cells have a characteristic which must be considered in their use: voltage delay. The delay is due to the lithium film formed on the lithium metal surface which provides a protective layer. When a load is switched on, the cell responds as if it contained a high resistance between electrodes. Thus, after the initial power surge, the cell or battery voltage drops well below the operating voltage for a short period, seconds to minutes depending on operating conditions, before the cell or battery reaches its normal operating voltage. This drop in voltage depends on the cells previous storage temperature, time, discharge rate, and temperature of the application. A number of corrective actions have been taken to avoid this problem including pulsing the battery before intended use and, more recently, the inclusion of additives of various types to the cell components.

Reversal of these cells, as with most electrochemical cells, produces undesirable reactions which not only result in loss of cell use but may result in a hazard condition. Diodes are used in parallel strings to avoid one of the strings forcing current through the other string resulting in a reversal. Diodes are also occasionally used across cell terminals to bypass the current when its voltage drops below a preset point.

The lithium soluble cathode cell group offers the user a power source that can be used in applications with a specific energy and energy density eight times greater than the present LeClanche cells over a wide range of temperatures, and, in addition, offers a long shelf life. Cells of bobbin construction (center electrode with thick surrounding opposing electrode) are considered low rate because of their limited electrode surface. However, high rate (spiral wound, large surface area) cells are used by the U.S. Army

Table 4.9 Lithium Soluble Cathode Cell Characteristics*

Name	Cell	Nominal Voltage (V)	OCV (V)	Specific Energy (Wh/kg)	Specific Energy (Wh/l)	Operating Temperature (°C)	Comments
Li-SO$_2$	Li // CH$_3$CN / LiBr // S0$_2$ / C	3.0	(2.9)	280	440	-55 to 60	Most advanced of soluble cathode type
Li-SOCl$_2$	Li // SOCl$_2$ / LiAlCl$_4$ // SOCl$_2$/C	3.6	(3.5)	600	900	-40 to 60	Highest energy density
Li-BCX	Li // SOCl$_2$ / LiAlCl$_4$ // BrCl //SOCl$_2$/C	3.9	(3.5)	430	960	-40 to 60	BrCl reported to prevent formation of S
Li-SO$_2$Cl$_2$	Li // S0$_2$Cl$_2$ // AlCl$_4$ // S0$_2$Cl$_2$ / C	3.9	(3.5)	500	1000	-40 to 60	In early stages of production

*Cylindrical spiral wound (jellyroll) construction.

for a number of field applications. Also, the lithium-sulfur dioxide cell was approved by the U.S. Federal Aviation Administration for use in electronic-locating transmitters (ELT) for private noncommercial aircraft, and the lithium-thionyl chloride cells are used in oil well logging tools. Almost all other applications are military.

The size and configuration of a cell determines its application. Lithium-thionyl chloride cells in flat disc-type (button) cells containing multiple plates have been used by the military in multi-cell batteries for several portable applications. These cells have been used in the Mars Rover and are planned for the DS-2 probe into the surface of Mars. They were also developed using flat plate electrodes into 250 Ah 28 V batteries for the Centaur launch vehicle. Their designs include button cell (very low rate applications), bobbin cylindrical cell construction to replace the alkaline cell applications (moderate rate), spiral wound cylindrical cells for higher rate applications, reserve types (high rate) where the acidic electrolyte increases the discharge rate capability, and in prismatic types to the 10,000 ampere hour size.

The cylindrical lithium-sulfur dioxide (spiral-wound) cells and lithium-thionyl chloride (spiral and bobbin) cells have been used in a wide variety of military applications. The lithium-sulfur dioxide cell has better low-temperature rate capability than the lithium-thionyl chloride cell, which has higher energy density (Wh/kg) and is capable of operating at higher temperatures. The spiral type is used in radios, transceivers, in sonobuoys, and a wide range of portable power equipment, as well as in the Galileo and Cassini probes. The lithium-thionyl chloride bobbin type (low rate) has been used for small instruments and computer memory applications. The reserve lithium-thionyl chloride cell offers an extremely high current within seconds of activation and is being considered for monitors and missile activation. The very large prismatic (10,000 ampere hour) size has been used as a standby power source in missile silos. There is also the flat disc cell which is available in 50 to 2000 ampere hour sizes. To date, soluble cathode cells that have been approved for NASA use in the Shuttle include the lithium-thionyl/chloride (bromine chloride) cell used in the astronaut helmets to power the TV camera and the EVA lights, lithium-sulfur dioxide cells for the Galileo Probe and Long Duration Exposure Facility (LDEF), and a few low rate M, 1/2 MA and C bobbin construction types in various applications. The safety of these spiral-wound "D" cells is assured by the use of a diode across each cell, a fuse in each cell, under-voltage battery cutout for the TV camera battery and thermostatic switches, and adequate heat sinking.

Lithium-Sulfur dioxide (Li-SO_2)

This cell is available in spiral wound construction. It utilizes a polypropylene separator between layers of lithium foil and porous carbon structures of roughly equivalent size. The carbon cathode is prepared to size by pressing or rolling a mixture of carbon and Teflon (as a binder) over a nickel screen with a lubricant (isopropyl alcohol). The result

CHEMICAL STORAGE AND GENERATION SYSTEMS 197

is an 80% porous structure that provides the necessary surface area for the discharge reaction to occur and volume for the discharge product (lithium dithionite) to deposit. The reactions are :

$$2\text{Li} \rightarrow 2\text{Li}^+ + 2\,e^- \quad (4.18)$$
$$\underline{2\,SO_2 + 2e^- \rightarrow S_2O_4^=} \quad (4.19)$$
$$2\,\text{Li} + 2\,SO_2 \rightarrow \text{Li}_2S_2O_4 \quad (4.20)$$

The electrolyte consists of a solvent acetonitrile (CH_3CN) to store the SO_2 and a salt LiBr to enhance the conductivity. The internal pressure of the cell at the start is in the range of 3-4 atmospheres (3-4 x 10^{-5} Pascals). A glass to metal seal is used for hermetic sealing and a pressure vent is used to release the pressure when it reaches a preset level of 26-30 atmospheres (at temperatures of 93-106°C). Cells of this type developed by Alliant Techsystems have been used in the Galileo probe to Jupiter. A photo of one of the three probe batteries is shown in Figure 4.12. A similar design was launched in 1997 on the Hugyens probe on the Cassini spacecraft to Saturn.

Lithium-Thionyl Chloride (Li-SOCl$_2$)

The lithium-thionyl chloride cell is composed of a lithium foil anode and a cathode current collector of carbon-Teflon composition similar to the Li-SO$_2$ cell. The liquid SOCl$_2$ serves the dual role as the cathode active material (catholyte) and the solvent. The conductivity is provided by the 0.5 m LiAlCl$_4$ dissolved in the electrolyte. Occasionally other salts or higher concentrations are added to enhance the performance. The separator is a nonwoven glass held together with an inert binder. The reactions are given as:

$$4\,\text{Li} \rightarrow 4\,\text{Li}+ + 4\,e- \quad (4.21)$$
$$2\,SOCl_2 + 4\,e- \rightarrow 4\,Cl- + S + SO_2 \quad (4.22)$$
$$4\,\text{Li} + 2\,SOCl_2 \rightarrow 4\,LiCl + S + SO_2 \quad (4.23)$$

The LiCl discharge product deposits in the pores of the cathode. It eventually clogs enough pores to render the cathode inactive (passivated) and thus causes the cell to fail. The SO$_2$ is evolved as a gas and reaches a pressure of 1 atmosphere at the end of discharge at 20°C. Because the solvent is consumed during discharge, an adequate volume of electrolyte is required when the cell is manufactured to assure that the SOCl$_2$ will not run out before discharge is completed. 250 Ah Li-SOCl$_2$ cells and batteries have been used recently in space for launch vehicle power on the Centaur launch vehicle to double the mission capability over the Ag-Zn batteries. Two types of cells and batteries were developed for this purpose. The flat pancake design of Saft (France) as seen in Figure

198 SPACECRAFT POWER TECHNOLOGIES

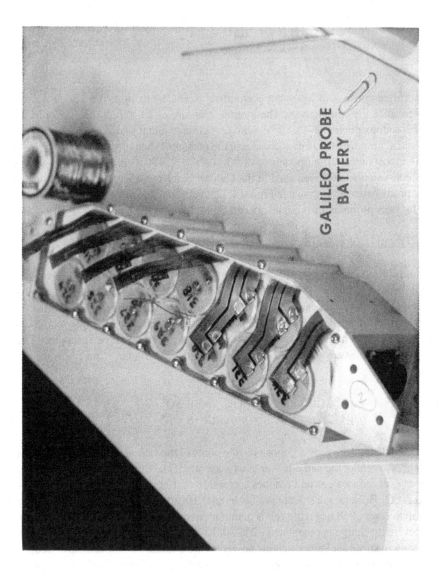

Figure 4.12 Galileo Probe Li-SO$_2$ Battery

CHEMICAL STORAGE AND GENERATION SYSTEMS 199

4.13, was selected for the first Centaur mission. Both designs passed a myriad of strenuous hot and cold temperature tests and several vibration and shock levels associated with a launch vehicle.

A more recent application is the Li-SOCl$_2$ Mars Pathfinder Rover battery, Figure 4.14, which used a 'D' cell design. It served as the power source for experiments on the Rover that could not be handled by the solar array on the top of the Rover.

Lithium-Thionyl/Chloride (Bromine Chloride) (Li-BCX)

This cell is produced almost exactly like the lithium-thionyl chloride cell except for the addition of BrCl. The BrCl additive is said to scavenge the free sulfur (S) formed during the early stages of discharge, thus minimizing potential hazardous reactions. As is the case with lithium-thionyl chloride cells, the SO$_2$ is gaseous and the clogging of the carbon by LiCl can limit the discharge. These cells have been successfully used in a number of astronaut applications in 'C', 'D', and 'DD' configurations.

The advantages include: 1) high storage capability including specific energy, energy density, and operating voltage, 2) lengthy storage life, 3) high rate capability for lithium cells, 4) good low-temperature performance of lithium-sulfur dioxide, and 5) relatively flat discharge voltage. The disadvantages include: 1) inherently higher hazard potential, 2) safety precautions necessary, 3) internal cell pressure increases rapidly as temperature increases (especially Li-SO$_2$), 4) reversal and charging can result in venting or explosion, 5) catholyte and electrolyte are toxic if cell is opened, 6) disposal, handling, and transportation procedures must meet safety requirements of U.S. Department of Transportation (DOT) and the U.S. Environmental Protection Agency (EPA), and (7) voltage delay is possible depending on temperature and time of storage and rate and temperature of application.

The potential hazards are based on the concern for the reactivity of lithium. The subject of safety immediately comes to mind when the use of a lithium cell is contemplated. What is it about these cells whose energy and performance is well beyond that of the present aqueous electrochemical systems? The conclusion is obvious: lithium is one of the most electrochemically energetic of the elements. In addition, the non-aqueous constituents used in the cell and the lithium are toxic and produce complex products whose reactivity, composition, and role in the cells is not well understood.

It is known that lithium cells are not as tolerant as aqueous cells to design flaws and abuse. The hazards associated with these cells were identified earlier in this Chapter as user induced and manufacturer-induced. Assuming a cell has been designed properly and has been manufactured under the quality control required to meet the specified applications, the safe use of the cell/battery is of primary concern.

The hazards include excessive temperature during storage and operation which can: 1) increase internal gas pressure, i.e., p = f(T), 2) increase vapor pressure of solvent or

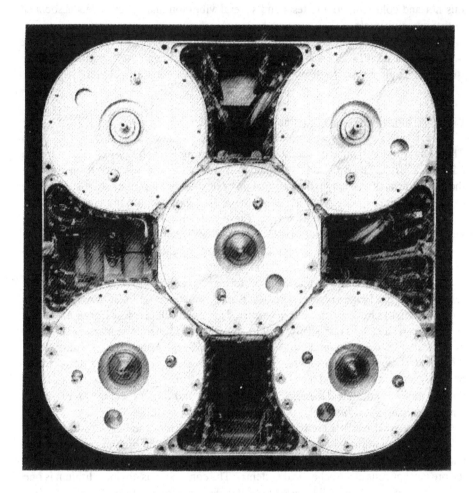

Figure 4.13 Top View of Saft 250 Ah, 28V Centaur Battery

other volatile constituents, thus further increasing internal pressure, 3) increase the rate of reactions, resulting in additional heat and pressure, 4) cause lithium to melt and react with other constituents and/or products, producing highly exothermic reactions and causing extremely rapid temperature and pressure increases, 5) result in thermal runaway leading to venting of gases (methane from lithium-sulfur dioxide cells) and explosion, 6) charge a cell, thus producing gases and other products while generating heat or causing unexpected reactions to occur which in themselves are heat initiated, 7) continue the

CHEMICAL STORAGE AND GENERATION SYSTEMS 201

Figure 4.14 The Mars Pathfinder Rover Battery with Li-SOCl$_2$ Cells

discharge beyond the point at which the cell's useful capacity is depleted (reversal or overdischarge) so that abnormal exothermic gas producing and other reactions take place, thus increasing the temperature or causing other reactions to occur, and 8) deplete the soluble cathode (SO$_2$, SOCl$_2$, SO$_2$Cl$_2$) at the end of discharge so that dry conditions are created, resulting in increased impedance and localized dry spots leading to dangerous localized intense heating. Note that the construction of the cell and battery determines the heat dissipation rate, and thus the rate at which the cell temperature increases; the merits of an anode-limited vs cathode- limited design are still being debated.

Numerous abuse tests have been performed, including exposure to flame, impact, penetration, and rapid high temperature heating. It is the consensus that cell venting is not objectionable as long as it does not occur in areas where personnel are affected, such as the Shuttle cabin. Explosion or detonation is totally unacceptable and all effort must be made to insure against this event. For example, the lithium thionyl/ chloride (bromine chloride) cell appears to offer a measure of added safety and performance over the lithium-thionyl chloride cell. Heat-tape tests performed recently on these cells, in which the temperature was increased at a high rate (10° C/sec), resulted in total detonation of the lithium-thionyl chloride (bromine chloride) cell whereas the lithium-thionyl chloride cell was found to vent its products, otherwise remaining intact.

Controversy remains over the subject of whether a cell that vents is safer than one without a vent. The latter cells can release internal products only by violation of the cell case, usually designed for up to 1000 psi. The selection of a cell type ultimately depends on the application and related environmental considerations.

Lithium Anode Soluble Cathode Cells/Batteries

Lithium-Manganese Dioxide (Li-MnO$_2$)

This cell utilizes an electrolyte composed of propylene carbonate (PC) and 1,2 dimethoxyethane (DME) with lithium perchlorate (LiClO$_4$) salt which enhances the conductivity. The reaction is given as:

$$Li + MnO_2 \rightarrow Mn^{3+}O_2(Li^+) \qquad (4.24)$$

where the lithium ion (Li$^+$) enters the MnO$_2$ lattice. The cathode is either pressed powder or a thin pasted electrode on a conductive support. It has been used for memory backup and small light loads.

Lithium-Carbon Monofluoride (Li-(CF)$_x$

The active components are a lithium anode and solid carbon polymonofluoride (CF) formed by reaction of carbon monoxide with fluorine gas. Several electrolytes have been used, including lithium hexafluorarsenate (LiAsF$_6$) in dimethylsulfide (DMSI). Other electrolytes have included lithium tetrafluoroborate (LiBF$_4$) salt in butyrlactone (BL) tetrahydrofuran (THF) or propylene carbonate (PC) and dimethoxyethane (DME). The simplified reaction is given as:

$$Li + (CF)_x \rightarrow x\,LiF + xC \qquad (4.25)$$

CHEMICAL STORAGE AND GENERATION SYSTEMS 203

where $(CF)_x$ acts as an intercalation compound. Li-(CF_x) cells were initially used in a launch vehicle safety application. However, since the Shuttle accident, use has been discontinued.

Lithium Solid Electrolyte Cells

There is only one cell of this type available today, the solid-state lithium-iodine cell.

Lithium-Iodine (Li-I_2)

The lithium-iodine cell uses a solid lithium anode and an iodine charge transfer complex as the cathode. The cathode consists of a mixture of the iodine and poly-2-vinylpyridine. The electrolyte is solid lithium iodide. The reaction is given as:

$$2Li + I_2 \rightarrow 2LiI \qquad (4.26)$$

Because of its very low discharge rate capability, the primary use of this cell is in pacemakers. However, these cells have applications in comparative circuits or computer memory retention, watches, and calculators, and have been used in several such applications in the Shuttle orbiter.

The advantages include: 1) excellent storage capability (~10 ears), 2) sealed (no leakage), 3) wide operating temperatures, and 4) safety. The disadvantages include: 1) low current drain only, and 2) low power capability. The potential hazards include placing the cell in a flame could result in a venting or deformation.

7.2 Rechargeable cells and batteries

Rechargeable or secondary cells and batteries differ from the primary cells and batteries in that the chemically stored energy used during the discharge can be returned to the chemical form by recharge. This is accomplished by causing a source of electrons from the solar array or power supply to flow in the reverse direction from that of the discharge direction. When this occurs the reactions are reversed and therefore the electrode that served as the anode on discharge becomes the cathode on charge. The cathode where reduction took place during discharge becomes the anode where oxidation occurs. Similarly the anode becomes the cathode. Even though the reactants and products are changing, the polarity remains the same as that during discharge because the quantities of active materials are primarily in the charged state. A diagram of charge and discharge configurations is shown in Figure 4.15 for the Ni-Cd cell.

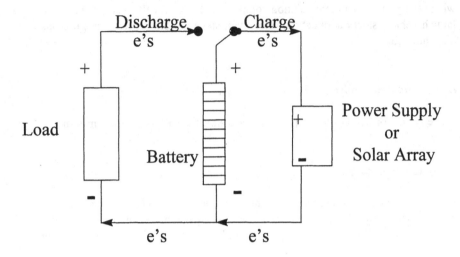

Figure 4.15 Charge and Discharge Configurations For Rechargeable Cells

Secondary cells/batteries require additional circuitry to provide the means for charging. Depending on type of electrochemical cell application and temperature, cells can be cycled (charged/discharged) hundreds of times to tens of thousands of times. Because of its reversibility and long life this battery type has been used in the majority of low Earth orbiting (LEO) and geosynchronous orbiting (GEO) spacecraft.

It is essential for long life missions that the cells in the battery be hermetically sealed to maintain their chemical and electrochemical balance. The rechargeable batteries that have seen the most use in space are those that consist of nickel hydroxide as the discharged active material. These are the 'nickel-cadmium (Ni-Cd)' and 'nickel-hydrogen (Ni-H_2)' types. They have demonstrated long cycle life in LEO and GEO mission applications. During eclipse, they provide power to the spacecraft, the instruments, and electronics. During Sun periods, the solar array is adequate enough to provide spacecraft energy requirements and recharge the battery.

A second type is based on the silver oxide electrode. These batteries have been used in more limited applications because of their limited cycle life. Specifically, these include silver-cadmium (AgO-Cd) and silver-zinc (AgO-Zn) batteries. The silver-hydrogen (AgO-H_2) battery has received some attention because of its higher specific energy but has not been used in space. The nickel and silver oxide based cells are similar in that they use a strongly-alkaline electrolyte.

Although the most well known rechargeable battery is the lead acid system, it is not truly a sealed system and therefore use in space is not practical for long term missions; its use in space has been limited to short term shuttle applications in a cylindrical configuration.

CHEMICAL STORAGE AND GENERATION SYSTEMS 205

Recent developments have resulted in cells based on lithium, with the initial designs using lithium foil. However, their use is a safety concern. Even though the primary cells use lithium foil, the replating of the lithium during charge was determined to be unsafe. A lithium-based cell that provides an adequate level of safety is the lithium-ion cell, in which lithium metal is absent.

Sodium-Sulfur (Na-S) cells operate at temperatures of 350°C where the constituents are in the molten condition. The electrodes are separated by a beta alumina separator which allows sodium ions to flow between electrodes. Although they have demonstrated long cycle life in ground tests, concern for the reliability of the high temperature system has limited its use to a single flight experiment. Sodium-Nickel Chloride (Na-NiCl$_2$) cells which operate at 250°C have also shown long life in commercial ground tests and are being considered for electric vehicle applications but have not had any space use.

A comparison of the characteristics of rechargeable cells is shown in Table 4.10. A comparison of the characteristics based on 28 V space batteries, given in Table 4.11, reflects the improvement in specific energy and energy density. The other factors that need to be considered in selecting a battery for a mission, such as cycle life and LEO or GEO, are also given.

Nickel hydroxide cells/batteries

There are three types of nickel hydroxide cells/batteries. These are identified as the cells containing Ni(OH)$_2$ as the discharged active material and NiOOH as the oxidized active material in the charged state. The Ni(OH)$_2$ exhibits a positive polarity in this type of cell. The reaction is:

$$Ni(OH)_2 + OH^- \rightarrow NiOOH + e^- \qquad (4.27)$$

The types of nickel hydroxide cells include Ni-Cd and Ni-H$_2$ and have been used extensively in space. Nickel-Metal Hydride (Ni-MH), a version of the Ni-H$_2$ type in which the hydrogen is bonded chemically as a hydride as compared to pressurized H$_2$, has until now received little attention as a space battery. Nickel-zinc (Ni-Zn) and nickel-iron (Ni-Fe) types are also in the nickel hydroxide family but with higher specific energies than Ni-Cd and Ni-H$_2$. However, they are not sealable. The latter two have been used, or considered for use, in electric vehicles, load leveling, and photovoltaic power generation systems and will not be discussed here.

Table 4.10 Secondary Cell Characteristics

Name	Design	Anode*	OCV (V)	Nominal Voltage (V)	Specific Energy (Wh/kg)	Specific Energy (Wh/l)	Operating Temp. C	Cycle Life 80% DOD
NiOOH-Cd		Cd // KOH // NiOOH	1.29	1.25	30	80	-10 to 35	-2000
NiOOH-H$_2$	(IPV)	H$_2$ // KOH // NiOOH	1.32	1.25	55	60	-10 to 35	-2000
NiOOH-H$_2$	(CPV)	H$_2$ // KOH // NiOOH	1.32	1.25	55	60	-10 to 35	-2000
NiOOH-H$_2$	(SPV)	H$_2$ // KOH // NiOOH	1.32	1.25	55	60	-10 to 35	-2000
AgO-Cd		Cd // KOH // Ag$_2$O / AgO	1.41	1.3/1.1	55	110	-25 to 60	400
AgO-Zn		Zn // KOH // Ag$_2$O/AgO	1.86	1.6/1.5	90	180	-20 to 60	100
PbO$_2$-Pb		Pb // H$_2$SO$_4$ // PbO$_2$	2.1	2.0	30	60	-40 to 60	100
Na-S		Na // b" alumina // S	2.1	1.65	186**	304**	350	1000
Na-NiCl$_2$		Na // b" alumina // NiCl$_2$	2.4	2.3	120	110	275	1000
Li-CoO$_2$		Li // EC / DMC / CoO$_2$	4.00	3.64	120	125	-20 to 50	1000

*On discharge **Projected

Note: The values for Specific Energy, Energy Density and Cycle Life refer to the same battery in each case. Different trade-offs can be made between energy and life with the result that, in most cases, these values can vary over a wide range.

CHEMICAL STORAGE AND GENERATION SYSTEMS

Table 4.11 A Comparison of 28V Battery Characteristics

CHEMISTRY	Ni-Cd	SUPER Ni-Cd	IPV Ni-H$_2$	2 CELL CPV Ni-H$_2$ (3.5")	22 CELL SPV	Li-ION	Ag-Zn
NUMBER OF CELLS	18	18	18	18	22	8	18
NOMINAL VOLTAGE	21.6	21.6	22.5	22.5	27.5	28.8	27
CAPACITY (Ah)	18	20	20	20	15	5	40
SPECIFIC ENERGY (Wh/Kg)	25	25	29	31	44	90	72
ENERGY DENSITY (Wh/L)	35	31	16	18	38	140	86
CYCLE LIFE	30000 LEO	30000	> 30000	~ 10000	N/A	~ 1000 (> 80% DOD)	< 100 (> 80% DOD)
WEIGHT (Kg)	16	17.9	13.8	13.1	9.6	1.6	15
VOLUME (L)	11.4	14.4	13.6	11.4	8.6	1.0	12.6
BATTERY DIMENSIONS (in)	12.3 x 7.3 x 7.7	13.5 x 8.1 x 8.4	25.0 x 17.6 x 3.5	18.0 x 14.5 x 4.5	21.0 x 5.0 dia.	N/A	12.0 x 8.0 x 8.0
HERITAGE	LEO / GEO	LEO / GEO	LEO / GEO	MARS GLOBAL SURVEYOR	MARS '98	NEW MILLENIUM	MARS PATHFINDER

Nickel-Cadmium (Ni-Cd)

In the Ni-Cd cell, cadmium hydroxide, $Cd(OH)_2$, is the active material of the negative electrode. The $Cd(OH)_2$ negative electrode active material like the $Ni(OH)_2$ material is contained within an 80 percent porous nickel plaque containing a perforated sheet or nickel screen which serves as the current collector. The active material is deposited into the pores by a series of steps filling approximately 40 percent of the remaining space in the plaque. A 31 percent aqueous solution of potassium hydroxide is used as the electrolyte, and the separator is a nonwoven nylon or polypropylene material. The cell is hermetically sealed with two ceramic/metal terminal seals. The electrode and net cell reaction shown as a charge reaction is:

Charge
$$2\,Ni(OH)_2 + 2OH^- \rightarrow 2NiOOH + 2H_2O + 2e^- \quad (4.28)$$
$$\underline{Cd(OH)_2 + 2e^- \rightarrow Cd + 2OH^- \quad (4.29)}$$
$$2\,Ni(OH)_2 + Cd(OH)_2 \rightarrow 2\,NiOOH + Cd + 2H_2O \quad (4.30)$$

In order to replace the energy removed from the cell during discharge, the cell must be recharged to the full state of charge and then must receive an overcharge which offsets the charge inefficiency. The inefficiency is a function of rate and temperature and is a result of the production of oxygen gas at the nickel hydroxide electrode produced primarily near the end of the discharge. The reaction resulting in the inefficiency at the positive electrode is:

$$4OH^- \rightarrow O_2 + H_2O + 2\,e^- \quad (4.31)$$

In the hermetically sealed (semi-dry or starved) type, the electrolyte is immobilized in the separator and plates, allowing the gas to distribute itself around and within the cell pack. The evolution of the oxygen gas is offset by its recombination at the negative electrode which contains a significant quantity of cadmium metal (Cd) because it is being charged. The oxygen recombines chemically according to the following reaction:

$$O_2 + 2Cd + 2\,H_2O \rightarrow 2Cd(OH)_2 \quad (4.32)$$

Thus, when the charging process has converted most of the active materials to NiOOH and Cd as given in Eqs. (4.28-4.30) and oxygen evolution occurs, the oxygen reacts with the charged Cd via Eq. (4.32) to discharge the cadmium electrode by the same amount resulting in no net change of reactants and products thus preserving the electrode balance.

CHEMICAL STORAGE AND GENERATION SYSTEMS

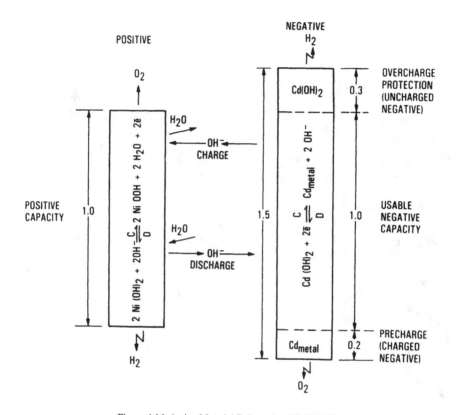

Figure 4.16 Active Material Balance in a Ni-Cd Cell

The aerospace cell is referred to as positive-limited on charge (and discharge). If it were to be negative-limited on charge, as in electrolysis, the negative electrode would evolve hydrogen gas which is not recombined in this cell. A high internal pressure and potential safety problem would result. The discharge is positive-limited because of more rapid degradation of the negative electrode. There is always excess negative capacity and therefore the cell capacity is limited by the positive electrode. The balance between the active materials in the Ni-Cd cell is given in Figure 4.16.

The most efficient charge scheme is to use a high current (>C/4) at the start of charge and convert to constant voltage (current taper) at a prescribed voltage limit until the current drops to a C/15 rate. The cells can generally be constantly charged at the C/10 rate without producing excessive pressure. However, this charge rate results in a temperature increase based on I^2R heating. The discharge process results in a thermal inefficiency of 16 to 18 percent which must be provided for in a system design. Because there is a rapid decrease of cell cycle life with high depth of discharge (DOD), operation has been limited in aerospace applications to 25 percent DOD for LEO applications and

Figure 4.17 The NASA Standard 20 Ah Battery used on the Solar Max Spacecraft

CHEMICAL STORAGE AND GENERATION SYSTEMS

65 percent for GEO applications. The NASA Standard 28 V battery, developed by McDonnell-Douglas and containing the NASA Standard 25 Ah G.E. prismatic hermetically sealed cells, was used on the Solar Max Mission. That mission lasted more than 8 years in LEO orbit (See Figure 4.17).

The advantages of this design are: 1) maintenance-free sealed cells, 2) long cycle life, 3) rugged/sealed, 4) high rate/power capability, 5) flat discharge, 6) long storage life and 7) a large operational data base. The disadvantages include: 1) a relatively high cost, 2) a memory effect/double plateau (may need reconditioning), 3) the need for charge control methods, 4) poor charge acceptance at high temperature or low charge rate, 5) poor capacity retention on storage, and 6) prismatic sealed cells require end-plates to prevent expansion due to internal pressure.

There are several potential hazards: 1) pressure buildup in sealed case on charge (adequate temperature, rate, and charge controls required because of an undesirable ratio, balance, and stability of active materials), 2) leakage of alkaline (white encrustation on seal or case is an indication of leak (inspection and welder certification, qualification, and calibration is necessary as is the use of nylon/cotton gloves in handling electrolyte), 3) leakage of oxygen can cause imbalance problems resulting in dangerous H buildup (use reliable ceramic/metal seals, perform helium leak test on each cell), 4) internal shorting can result in hot spots, arcs, etc. leading to potential explosion (X-ray cell, use "clean" methods in cell assembly, inspect plates for sharp edges, burns, etc.), 5) cell reversal results in gas evolution and irreversible electrode imbalance, making it necessary to monitor each cell during operation (either the undervoltage cutout can be utilized, or one can use a 1/2 battery voltage to monitor changes in individual cell, and 6) external shorts can cause very high-rate, high temperature excursions which can lead to a violent venting (use tools and equipment that are insulated to prevent bridging, coat cell terminals to prevent inadvertent metal contact; use fuses and protective circuitry). Please note that although the potential hazards are concerns to be considered, the Ni-Cd cell and battery has been the battery most often used in space from the 1960s to 1990s.

Nickel-Hydrogen (Ni-H_2)

The Ni-H_2 battery has been selected for GEO and some planetary missions. The Hubble Space Telescope mission also uses these batteries. Three types of NiOOH-H_2 cells have been demonstrated for space missions. These include the Individual Pressure Vessel (IPV), the Common Pressure Vessel (CPV), and the Single Pressure Vessel (SPV). The Ni-H_2 battery has a higher specific energy, greater depth of discharge capability, and longer life than the Ni-Cd battery because the cadmium electrode, which degrades with life and depth of discharge, is replaced by the H_2 electrode. However, its energy density is generally less than the Ni-Cd battery because the Ni-H_2 cylindrical cell configuration has a significantly lower packing density.

The single IPV Ni-H$_2$ cell was the first configuration used in space in a GEO application. Eagle-Picher, one of the manufacturers of CPV Ni-H$_2$, found that they could install two cells in the same pressure cylinder thus increasing the specific energy and energy density. The SPV involved containment of all 22 cells in a single pressure cylinder thus enhancing the specific energy and energy density. Since its first use by the Naval Research Laboratory, the IPV has been used for commercial purposes as well.

This cell, like the Ni-Cd cell, is quite reversible. The positive NI(OH)$_2$ plates are also similar in construction to those described above. The hydrogen electrode comprises a Teflon bonded platinum back on a nickel mesh screen to allow the reduction and oxidation of hydrogen. A gas diffusion screen is used to facilitate hydrogen diffusion. A plasma sprayed zirconia oxide is sometimes used to coat the inside of the cylindrical pressure vessel to enhance electrolyte distribution. It has the potential for greater depth of discharge and higher rate capability than the nickel-cadmium. The electrolyte is 30 to 35 percent aqueous potassium hydroxide by weight. An asbestos fuel cell bidirectional separator is used.

The IPV, CPV, and SPV cell configurations use a pressure cylinder to contain 600 to 1000 psi of H$_2$ at full charge. The cell discharge reaction at the NiOOH electrode is the same as that in the Ni-Cd cell. The H$_2$ recombination reaction with NiOOH at the negative during discharge is given in Eq. (4.34). The net reaction is given in Eq. (4.35). The reaction on discharge shows that the H$_2$ built up on charge decreases during the discharge.

$$\begin{array}{lr} \text{Discharge} & \\ 2\text{NiOOH} + \text{H}_2\text{O} + 2\,e^- \rightarrow 2\text{Ni(OH)}_2 + 2\text{OH}^- & (4.33) \\ \underline{\text{H}_2 + 2\text{OH}^- \rightarrow 2\text{H}_2\text{O} + 2\,e^-} & (4.34) \\ 2\,\text{NiOOH} + \text{H}_2 \rightarrow \text{Ni(OH)}_2 & (4.35) \end{array}$$

During charge, H$_2$ Pressure during charge builds to the design level of 600 - 800 psi. During the overcharge period O$_2$ is also evolved as shown in reaction at the positive Nickel electrode Eq. (4.35). The O$_2$ produced reacts with the water to produce OH$^-$ according to Eq. (4.36) resulting in no net reaction (Hydrogen evolution reaction is suppressed) and thus the cell pressure remains constant once full charge is reached.

$$\begin{array}{lr} 4\text{OH}^- \rightarrow 2\text{H}_2\text{O} + \text{O}_2 + 4\,e^- & (4.36) \\ \underline{\text{O}_2 + 2\text{H}_2\text{O} + 4\,e^- \rightarrow 4\text{OH}^-} & (4.37) \\ \text{Equilibrium maintained} & (4.38) \end{array}$$

The two gases are combined at the fuel cell electrode at the same rate as they are being produced and thus electrode balance is maintained. The pressure remains at the maximum until the start of discharge when the O$_2$ reaction subsides. The H$_2$ then reacts until the end of discharge when pressure is at the minimum as shown in Eq. (4.34). The

CHEMICAL STORAGE AND GENERATION SYSTEMS 213

cell can be charged at a relatively high rate constant current (~ C/2) similar to the Ni-Cd cell.

This type of cell has had one major application, that of replacing the Ni-Cd in GEO spacecraft. The enhanced specific energy (Wh/kg) over Ni-Cd cells offers a greater opportunity for increasing the spacecraft capability while reducing total mass. GEO orbit (22,000 miles above the Earth) is one in which there are only two eclipse periods per year, thus requiring ~100 cycles/year. The Ni-H_2 cells are projected to reach 85 percent DOD during the longest eclipse period (about 75 minutes) which occurs twice each Earth year. There is also interest in using this cell in LEO where its capability for greater DOD would provide a distinct advantage. While the specific energy of Ni-H_2 cells is greater than Ni-Cd cells, the volume energy density is less because of the cylindrical gas canister design.

Two types of Ni-H_2 cells were developed initially: the Comsat 35 Ah design (Figure 4.18) used in the U.S. Navy NTS-2 spacecraft and the U.S. Air Force 50 Ah baseline design developed by Hughes Aircraft. The second generation design combined the best of both designs. Eagle-Picher produced an 80 Ah design and Gates Energy Products 3 Ah cell used for Intelsat. Figure 4.19 is a photo of the Intelsat Ni-H_2 battery containing IPV cells produced by Ford Aerospace.

Figure 4.20 shows the concept of the two-cell common pressure vessel cell used for the Mars Global Surveyor Mission and the 22-cell single pressure cell used in the Clementine Mission. The SPV design has been used on more than 40 Iridium™ communications spacecraft developed by Motorola.

This cell offers several advantages, including: 1) state of charge directly related to pressure, 2) longer cycle life at higher DOD than Ni-Cd, 3) specific energy/specific power greater than Ni-Cd, 4) higher depth of discharge capability than Ni-Cd, and 5) the ability to tolerate overcharge and reversal at controlled rates.

The disadvantages include: 1) high cost, 2) self-discharge proportional to decrease H_2 pressure, 3) higher cost than Ni-Cd, and 4) volume energy density and power density less than Ni-Cd. The potential hazards are: 1) high rate charging and heat buildup (use charge control and thermostatic devices), 2) high pressure buildup (use strain gage to monitor pressure), 3) leakage of strong alkaline electrolyte, 4) gas (H_2 or O_2) leakage creates cell imbalance leading to excessive pressure (perform helium leak testing on each cell and use high reliability well tested seals), and 5) external shorting can result in very high-rate, high temperature excursion leading to a violent venting (use fuse and other protective circuitry).

Silver oxide cells/batteries

Three types comprise this group: silver-cadmium (Ag-Cd), silver-zinc (Ag-Zn), and silver-hydrogen (Ag-H_2). The first two have found applications in space for limited

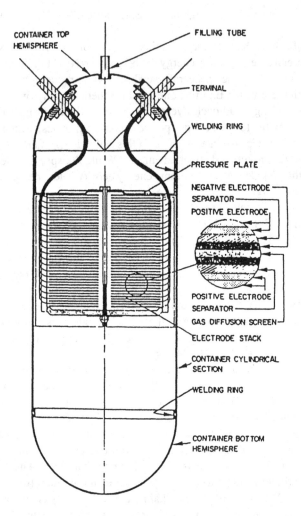

Figure 4.18 Comsat IPV Ni-H$_2$ Cell Design

mission life. The latter has not been developed for space. The Ag-Cd cell is nonmagnetic and, although not extensively used, was favored in applications in which magnetometers are among the spacecraft instruments. Ag-Zn (also known as silver oxide-zinc and zinc-silver oxide) cells have been used in mid-altitude orbits with relatively short-life spacecraft where the high specific energy plays a part in helping to reduce spacecraft weight. Ag-Zn batteries have been used in launch vehicles for communication and control until the spacecraft is placed in orbit (i.e., during the insertion phase). The cell is similar to that described above under zinc anode primary cells except that it can be

CHEMICAL STORAGE AND GENERATION SYSTEMS 215

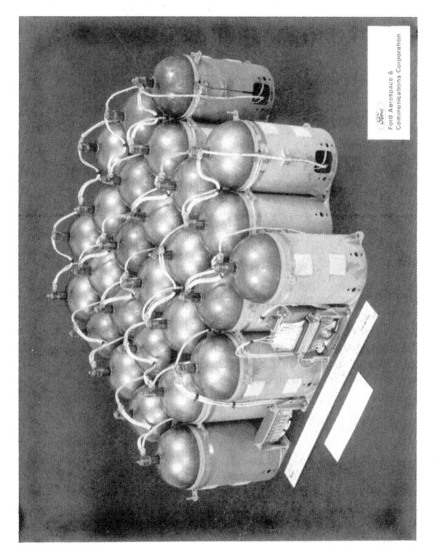

Figure 4.19 Intelsat-V Battery with 30 Ah IPV Cells

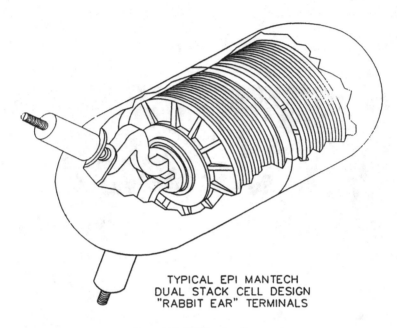

Figure 4.20 CPV Design Concept

recharged. However, because of the inefficiencies caused by gaseous reactions at the electrodes pressure relief vents are used in these cells.

As described earlier, the silver oxide electrode is the cathode on discharge and anode on charge (polarity + on both charge and discharge). The zinc, cadmium, and hydrogen are the anodes on discharge and cathode on charge (polarity - on both charge and discharge). See Table 4.10 for the cell characteristics.

Silver-Cadmium (AgO-Cd)

The silver electrode is prepared by sintering silver powder on a silver grid after molding or continuous rolling. Pasted or pressed plates have also been used. The cadmium electrodes are produced by the same process using cadmium oxide or hydroxide on a silver or nickel grid. The electrolyte is 40 percent KOH. The plates are wrapped with multiple layers of separators to prevent silver migration, which is a life-limiting process. The discharge reaction is given as:

$$AgO + Cd + H_2O \xrightarrow{Discharge} Cd(OH)_2 + Ag \qquad (4.39)$$

CHEMICAL STORAGE AND GENERATION SYSTEMS

A second discharge plateau noted in silver cells is due to the discharge of Ag_2O according to:

$$Ag_2O + H_2O + Cd \xrightarrow{Discharge} Cd(OH)_2 + 2Ag \qquad (4.40)$$

The charging of Ag0-Cd cells occurs at two voltage levels. Conversion of Ag to Ag_2O and AgO charge is best performed at constant current at the C/20 - C/10 rate to 1.6V/cell. Thereafter, O_2 is generated at the silver electrode (anode on charge) as with the nickel hydroxide electrode (anode on charge). H_2 is also generated on the Cd electrode via electrolysis of water during the overcharge if not prevented by cutoff voltage. There is little or no gas recombination so that the overcharge capability of the silver-cadmium is not possible. H_2 evolution appears to be less of a problem with these cells than the Ag-Zn cells. The cells are usually available in polymeric cases in the prismatic configuration and their use has been primarily on magnetic spacecraft and in military applications.

The advantages of this cell are: 1) higher specific energy than Ni-Cd, 2) higher volume energy density than Ni-Cd and Ni-H_2, 3) nonmagnetic, and 4) less sensitive to overcharge than silver-zinc, while the disadvantages include: 1) short cycle life (better than Ag-Zn), 2) operation strongly dependent on temperature, and 3) lower voltage than Ag-Zn.

The potential hazards and suggested cautions are: 1) high rate charging and heat buildup (use charge control and thermostat devices), 2) high-pressure buildup (some cell pressure relieved, use strain gage to monitor pressure), 3) leakage of strong alkaline electrolyte (prevented by performing seal and cell testing; white encrustation on seal or case is an indicator of a leak), 4) inspection and welder qualification and calibration are necessary (use rubber gloves in handling electrolyte), and 5) gas (H_2 or O_2) leakage creates cell imbalance leading to excessive pressure (perform helium leak testing on each cell; use high reliability, well-tested seals; external shorting can result in very high-rate, high temperature excursion, leading to a violent venting; use fuse and other protective circuitry).

Silver-Zinc (Ag-Zn)

The cell is a modified version of the primary Ag-Zn cell. The silver electrode is the same type as previously described. The zinc electrode is prepared by pressing a paste or slurry of zinc oxide binder, or by electrodeposition in plating tanks onto metallic grids. As in the Ag-Cd cell, the plates are also wrapped with layers of separator to prevent silver migration. Zinc dendrite formation is the major cause of life limitation. The inner separator serves as an electrolyte reservoir and barrier to minimize AgO oxidation of the separator. The outer separator stabilizes the zinc electrode and retards zinc penetration. The outer separator is cellophane but is being replaced by radiation-grafted polyethylene in some applications. The electrolyte is 40 percent KOH. The two-step discharge reaction is given by:

$$AgO + Zn + H_2O \xrightarrow{Discharge} Zn(OH)_2 + Ag \qquad (4.41)$$

$$Ag_2O + Zn + H_2O \xrightarrow{Discharge} Zn(OH)_2 + 2Ag \qquad (4.42)$$

The charge and discharge are similar to that of the Ag-Cd cell. In this case charge is terminated at 2.0 V to avoid the generation of gas which cannot be recombined and to prevent the formation of zinc dendrites. The high rate capability is due to the electrical conductivity of the silver grid and the conductivity of the positive electrode.

These cells are generally prismatic in configurations. The capacities available range from very low to thousands of ampere hours as in military equipment. High rate (HR) and low rate (LR) versions exist. These cells are usually specially ordered to meet the requirements. HR cells were used in the tools and MMU on the SMM repair mission in April 1984. The Ag-Zn battery is also available in reserve configuration for very long storage periods prior to use. Cells with active electrolyte will lose capacity and therefore storing the electrolyte outside the cell will prevent the high self-discharge rate. Several planetary missions have used batteries of this type to meet long term inactive cruise requirements. Figure 4.11 is a photo of the Mars Pathfinder Lander Ag-Zn battery that landed on Mars, July 4, 1997.

Again, the advantages are: 1) highest specific energy and power of presently available secondary batteries, and 2) highest energy density and power of presently available secondary batteries. The disadvantages: 1) high cost, 2) limited cycle life, 3) poor low-temperature performance, and 4) a two-step voltage plateau. The potential hazards and mitigating actions are: 1) high rate charging and heat buildup (use charge control and thermostat devices), 2) high pressure buildup possible (cells may be pressure relieved, use strain gauge to monitor pressure), 3) leakage of strong alkaline electrolyte (prevent by performing seal and cell testing; in vented designs, absorb electrolyte before it escapes from the battery; search for white encrustation leak, which requires inspection and cell assembler qualification and calibration; use rubber gloves in handling electrolyte), and 4) external shorting can result in very high rate, high temperature excursion leading to a violent venting (use fuse or other protective circuitry).

CHEMICAL STORAGE AND GENERATION SYSTEMS 219

Lithium ion cells/batteries

Initially, rechargeable lithium cells were comprised of lithium foil as the anode material. A number of cathode materials were also investigated. Limited cycle life cells were demonstrated with $Li-TiS_2$ and $Li-V_6O_{13}$, however, concern for the safety of the lithium foil as a rechargeable electrode led to the conclusion that further development for space was not advisable.

An increase in power requirements for space missions has necessitated the development of high energy density rechargeable batteries. Among the various electrochemical systems, batteries utilizing lithium have received widespread attention because of the high electropositive nature of lithium and its low equivalent weight. A number of soluble and insoluble materials have been examined as candidates for cathode materials. Even though soluble cathode materials offer a number of advantages (such as rate capability, low sensitivity to overcharge, etc.), these cells exhibit high rates of self-discharge.

The development of a lithium ion cell, absent of lithium metal, offers an opportunity to significantly increase the specific energy and energy density, thus reducing battery mass and volume. The three types presently under development are the Lithium-Cobalt Oxide ($Li-CoO_2$), Lithium-Manganese Oxide ($Li-Mn_2O_4$), and Lithium-Nickel Oxide ($Li-NiO_2$). Of the three, only the first has demonstrated the potential for cycle life required for space. A major advantage of this system is the cell voltage of near 4 V per cell, a specific energy of 100 Wh/kg, and energy density in excess of 250 Wh/l.

The cell potential is based on the difference in potential between a cathode material containing lithium ion, such as Lithium Cobalt Oxide ($LiCoO_2$) and a carbon electrode that can intercalate lithium ion. When the lithium ions are in the carbon anode the cell is charged, and when the lithium ions return to the cathode matrix the cell is discharged. The most familiar of these is the Lithium-Cobalt Oxide ($Li-CoO_2$) cell. Although the lithium -ion cell has not been used in space to date, it has been added to the manifest of upcoming planetary missions, e.g., the Mars missions.

The basis for the Lithium-Ion cell concept appears in Figure 4.21. The cell is assembled with the Li^+ intercalated in the cathode material, e.g., $LiCoO_2$, and a layered carbon or graphite anode. The cell is activated by charging, i.e., forcing the Li^+ to flow into the anode structure. When the charging step is completed the potential difference is approximately 4.0 V. During discharge the Li^+ flows back into the cathode during which time the cell voltage decreases. During the initial charge process there is some loss of the Li^+ in the structure of the anode.

$$LiC_6 + MO_2 \xrightarrow{Discharge} 6C + LiMO_2 \quad (4.43)$$
$$(M = Co, Ni, \text{ and } MO_2 = Mn_2O_4)$$

220   SPACECRAFT POWER TECHNOLOGIES

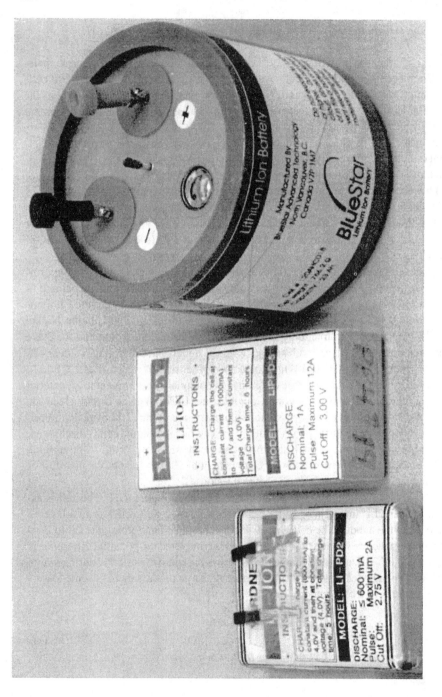

Figure 4.21 Examples of Li-Ion Cells

CHEMICAL STORAGE AND GENERATION SYSTEMS

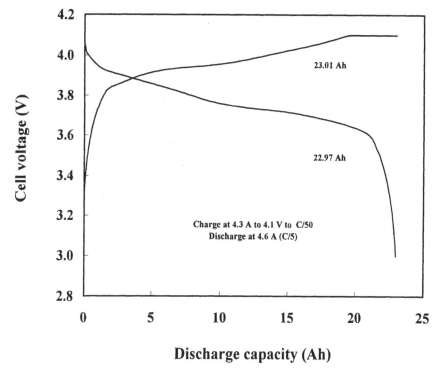

Figure 4.22 Charge and Discharge Curves of the Blue Star Li-CoO₂ Cell at the C/5 Rate

The electrolyte is generally a solvent mixture of Ethylene Carbonate (EC), Diethyl Carbonate (DE), and Dimethyl Carbonate (DMC), with a salt such as $LiPF_6$ or $LiBF_6$. Examples of flat plate, cylindrical, and prismatic design Li-Ion cells appear in Figure 4.21. The prismatic cells are from Yardney Technical Products and the cylindrical cell from Blue Star Batteries. The charge and discharge curves for the Blue Star $LiCoO_2$ battery at a C/5 rate are shown in Figure 4.22.

The advantages include: 1) cell voltage of 4.0 V, 2) high specific energy and energy density (4 x Ni-Cd and 2 x Ni-H₂), 3) 99% charge efficiency, 4) available for commercial products, 5) operation demonstrated at -20°C, and 6) no lithium metal in the cell. The disadvantages include: 1) relatively short cycle life (compared to Ni-Cd), 2) only small size cells available for commercial products, and 3) individual cell control to maintain cell balance.

There are two noteworthy potential hazards: 1) excessive overcharge and overdischarge can result in undesirable products, and 2) cell balance must be maintained to avoid cell imbalance, resulting in undesirable reactions.

Sodium cells and batteries

The sodium cell is a high-temperature technology that has generated significant interest in recent years. The Sodium-Sulfur (Na-S) system (350°C) continues to be tested by the U.S. Air Force for a potential space application. Lengthy cycle life has been demonstrated. However, the closest to space use is in an Air Force flight experiment launched in November 1996. Significant resources were allocated for this work for both electric vehicles and load leveling, and thus transfer of the technology for flight use appeared worthwhile for large spacecraft. However, the effect on materials of the high temperature, together with the change in NASA direction to 'smaller, better, cheaper' spacecraft, resulted in reduced interest in this technology. The Sodium-Nickel Chloride (Na-NiCl$_2$) system with lower operating temperature (250°C) and absent of molten sulfur appeared to be worthwhile. However, with the advent of the new NASA approach, this too was dropped from consideration for space use. The Na-S system is discussed below in view of the continued, albeit low, interest.

The system requires a high temperature (350°C) to maintain its anodic material, sodium, and cathodic material, sulfur, in the molten state. Solid beta alumina or glass acts as both the separator and the electrolyte in which the sodium ions produced on discharge diffuse through the ion selective material to produce the product sodium sulfide according to:

$$\text{Discharge}$$
$$2Na + xS \rightarrow Na_2S_x \qquad (4.44)$$

The depth of discharge and operation determines the value of x which ranges from five to two.

The projected operation appears to be rather straightforward. The cell is designed to operate at 90 percent energy efficiency. However, the impedance, and therefore open circuit voltage, varies with state of charge/discharge. Charging is terminated once the cell resistance equals twice the discharge endpoint resistance. Interest in this projected high specific energy system makes it a candidate for several applications, including electric vehicles and space. The major difficulty is the weight penalty that is associated with the thermal management of the system at 350°C and as such it is geared for large systems. The materials problems associated with this temperature have not been solved and the impact on safety also represents an unresolved issue.

The advantages of sodium cells include: 1) adaptibility to large energy systems, 2) high peak power, 3) high level of support for development is available, and 4) large specific energy and energy density ideal for a secondary system. There are several disadvantages: 1) operation at 350°C, 2) materials and safety problems associated with high temperatures, 3) beta alumina technology refinement, 4) heat management, 5) high cost, and 6) the weight penalty for thermal management.

CHEMICAL STORAGE AND GENERATION SYSTEMS 223

The potential hazard includes operation at a high temperature where molten sulfur and sodium are present separated only by a ceramic separator.

8. Fuel cell systems

A fuel cell system consists of a group of cells connected in series (fuel cell stack), the fuel and oxidant stored external to the stack, and ancillaries including pumps, plumbing, sensors, and controls to process the products and reactant product. The fuel cell system differs from a battery in that the reactants are stored outside the cell in cylinders. This arrangement infers that the more fuel and oxidant available the larger the energy storage capacity and the longer the fuel cell system will operate. Battery storage capacity, however, is limited by the quantity of active material contained within the case. Fuel cell stack power is the sum of the voltages of each cell times the current. The fuel cell system uses some of the stack power for operation of the ancillaries and thus will be less than the stack power. The capacity of a fuel cell system is determined by the quantity of fuel and oxidant stored. This is a convenient system for manned flight because the fuel cell system can be supplied with enough fuel and oxidant to meet the relatively short mission length and provide water and heat for life support. For longer missions, such as LEO and GEO, rechargeable batteries are used.

The applicability of a fuel cell power plant for space use has, besides the attractive features of being a pollution-free power source based on direct conversion with immovable parts, the primary advantage of its ability to be incorporated into the ecological cycle of the space crew. Further, liquid oxygen and hydrogen are also available for propulsion on board space vehicles because of their high specific impulse. These characteristics have extended the utility of fuel cell power sources for space use in the Gemini, Apollo, Shuttle, and Spacelab programs. The H_2-O_2 alkaline fuel cell has emerged as the most attractive candidate for space use.

While making a choice among suitable alternatives to meet the requirements of a spacecraft, the factors to be considered are: 1) reliability, 2) efficiency, 3) life, 4) environmental compatibility, 5) endurance to environmental conditions in space (zero gravity, vibration, shock, acceleration, acoustic noise, etc.), 6) energy densities, 7) storage, 8) heritage, and 9) cost. The reliability of the hardware, once a choice is made, will have to consider proper choice of materials, chemicals, and components of space grade and incorporate a detailed and stringent qualification test plan.

Fuel cells have been known from the time of Grove in 1839. Since then several types of fuel cells have been developed for various stationary, vehicular, and other applications. The types of fuel cells that have been or are being considered for space use include:

Alkaline Electrolyte H_2-O_2 fuel cells (AFC) which operate at 40 - 60°C, Proton Exchange Membrane Electrolyte H_2-O_2 Fuel Cells (PEMFC) which operate at 60-80°C, and Direct Methanol/O_2, Liquid-Feed Proton Exchange Membrane Fuel Cells (DMLFFC/PEM) operating between 20°C and 90°C.

Regenerative fuel cells using the PAFC, AFC, or PEMFC, together with an electrolyzer, are being considered for applications where charge and discharge operations are required. The electrolyzer, powered by solar cells, produces the fuel and oxidant to the fuel cell when solar energy is not available. This system has yet to prove feasible for space applications.

The most common fuel cell system used in terrestrial applications is the Phosphoric Acid fuel cell (PAFC) that operates at 250°C. Higher temperature fuel cells include the Molten Carbonate Fuel Cells (MCFC) which operate at 650°C, and Solid Oxide fuel cells (SOFC) which operate near 1000°C. These have not been used in a space application.

Molecular oxygen is the oxidant in these six types of systems, and molecular hydrogen is the fuel used in the AFC and PEMFC designs. In the latest innovation, DMLFFC/PEM, aqueous methanol in the form of a liquid is the fuel. The higher temperatures of the MCFC and SOFC allow a greater variety of fuels including hydrocarbons and diesel fuel. Except for the Biosatellite missions, only the alkaline and PEMFC have been used in space, specifically for manned space applications, i.e., Gemini, Apollo, and Shuttle spacecraft.

8.1 History

The first use of a fuel cell system in space was in the Gemini program, August 21, 1962. This was the first of the seven Gemini Earth-oriented manned missions from 1962-1965. A Proton Exchange Membrane Electrolyte Fuel Cell (PEMFC), known at that time as the Solid Polymer Electrolyte Ion Exchange Membrane Fuel Cell, was used. The solid ion exchange membrane (sulfonated polystyrene resin) served as an alternative to the aqueous liquid electrolyte providing a path for the H^+ to move from anode to cathode. Providing a mechanism for the H_2 and O_2 required the use of a platinum catalyst.

The system produced by General Electric provided 350 W for each fuel cell module. Three stacks of 32 cells were used in parallel to provide a total 1 kW for the spacecraft fuel cell system. Cryogenic H_2 and O_2 were used. The Platinum (Pt) catalyst loading was 28 mg/cm^2. The H_2 supply pressure was 1.6 psi above water pressure and O_2 was 0.5 psi above H_2. The efficiency was reported as 50-60%. Although the mass of the system was lower than the others, limitations of this fuel cell technology were the voltage losses due to ohmic drops in the solid electrolyte and sensitivity to water content.

The Biosatellite 2 launched September 7, 1967, utilized a similar PEMFC system, with an important change. The PEM was Nafion (perfluorosulfonic acid), a registered

CHEMICAL STORAGE AND GENERATION SYSTEMS 225

trademark of the Dupont Company. Since then, Nafion has been the membrane of choice for all PEM fuel cells, including the DMLFFC/PEM.

To overcome the limitations of the Gemini PEMFC, Apollo manned flights (1968 - 72) utilized the alkaline electrolyte fuel cell (AFC). The electrolyte was potassium hydroxide (KOH). The fuel and oxidant were H_2 and O_2. These fuel cells used an asbestos separator. The reactions for the alkaline fuel cell system are addressed in Section 8.3.

The Apollo fuel cell plant developed air and particulate contamination problems. These were trapped in the coolant system and resulted in pump cavitation. This was solved by improving servicing procedures. Another problem observed onboard was temperature oscillations due to low gravity and flow instability under certain operating conditions. This was solved by valve schedule changes. Problems in two-phase fluid handling (caused by the microgravity environment) were avoided by the use of a supercritical stage. Problems faced with insulation, heaters, pressure vessels, fans etc., were all solved by improved system design, quality control, and manufacturing/ maintenance procedures, although these failures were observed in Apollo 13.

The 1.5 kW, 26 V power plant was developed by Pratt and Whitney, a Division of United Aircraft Technologies. The total fuel cell utilized three fuel cell modules connected in parallel. The fuel cell operated at 260°C increased from the original temperature of 205°C to improve performance. A platinum catalyst was not needed. The pressure of the fuel and reactant were 60 psia. The system operated at 150 mA/cm^2 and the voltage was 0.72 V/cell. The peak power was 2.3 kW at 20.5 V, and the fuel cell weighed 100 kg. It operated for 690 hours without failure.

The Shuttle orbiter, developed by United Technologies Corporation and in use today, contains three H_2-O_2 alkaline fuel cell power plants supplying 12 kW at peak and 6 W average power in performance. Asbestos is also used as the separator. The operating temperature is 83-105°C. The current density is 66 - 450 mA/cm^2. The system is capable of 2000 hours of operation. The Shuttle orbiter fuel cell power plant was 23 kg lighter and delivered eight times the power of the Apollo fuel cell system. For additional details see Table 4.12.

In the Shuttle fuel cell plant there was a H_2 pump seizure problem which was solved by appropriate design changes followed by a 2000 hour qualification test. The magnesium separator plates in contact with water corroded, and this was solved by plating them with nickel. Cell-to-cell voltage variations resulted in a significant drop in the total voltage. These were solved by coating the electrode surfaces with special materials and leaching the asbestos matrix to remove cadmium impurities. In STS-2, water contamination resulted in non-nominal operation of the H_2 pump. By incorporating suitable filters and making changes in the recirculatory system, the problems were solved.

NASA is considering upgrades in the fuel cell system. Among these is the use of the Proton Exchange Membrane technology. Significant improvements have been made since their use in the Gemini mission. The PEMFC and the DMLFFC/PEM are also being considered for use in lunar, Mars colony, and Rover applications. Another development under consideration is a rechargeable fuel cell system known as a Regenerative Fuel Cell System. It combines an electrolyzer that can produce the H_2 and O_2 during the sunlight portion of a mission using solar cells and a fuel cell stack that can provide the energy during the eclipse period. To date, the system has not met the performance requirements required for a mission.

8.2 Fuel cell system basics

In the common acid electrolyte fuel cell, fuel (H_2) reacts electrochemically at a catalyzed anode electrode/electrolyte surface to produce protons (H^+) and electrons which flow through the load to the cathode. The protons diffuse through the electrolyte and react with O_2 to produce water at the negatively charged cathode electrode/electrolyte interface. The fuel cell useful in space applications can generally be described as consisting of two electrodes separated by and in direct contact with an electrolyte. The electrolyte can be acidic, either liquid (phosphoric acid) or solid (Nafion), or alkaline. Conductive biplates are in contact with the anode of one cell and the cathode of the adjacent cell. Thus, they separate the cells and conduct the electrons through the stack resulting in a positive terminal at one end of the stack and a negative terminal at the other end. On both sides of the biplate are flowfields that allow the fuel and oxidant to come in contact with the anode on one side and cathode on the other maximizing the contact area. The fuel and oxidant flow to the cells is provided by an internal or external manifold so that they flow into the flowfield of all cells at the same time. The fuel cell concept is given in Figure 4.23.

The reactions for the PAFC are:

Anode:	$H_2 \rightarrow 2H^+ + 2\,e^-$	0.000	(4.45)
Cathode:	$1/2 O_2 + 2H^+ + 2\,e^- \rightarrow H_2O$	1.229	(4.46)
Overall:	$H_2 + 1/2 O_2 \rightarrow H_2O$	1.229	(4.47)

The electrochemistry fundamentals and thermodynamics of this system are the same as those discussed in Section 4.4. However, there are important differences between the fuel cell parameters and those of a battery. While capacity is the parameter that best characterizes the battery cell capability, power best describes the fuel cell capability. The capacity is increased or decreased by the volume of stored fuel and oxidant. The fuel cell polarization curve is given in Figure 4.24.

CHEMICAL STORAGE AND GENERATION SYSTEMS

Table 4.12 Comparison of Gemini, Apollo, Shuttle Fuel Cell Characteristics

	Gemini	Apollo	Shuttle
No. Flights	7 (#5-12)	All	All
Manufacturer	General Electric	P & W	United Tech. United Aircraft
Type PEMFC	AFC	AFC	
Power Module continuous	500-620 W s 3 -350 W modules	1.5 kW 3 modules	14 kW 3 modules
1 kW Peak	Peak 2.3 kW at 20.5 V	36 kW peak	
Cell temperature	40 to 60°C	200-250°C	83 - 105 °C
Voltage	23.3 - 26.5V	26 - 31 V	26.5 -32.5V
No. of sections per s/c	2	3	3
No. stacks/section	3	2	
No. cells/stack	32	31	32
Stack Size, cm	66 x 33 diam	110 x 56 diam	35 (h)x 30(w)x 01cm(l)
Battery weight, Kg	31	110	91
H_2 / O_2 Oper Pressure	22 /23 psia	60 / 60 psia	60 /60 psia
H_2 Storage pressure:	210 to 250 psi	245 psi	290-290 psi
O_2 Storage pressure	800-psi	900 psi	850-950 psi
Current density (ma/cm^2)	15	92	67 - 450
Electrolyte	Sulfonated polystyrene	85% KOH	30 - 40 % KOH
Efficiency	50 - 60%	60%	61.8% @6 kW
Service life	400 - 800 Hrs @ 0.5kW	400 - 1500 Hrs @ 1kW	2000 Hrs @ 4.5 kW
Reactant used / 3 day Mission	0.41 kg/kWh Composition Rate	0.36 kg / kWh 0.55 kg / Hr	450 kg / 600 L
Time in Space	840 Hrs	1995 Hrs	Serviced 2000Hrs

Power

Power, the product of current and voltage, describes the capability of the fuel cell to convert the fuel and oxidant into the product water. The larger the electrode surface and the more active the surface, the greater is the current capability. In addition, the voltage is affected by ohmic, activation, and concentration polarization, as well as temperature. Therefore, power as a function of temperature and rate is used to describe the capability of a fuel cell or fuel cell stack.

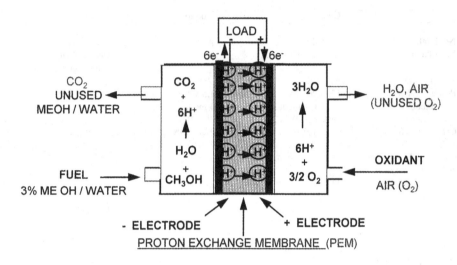

Figure 4.23 Fuel Cell Concept

Efficiency

The theoretical thermal efficiency, ε° at 25°C is 0.95 as determined by:

$$\varepsilon_{th}^\circ = \Delta G_{298} / \Delta H_{298} \tag{4.48}$$

where ΔG_{298}, the change in free energy, is -54.63 and ΔH_{298}, the change in enthalpy, is -57.80. This equation can also be shown to be:

$$\varepsilon_{th}^\circ = nFE^\circ / \Delta H_{298} \tag{4.49}$$

and thus, the actual thermal efficiency of a fuel cell can be calculated from:

$$\varepsilon_{th} = nFE / \Delta H_{298} \tag{4.50}$$

where n is the number of electrons, F is the Faraday constant, E is defined as kinetic losses minus resistive losses, and ΔH_{298} is the change in enthalpy at 25°C.

The voltage efficiency can be shown as:

$$\varepsilon_v = E_{meas} / E^\circ \tag{4.51}$$

CHEMICAL STORAGE AND GENERATION SYSTEMS 229

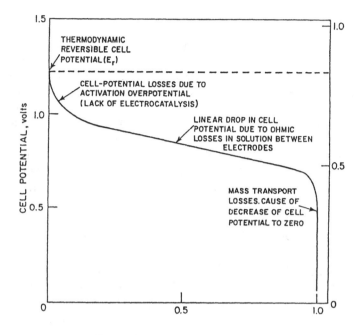

Figure 4.24 Fuel Cell Polarization Curve

The electrochemical efficiency, ε_{ec}, can be shown to be the product of the thermal, voltage, and current efficiencies.

8.3 Alkaline fuel cells

The reactions of the alkaline fuel cells are somewhat different from that of the acid fuel cells because of the pH. However, the E° is the same for both. The reactions are:

Anode:	$H_2 + 2OH^- \rightarrow 2H_2O + 2e^-$	0.828	(4.52)
Cathode:	$1/2 O_2 + H_2O + 2e^- \rightarrow 2OH^-$	0.401	(4.53)
Overall:	$H_2 + 1/2 O_2 \rightarrow 2H_2O$	1.229	(4.54)

The fuel cell system (power plant) is comprised of several elements consisting of the fuel cell stack, H_2 and O_2 storage tanks, condenser, gas flow control system, power conversion equipment, and thermal control equipment. The most important element of this complete system is the fuel cell stack in which the power-producing electrochemical reactions take place. The stack is comprised of a group of thin bipolar plates with electrodes on either side and a layer of electrolyte between each. The plates are machined with holes on their outer edges along with grooves on their surfaces so as to form a manifold and gas distribution system when assembled into the stack. Hydrogen is fed to one side and oxygen to the other side of each plate through this manifold system. Upon entering their respective compartments, the gases come in contact with the catalyzed electrode/electrolyte interfaces where the electrochemical reactions take place.

In both the acid and alkaline types the catalyst employed within both anodes and cathodes is a finely divided form of platinum black. Different amounts are incorporated within each electrode of each type. Other materials of construction for the acid type consist of carbon black and Teflon for the electrodes and a solid polymeric electrolyte (an acid type ion exchange membrane material). Other materials of construction for the alkaline type consist of carbon and nickel for the electrodes and a porous asbestos layer containing aqueous KOH solution as the electrolyte. Operating temperatures of both types of fuel cells are near 100°C.

The advantages include: 1) very high energy density, 2) few moving parts, 3) moderate operating temperature, 4) dual use for oxygen (can be used for life support), and 5) high thermal efficiency. The disadvantages include: 1) high cost, 2) complex assembly (numerous interconnecting parts), and 3) complex operation (many controls are required).

There are a number of potential hazards accompanying the use of fuel cells: 1) external H_2 leaks can produce explosive conditions, 2) internal H_2 and O_2 leaks cause local hot spots and also explosive conditions, 3) external O_2 leaks can cause fires, 4) shunt currents can cause internal generation and mixing of gases with resultant heating, 5) age of manifolds can cause cell reversal and subsequent generation and mixing of gases with resultant heating, 6) inadequate cooling can cause thermal runaway (thin bipolar cell stacks can be readily shorted without proper insulation and handling), and 7) the alkaline electrolyte is corrosive and toxic.

8.4 Proton exchange membrane fuel cells

The simplest of the fuel cell systems is the proton exchange membrane type. The electrolyte is a polymer film, Nafion (perfluorosulfonic acid), and thus the complexity of requirement for a liquid electrolyte is eliminated. The reactions are the same as in the phosphoric acid fuel cell system as is the theoretical $E°$ for the reaction.

This system operates at a temperature to 80°C, a significantly lower temperature than the alkaline system. It was first used in the Gemini and Biosatellite spacecraft. However, it lacked the power of the higher temperature alkaline system. The operating temperature of the PEMFC is limited because of the 120°C melting point of the polymer membrane. In addition, the fuel and oxidant had to be humidified before entering the

CHEMICAL STORAGE AND GENERATION SYSTEMS   231

fuel cell stack to avoid dryout and degradation of the Nafion membrane. The Gemini system also had a water control problem. For these reasons, the alkaline system was selected for the Apollo and Shuttle orbiter applications.

Because of the advantage of system simplicity, due to the solid electrolyte, this system is receiving attention as an upgrade to the orbiter and additional NASA applications.

The Gemini fuel cell stack assembly consisted of the ion exchange membrane and platinum catalyst. Each side of the membrane was covered by a titanium screen which was coated with platinum. One side of the membrane was bonded at its edges to a titanium sheet to form the cavity for the hydrogen gas. Hydrogen was introduced into the cavity through a tube connected to the edge seal. Two loops of coolant tubing were placed against the other face of the titanium sheet and wicks were placed between each pass of the tubing to remove the water produced at the cathode. These were necessary to prevent accumulation of water at undesirable places in a zero gravity environment. A total of 32 such assemblies were bolted together between end plates to form the stack assembly. Each stack contained its own hydrogen and coolant manifolding and water-oxygen separator. These stacks were installed in a cylindrical container to form the fuel cell section. Two such sections were used onboard each Gemini spacecraft.

8.5 Regenerative fuel cells

Regenerative fuel cells are comprised of the primary fuel cell power plant in conjunction with a water electrolysis unit. These two elements are integrated into a completely rechargeable type system. The fuel cell portions of these are essentially the same as described above. The only major difference is that the gases are stored under pressure in steel containers. The electrolyzers are in both cases very similar in configuration to the fuel cells. These are comprised of bipolar stacks of cells with gas manifolds. Product water from the fuel cells (collected in a reservoir tank) is fed to the electrolyzer where it is decomposed into H_2 and O_2 gases which, in turn, are stored in the high-pressure tanks.

In the primary fuel cells, the H_2 and O_2 gases from the storage reservoirs are fed via manifolds to the anodes and cathodes of the cell stacks. Here the gases react electrochemically to produce electrical energy. The specific electrochemical reactions that take place within the cells depend on whether they are of the acid or alkaline type.

The product water formed within the cells is removed by a gas recirculation scheme wherein water vapor is transferred to the hydrogen stream and then removed externally in a condenser. Thermal management is provided by circulating a coolant around the cell stack to absorb heat and then removing the heat in a radiator. Small heaters are incorporated to heat the stack up to operating temperature during startup. Fans and pumps are used to circulate the gases and coolants. Hydrogen and oxygen are stored as cryogenic liquids in metal storage tanks (titanium, Inconel, or aluminum).

In the regenerative fuel cells, the product water from the fuel cell stack is fed to an electrolysis unit where it undergoes electrochemical decomposition into H_2 and O_2 gases. The specific reactions that take place within the electrolyzer again depend upon whether it is of the acid or alkaline type. In the acid electrolyzer:

Anode:	$2H_2O$	$\rightarrow O_2 + 4H^+ + 4e^-$	(4.55)
Cathode:	$4H^+ + 4e^-$	$\rightarrow 2H_2$	(4.56)
Overall:	$2H_2O +$	$\rightarrow 2H_2 + O_2$	(4.57)

In the alkaline electrolyzer:

Anode:	$4OH^-$	$\rightarrow O_2 + 2H_2O + 4e^-$	(4.58)
Cathode:	$4H_2O + 4e^-$	$\rightarrow 4(OH)^- + 2H_2$	(4.59)
Overall:	$2H_2O$	$\rightarrow 2H_2 + O_2$	(4.60)

The gases from the electrolyzers are first passed through a condenser to remove water and then transferred to their respective storage tanks. The water is transferred to the water reservoir tank. Both the fuel cell and electrolyzer portions generate heat during operation and must be cooled. This cooling is provided by circulating a coolant through the outer shells of each to a heat exchanger and then to a radiator. During some conditions it is necessary to supply heat to one unity or the other to maintain operating temperatures. This is provided by circulating the hot coolant from one to the other via the heat exchanger.

The regenerative fuel cells are intended for use on long term orbital missions requiring very high power levels up to 500 kW. These missions include a wide range, from low Earth orbit to geosynchronous orbits. In these applications, the cells serve the same function as rechargeable batteries, i.e., to provide power during eclipse periods to all spacecraft loads. The cells are recharged during the light period.

The advantages include: 1) high energy density (mass), 2) projected high cycle life, and 3) built-in state of charge indicator. The disadvantages include: l) low volumetric energy density, 2) high cost, 3) relatively low energy efficiency, 4) complex assembly (numerous interconnecting parts, and 5) complex operation (many controls required). The potential hazards are: 1) all of the same hazards of primary fuel cells above, and 2) external leakage problems are enhanced by high pressure storage and numerous gas plumbing connections.

8.6 Direct methanol liquid-feed fuel cell/PEM

In 1991, a breakthrough in fuel cell technology was demonstrated by the Jet Propulsion Laboratory and the University of Southern California on a U.S. DARPA-sponsored pro-

CHEMICAL STORAGE AND GENERATION SYSTEMS 233

gram. This fuel cell is similar to the PEM cell described above that uses H_2 as the fuel. This technology utilizes an aqueous liquid methanol solution (presently only 3%) directly without reformer as the fuel. Methanol has a theoretical capacity of 5 kW/liter. The Direct Methanol, Liquid-Feed, Fuel Cell with Proton Exchange Membrane (DMLFFC/ PEM) uses either compressed air or oxygen as the oxidant. The only products of reaction are water and carbon dioxide. The system is being considered for use on Mars where the concentration of CO_2 is high and the product would not have an effect. The reactions of the DMLFFC/PEM are:

$$CH_3OH + H_2O \rightarrow 6H^+ + CO_2 + 6e^- \quad (4.61)$$
$$\underline{3/2O_2 + 6H^+ + 6e^- \rightarrow 3H_2O} \quad (4.62)$$
$$CH_3OH + 3/2O_2 \rightarrow CO_2 + 2H_2O \quad (4.63)$$

Laboratory versions of the fuel cell operating directly on a 3% methanol/water solution at a temperature of 90°C have delivered a continuous output of 50 A on a 4" x 6" electrode (300 mA/cm^2) at 0.55 V with oxygen. The peak power of >320 mW/cm^2 occurs at 100 A. The size of a 5 kW stack is projected to be an 8-inch cube with a weight of less than 1 kg.

A 50 W DMLFFC/PEM full system has been demonstrated using a recently developed methanol sensor that controls the flow of methanol to the aqueous mixing tank. Two twelve-cell DMLFFC/PEM stacks (eleven cells per inch) in parallel were integrated into a complete system and continuous operation was demonstrated.

The DMLFFC/PEM concept is shown in Figure 4.25. The heart of the technology is the central section that comprises the .020-inch thick membrane electrode assembly (MEA). The MEA consists of the proton exchange membrane sealed between two carbon electrodes containing the catalyst. The 3% aqueous methanol enters the anode chamber and is converted at the electrode interface to six protons (H^+), six electrons (e^-), and carbon dioxide (CO_2). The protons diffuse through the membrane and react at the Cathode interface with air and the returning electrons to produce water. The theoretical energy density of methanol is five kWh/liter compared with cryogenic hydrogen of 2.7 kWh/liter. An efficiency of 34% or 1.7 kWh/liter has been achieved in a five cell stack.

The Nafion, in addition to serving as a good proton conductor, also allows methanol to diffuse from anode to cathode. The reaction at the cathode is equivalent to oxidizing methanol, with CO_2 and water as the same products as the electrochemical reaction. However, the result is a loss in efficiency of approximately 20%. A new membrane, with the same conductivity as Nafion and reduced methanol crossover to <5%, will be implemented to increase the stack efficiency from 34 to 45%.

Figure 4.26 is a closed system conceptual design. Its simplicity, compared with hydrogen or methanol-reformed systems, is due to liquid aqueous fuel circulation thus avoiding humidifiers and thermal fins (estimated at every fifth biplate) used to remove

234   SPACECRAFT POWER TECHNOLOGIES

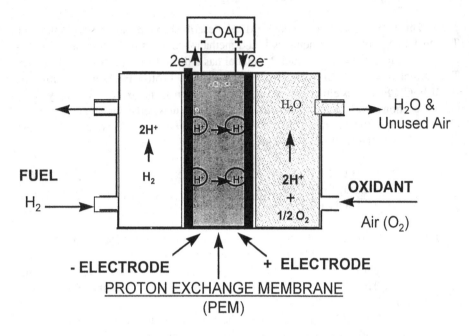

Figure 4.25 Schematic of a DMLFFC/PEM

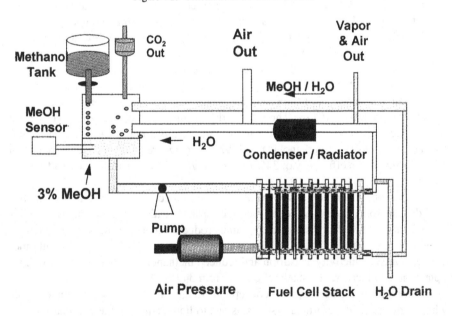

Figure 4.26 Direct Methanol, Liquid-Feed Fuel Cell Schematic

CHEMICAL STORAGE AND GENERATION SYSTEMS

heat in the stack. Methanol from the fuel tank is mixed with recirculating water from the stack output to produce the desired concentration. The aqueous fuel is fed into a manifold that allows the fluid to flow past the anode in each cell. The aqueous fluid, less the reacted methanol, returns to the mixing tank for methanol addition. The carbon dioxide is released as a gas. Likewise, pressurized or unpressurized air flows past the cathode of each cell. The water produced is picked up in the air stream and returned to the mixing tank. Depending on temperature and application, the water can be released as a vapor or liquid. For obvious reasons, this new technology is also being considered for a number of commercial applications including electric vehicles, marine and RV use, electronic and consumer devices, and remote and emergency power.

The advantages include: 1) no harmful emissions (important for use on Mars), 2) elimination of fuel vaporizer, 3) elimination of complex and voluminous humidification systems, 4) more efficient thermal management systems, 5) significantly lower system complexity, size, and weight, and 6) a projected 45% efficiency.

There are three main disadvantages: 1) limited to use on Mars or other planetary missions, 2) lower performance compared to the AFC, and 3) the Nafion membrane allows methanol to diffuse from anode to cathode reducing efficiency

9. Definitions and terminology

Anion: A positively charged ion.
Anode: The electrode at which oxidation takes place releasing electrons to the load. The anode can consist of many plates of the same polarity strapped together and tied to a current collector.
Battery: Two or more cells connected in series, parallel, or a combination of both, or a single cell used as a single cell battery.
C Rate: A method for describing the charge and /or discharge current relative to the cell capacity. 'C' is the manufacturer's or nominal cell capacity. Therefore, the 'C' rate is the current used to discharge a cell or battery in 1 hour, i.e., 'C'/1. To discharge the battery in 2 hours, the C/2 rate or 2 hour rate, etc. The same designation is used for charge as well.
Capacity: This is the term used to describe the electrical storage capability of an electrochemical cell or battery. It equates to the integral of the number of amperes provided by the battery over the number of hours the battery is discharged. The three terms below are used to describe capacity:
  a) Theoretical Capacity (Ah_T) - The capacity of a cell as derived from the theoretical conversion of the quantity of stored electrochemical active materials to electrical energy. For example, in Eq. (4.4) above, if the mass of zinc was 65.4 g (the atomic weight), the Ah_T for the two electron reaction would be 2 x 26.8 Ah = 53.6 Ah.

b) Actual or Measured Capacity (Ah_A) - This is the capacity measured from the state of full charge until the cell or battery energy is depleted as determined by a severe voltage drop when discharging through a load similar to that for planned use.

c) Manufacturer's Rated or Nominal Capacity (Ah_R) - This is the value the manufacturer labels the cell or battery as its design capability achievable over a range of operating conditions, e.g., rate and temperature. For ambient conditions and nominal rates, the Ah_A is generally 5-20% higher than the Ah_R.

Cathode: The electrode in the cell at which reduction takes place by accepting the electrons released at the anode. The cathode can consist of many plates of the same polarity strapped together and tied to a current collector.

Catholyte: Used in cells where the electrolyte also serves as the reactant at the cathode (as in lithium-thionyl chloride cells where the solvent and the reactant at the cathode are the same - Thionyl Chloride).

Cation: A positively charged ion that reacts at the cathode.

Cell Reversal: This occurs when the polarity of the cell is reversed. The cell is driven into a negative voltage by continuing the discharge beyond the cell's useful capacity. (see *Overdischarge*).

C/D Ratio: The ratio of Ah_{in}/Ah_{out} with rechargeable cells (a way of defining inefficiency, the inverse of coulombic efficiency).

Charge Rate: This is the rate or current used during charge. It can be described in amperes or as fractions of C' rate in relation to the cell capacity, low rates are considered to be < C/50, moderate rates C/50 to C/10, and high rates > C/10. However, bear in mind that high rate for one type of cell, e.g., lithium anode types, may be a moderate rate for another type, i.e., aqueous secondary cells.

Charge Retention: The capacity remaining in a cell or battery after being on open circuit stand where it can experience self-discharge.

Charging: The process in which electrical energy removed from the battery is returned to stored chemical energy by reversing the electron flow.

Corrosion: The wasteful consumption of cell components by conversion to non active materials by the galvanic action.

Coulombic Efficiency: The ratio of the measured capacity divided by the ampere hours required to return the cell or battery to full charge (Ah_{out}/Ah_{in}). Also, known as electrochemical efficiency.

Current Collector: The low resistance metallic portion of the anode or cathode at which electrons collect and are transferred to the terminal. This is sometimes referred to as the busbar.

Current Density: Current per unit of active electrode area (Amps /cm^2).

Cycle Life: The number of charge / discharge cycles the cell or battery has experienced in the regime in which it has been operated. It is occasionally reported in term of years or months.

Dendrite: The deposition of active material between anode and cathode which can serve as a means for internal discharge of a cell either temporarily or permanently.

Depth of Discharge: The 'DOD' is usually given as a percent of the total capacity removed from the cell or battery during a given cycle.

Discharge: The spontaneous removal of current from a cell or battery through a load resulting from the potential difference between electrodes.

Discharge Plateau: The relatively flat portion of a discharge curve occurring at the middle of the discharge period.

Discharge Rate: The same as charge rate except applied to discharge.

Electrochemical Efficiency: See *Coulombic Efficiency.*

Electrode: One or more electrochemically active plates in a cell containing stored chemical energy that can be converted to electrical energy and vice versa.

Electrolyte: A liquid or solid solution consisting of a solvent containing a dissolved ionic salt, thus forming an ionically conductive solution.

Energy Density: A figure of merit expressing the stored energy as a function of volume (Wh/Liter or Wh/in^3).

Energy Efficiency: The ratio of output energy to input energy (Wh_{out}/Wh_{in}).

Energy Storage: The energy storage capability of a cell or battery is given in terms of Watt hours = the integral of the voltage capacity product.

Float Charge: The process of using continuous voltage controlled low-charge current to offset self-discharge.

Flooding: Complete or almost complete filling of a cell with electrolyte to maximize the electrode/electrolyte contact. In some cell types this is detrimental because it minimizes the gas/electrode/electrolyte contact.

Formation: A series of charge and/or discharge operations performed on a newly manufactured cell or battery to condition it for service.

Grid: A metallic screen or perforated sheet that serves as a mechanical support and/or current collector within the anode or cathode structure.

Intercalation: The process in Lithium-Ion rechargeable cells wherein active species are inserted into the layers of a host compound such as Li$^+$ into graphite.

Ion: A species in which there is an excess or deficiency of electrons when compared with the number of protons in the nucleus. Negative ions (Anions) are formed by an excess or addition of electrons; positive ions (Cations) have less electrons than protons.

Overcharge: Continued charge of the cell after it has been fully charged.

Overdischarge: Forcing current through the cell in the discharge direction after all of the active materials have been exhausted. This results in cell reversal.

Oxidation: The process wherein the reactant releases electrons that can support a reduction reaction; one-half of a redox reaction.

Plate: One of many structures containing similar anode or cathode active material. The anode plates are strapped to a current collector to comprise the anode electrode; the same as for the cathode plates.

Packaging Factor: The ratio of battery mass or volume to mass or volume of all the cells in the battery. Prismatic cells have low packaging factors, cylindrical cells have high packaging factors.

Passivation: A process by which poorly soluble reaction products deposit on the electrode surface reducing electrode capability either by voltage or capacity loss.

Plaque: A porous body of conductive inert metal part of an electrode plate that is used as a container for active material and current collector in some cells, e.g., Ni-Cd.

Polarization: The voltage offset due to electrical, ionic, or kinetic impedance. The three types of polarization are: Resistance, Activation, and Concentration.

Reduction: The process wherein the reactant uses electrons that are released by the oxidation reaction; one-half of a redox reaction.

Reserve Battery: A dry cell or battery activated with electrolyte stored external to the cell. It is used in cases where the cell or battery exhibits a high self-discharge rate.

Sealed Cell: A cell that is sealed and isolated from the atmosphere.

Self-discharge: The spontaneous discharge of a battery while standing on open circuit (without load) due to reactions between the cell components.

Separator: An electronically insulating material which provides mechanical separation of anode and cathode plates and a path for ionic conduction. The separator can also serve as the reservoir for electrolyte in cells which contain minimum electrolyte.

Sintered Plate: An electrode formed by sintering metallic powders to form a porous structure which can contain the electrode material and current collector.

Specific Energy: The Specific Energy (SE) in Watt-hrs/kg is a parameter which describes the energy storage capability of a cell or battery in terms of terms of mass. The SE of a battery must take into account the battery mass including all electrical and mechanical parts.

Starved Cell: A cell containing little or no free electrolyte which enables gases to reach the electrode surfaces readily and permits relatively high rates of gas recombination. This type of cell is sometimes referred to a 'semi-dry' or 'semi-wet' cell.

State of Charge: That percentage of the total available active material of the cell that is in the charged state. (If the cell is unbalanced, this refers to the active materials of the limiting electrode.)

Substrate: A conductive support used as a plate support and as a current collector. It is used as a base on which the active material is placed.

Taper Charge: A charging method wherein the charge current is reduced as the cell or battery approaches the fully charged state. Usually similar or identical to modified constant potential charging.

Thermal Runaway: In batteries, the process by which a cell or battery will undergo a progressive and/or uncontrolled temperature increase which results in abbreviated or unexpected reaction.

CHEMICAL STORAGE AND GENERATION SYSTEMS 239

Trickle Charge: A low-level constant charging current used to maintain the battery at the fully charged level with a minimum of damage due to overcharging. (countering the self-discharge)

Utilization: That fraction of electrode active material mass that can be electrochemically converted from chemical to electrical energy or the reverse.

Voltage Limit: During charge, the limit above which the battery potential is not permitted to rise.

10. References

Bis, Richard F. and Murphy, Robert M., *Safety Characteristics of Non-Lithium Battery Systems,* Report Number NSWC TR 84302, (Naval Surface Weapons Center, Dahlgren, Virginia, July 1984).

Bockris, J. O'M., Conway, B. E., Yeager, E., and White, R. E., *Comprehensive Treatise of Electrochemistry,* Volume 3: *Electrochemical Energy Conversion and Storage,* (Plenum Press, 1981).

Bode, H., *Lead Acid Batteries,* (John Wiley & Sons, New York, 1966).

Falk, S. H. and Salkind, A. J., *Alkaline Storage Batteries,* (John Wiley & Sons, New York, 1969).

Fleischer, A. and Lander, J. J., *Zinc Silver Oxide Batteries,* (John Wiley & Sons, New York, 1961).

Gabano, J. P., *Lithium Batteries,* (Academic Press, [CITY], 1983).

Halpert, Gerald and Anderson, Art, *Lithium/Sulfur Dioxide Cell and Battery Safety,* NASA Reference Publication 1099, (November 1982).

Linden, D., *Handbook of Batteries and Fuel Cells,* (McGraw-Hill, New York, 1984).

Linden, D., *Handbook of Batteries,* Second Edition, (McGraw Hill, New York, 1995).

Nebergall, W.H., Schmidt, F.C., Holtzclaw, H.F. Jr., *General Chemistry,* (D.C. Heath and Co., 5th Edition, 1976).

Scott, W. R. and Rusta, D., *Sealed Cell Nickel Cadmium Battery Application Manual,* (NASA Reference Publication 1952, 1969).

Subbarao, S., Halpert, G., and Stein, I., *Safety Considerations of Lithium Thionyl Chloride Cells,* (Jet Propulsion Laboratory, Pasadena, California).

Trout, J. B./NSI (prepared by), *Manned Space Vehicle Battery Safety Handbook,* Report Number JSC-20693, (Lyndon B. Johnson Space Center, Houston, Texas, September 1985).

Venkatsetty, H. V., *Lithium Battery Technology,* (John Wiley & Sons, New York,1984).

Vincent, C. A., Bonino F., Lazzari, M., and Scrosati, B., *Modern Batteries,* (Edward Arnold Publications, 1984).

241

CHAPTER 5

NUCLEAR SYSTEMS

1. Introduction

From the early days of the space program nuclear power has played an important role and will continue to do so in the future. Nuclear power greatly enhances, and in some cases enables, solar system exploration. Its main advantage relative to solar power is the ability to operate independent of sunlight. In particular, the low solar flux available at the orbits of outer planets makes nuclear power the only practical choice. Another application of great interest is surface power on the Moon due to the long (two-week) lunar night. Some future missions may well require electrical power levels from thousands to millions of watts, and nuclear power will likely be necessary to fill that need. A few Earth orbital satellites also use nuclear power, one advantage being that energy storage is not required during eclipse periods. However, as discussed in Chapters 3 and 4, solar/battery power systems have been very successful and are by far the preferred technology for Earth orbital satellites. Cost and safety issues (real and perceived) are the main obstacles to the use of nuclear power in space.

Defined by the thermal energy source, space nuclear power systems fall into two general categories. Radioisotope systems generate heat by the natural decay of radioisotopes, and the heat can then be converted into electrical energy. Radioisotope Thermoelectric Generators (RTGs) have played a very significant role in the U.S. space program. The other category is nuclear reactor systems in which the heat is generated by nuclear fission. Both the U.S. and the Former Soviet Union (FSU) have conducted major R&D programs for space reactors, but only the FSU has applied this technology to actual space missions.

This chapter provides a brief history of U.S. and Soviet use of nuclear power in space, descriptions of radioisotope and reactor systems, and a discussion of the safety issues along with a summary of the U.S. safety review and launch approval process. Chapters 6 and 7 contain additional information on analysis and principles of operation of energy conversion options. In addition to electrical power, substantial work has been devoted to nuclear electric and nuclear thermal propulsion systems for upper stages and interplanetary rockets. A nuclear electric rocket uses propulsion thrusters (e.g., arcjet thrusters, ion engines, or magnetoplasmadynamic thrusters) driven by the nuclear electric power system, while a nuclear thermal rocket transfers the fission heat to hydrogen or another gas that is then expanded through a nozzle to produce thrust. Nuclear thermal propulsion is not covered in this text; the reader is referred to other references (e.g., Angelo and Buden, 1985).

2. History of the U.S. space nuclear program

The U.S. space nuclear program has its roots in an early Air Force investigation of the possible use of reconnaissance satellites (Voss, 1984). Initiated in 1948 by the Rand Corporation, Project Feedback identified the need for reliable power systems for such satellites. In 1954, the Air Force examined nuclear power systems under the Pied Piper Program and incorporated these results into the larger Weapons Systems 117–L study of satellite systems and power options. From 1952 to 1955, the Atomic Energy Commission (AEC) conducted studies of nuclear power for space systems.

In 1955, a joint Air Force–AEC committee established specifications for space nuclear power and renamed the Pied Piper program the Systems for Nuclear Auxiliary Power (SNAP) program to be administered by the AEC. The odd-numbered SNAP programs/systems involved radioisotopes for the heat source, while the even-numbered programs involved reactors. From 1965 to 1973, the AEC and NASA managed SNAP systems development through a joint program office. The formal AEC SNAP RTG program dissolved when the agency was disbanded in 1975. The AEC's regulatory functions shifted to the Nuclear Regulatory Commission, and the R&D work was assumed by the new Energy Research and Development Administration (ERDA). In 1977, the newly-formed Department of Energy (DOE) absorbed the ERDA functions and development of nuclear power for space became the responsibility of that department. Although funding levels have varied, RTG development has been a continuous process since it began in the 1950s. Development of U.S. space reactors has occurred periodically in the 1960s and 1980s.

Table 5.1 lists the nuclear power systems launched by the United States. Except for one experimental flight of a reactor power system (SNAP–10A), all U.S. nuclear systems have been RTGs. These devices are extremely reliable, and all have operated well beyond their design lifetimes. In nearly all cases, the satellite was either shut down intentionally or a failure not related to the RTG terminated the mission. RTGs have provided power for the Navy Transit navigational satellites, Nimbus weather satellites, two Air Force experimental communication satellites (LES 8 and LES 9), five Apollo Lunar Surface Experimental Packages (ALSEPs) deployed by astronauts on the moon, and several unmanned NASA planetary exploration satellites. Several RTG systems launched in the 1970s are still functional.

Table 5.1 Summary of Space Nuclear Power Systems Launched by the United States (after Bennett, *et al.*, 1996; Angelo and Buden, 1985)

Date	Spacecraft	Power Source[†]	Mission Type	Status
29 Jun 61	Transit 4A	SNAP-3B7	Navigational	RTG operated for 15 years. Satellite now shut down but operational.
15 Nov 61	Transit 4B	SNAP-3B8	Navigational	RTG operated for 9 years. Satellite operation was intermittent after 1962 high altitude test. Last reported signal in 1971.
28 Sep 63	Transit 5-BN-1	SNAP-9A	Navigational	RTG operated as planned. Non-RTG electrical problems on satellite caused satellite to fail after 9 months.
5 Dec 63	Transit 5-BN-2	SNAP-9A	Navigational	RTG operated for over 6 years. Satellite lost navigational capability after 1.5 years.
21 Apr 64	Transit 5-BN-3	SNAP-9A	Navigational	Mission was aborted because of launch vehicle failure. RTG burned up on reentry as designed.
3 Apr 65	Snapshot	SNAP-10A	Experimental	Successfully achieved orbit.
18 May 68	Nimbus-B-1	SNAP-19B2	Meteorological	Mission was aborted because of range safety destruct. RTG heat sources recovered and recycled.
14 Apr 69	Nimbus III	SNAP-19B3	Meteorological	RTGs operated for over 2.5 years (no data taken after that).
14 Nov 69	Apollo 12	SNAP-27	Lunar Surface	RTG operated for about 8 years (until station was shut down).
11 Apr 70	Apollo 13	SNAP-27	Lunar Surface	Mission aborted on way to moon. Heat source returned to South Pacific Ocean.
31 Jan 71	Apollo 14	SNAP-27	Lunar Surface	RTG operated for over 6.5 years (until station was shut down).
26 Jul 71	Apollo 15	SNAP-27	Lunar Surface	RTG operated for over 6 years (until station was shut down).
2 Mar 72	Pioneer 10	SNAP-19	Planetary	RTGs still operating. Spacecraft is beyond solar system, 6.2 billion miles from Earth. NASA officially ended mission on 31 March 1997.
16 Apr 72	Apollo 16	SNAP-27	Lunar Surface	RTG operated for about 5.5 years (until station was shut down).
2 Sep 72	"Transit"	Transit-RTG	Navigational	RTG still operating (Triad-01-1X).
7 Dec 72	Apollo 17	SNAP-27	Lunar Surface	RTG operated for almost 5 years (until station was shut down).
5 Apr 73	Pioneer 11	SNAP-19	Planetary	RTGs still operating. Spacecraft successfully operated to Jupiter, Saturn, and beyond.
20 Aug 75	Viking 1	SNAP-19	Mars Surface	RTGs operated for over 6 years (until lander was shut down).
9 Sep 75	Viking 2	SNAP-19	Mars Surface	RTGs operated for over 4 years until relay link was lost.
14 Mar 76	LES 8[*]	MHW-RTG	Communications	RTGs still operating.
14 Mar 76	LES 9[*]	MHW-RTG	Communications	RTGs still operating.
20 Aug 77	Voyager 2	MHW-RTG	Planetary	RTGs still operating. Spacecraft successful to Jupiter, Saturn, Uranus, Neptune, and beyond.
5 Sep 77	Voyager 1	MHW-RTG	Planetary	RTGs still operating. Spacecraft successfully operated to Jupiter, Saturn, and beyond.
18 Oct 89	Galileo	GPHS-RTG	Planetary	RTGs still operating. Spacecraft orbiting Jupiter.
6 Oct 90	Ulysses	GPHS-RTG	Planetary/Solar	RTG still operating. Spacecraft successfully measured environment over Sun's poles.
Oct 97	Cassini	GPHS-RTG	Planetary	RTGs still operating. Mission to Saturn

[†] All power sources are RTGs, except SNAP 10-A (reactor).
[*] Single launch vehicle with double payload.

2.1 Radioisotope space power development

Table 5.2 summarizes the history of several SNAP and other radioisotope systems developed by the United States. As seen, the RTGs include a variety of power levels and design lifetimes. Often, more than one version of a particular RTG was developed, either incorporating design improvements or tailored to a specific mission. As is the case in many technology programs of this nature, several RTGs were not fully developed, usually because their intended missions did not materialize or other RTGs proved superior. Table 5.3 lists some of the key engineering design features of U.S. RTGs. Although design specifics vary, all RTGs consist of two primary subsystems: a radioisotope heat source and a thermoelectric converter/radiator assembly.

Some of the early SNAP systems were proof-of-concept designs, initially of the radioisotope and later of the thermoelectrics. Although never launched, the SNAP–3 was a pivotal RTG. On the technical side, static energy conversion in the form of thermoelectrics replaced the dynamic mercury–Rankine cycle used for the SNAP–1 design. More importantly, a proof-of-principle SNAP–3 was exhibited to President Eisenhower in January, 1959. The President saw this new technology as another way to promote the peaceful use of nuclear energy and serve as an opportunity to emphasize space missions for non-military purposes under the newly-formed National Aeronautics and Space Administration (NASA). With the President's public endorsement, both the military and NASA proceeded with identifying applications for the new technology (U.S. DOE, 1987).

The first RTG flights were aboard Navy Transit navigational satellites. Although power levels and mission durations were modest in those days, RTGs were competitive in large part because of the poor reliability of batteries needed for solar conversion systems. In 1961, SNAP–3B RTGs were launched on the Transit 4A and 4B satellites. These units not only proved the application of RTG technology, but also established their excellent reliability and operating life potential. The SNAP–9A RTGs launched aboard the Transit 5BN satellites in 1963 also exceeded their minimum design life. The last Navy navigational satellite powered by an RTG was the Transit TRIAD–01–1X launched in 1972. Plutonia–molybdenum–cermet (PMC) fuel was still used, but the Transit RTG employed a new configuration for the thermoelectric elements (Angelo and Buden, 1985). Named ISOTEC, the PbTe thermoelectrics were encased in panels such that an inert cover gas was no longer required in the full RTG to inhibit sublimation.

A failure occurred during the 1964 launch of Transit 5BN–3. As part of the SNAP–9A safety philosophy, its ^{238}Pu metal fuel was designed to burn up and disperse at high altitude if a launch or other failure resulted in an inadvertent reentry. Burnup occurred when the spacecraft failed to achieve a stable long-lived orbit and the spacecraft with its on-board SNAP–9A RTG reentered the Earth's atmosphere. Although no evidence exists that this particular high altitude burnup resulted in any health consequences, subse-

NUCLEAR SYSTEMS 245

quent RTGs were designed to contain the radioisotope fuel through atmospheric reentry. The high altitude dispersion philosophy was based on the relatively small amount of radioisotopes in the first RTGs and was established at a time when background radiation was relatively high due to the atmospheric testing of nuclear weapons. RTGs were planned for missions requiring more power and, therefore, more radioisotope fuel per flight, thus dispersion as a policy and design philosophy became more difficult. Another consideration was that safety is less difficult to verify for systems designed to contain fuel rather than to guarantee adequate dispersion. As a result, safety philosophy evolved quickly to that of "intact reentry/intact on impact" under all credible accident conditions.

By the mid 1960s, several new RTGs (SNAPs 11, 13, 15, 17, and 19A) were designed and constructed in varying degrees up to a full system level, but were not flown. SNAP–11 used Curium–242 fuel for a short duration mission of the NASA Surveyor lunar lander, but the requirement was canceled in 1965. SNAP–13 was a demonstration device for thermionic conversion instead of thermoelectrics. SNAP–15 was a very small power source developed for a classified application. SNAP–17 evaluated the use of Strontium–90 for the heat source and was designed for the Air Force Medium Altitude Communications Satellite. SNAP–19A was developed in part for the NASA Interplanetary Monitoring Probe intended to chart the magnetic field between the Earth and the Moon. Although SNAP–19A was not flown, its derivatives became the workhorses for several missions to come.

The next application of RTG power was for the Nimbus series of weather satellites. These satellites used solar panels as their primary source of power, but two SNAP–19B RTGs provided auxiliary power for each satellite. Unfortunately, the May 1968 launch of the Nimbus–B–1 satellite was unsuccessful because an erroneous setting of a guidance gyroscope forced range safety destruction about 2 minutes into the flight. At the time of destruct, the vehicle's altitude was about 100,000 feet. The RTGs landed safely in the Santa Barbara Channel near San Miguel Island. Both heat sources were located and recovered from the ocean floor. No plutonium release was detected, and the fuel was reused in later systems. In April 1969, the Nimbus III carrying two SNAP–19Bs was launched successfully. The RTGs operated for more than twice their one-year design life until power telemetry was discontinued. At about 1.5 years, both RTGs experienced an accelerated rate of degradation. This change was attributed to depletion of the inert cover gas, which led to accelerated sublimation of the thermoelectric material and loss of the hot junction bond (Angelo and Buden, 1985).

NASA began planning missions to explore the planets, and the longer duration of those missions required radioisotopes with long half-lives. To satisfy these requirements, higher temperature and long-life fuel forms were developed, including PMC and pressed plutonium oxide (PPO) fuel forms, both using ^{238}Pu; the latter is used almost exclusively today. NASA exploration missions surged in the 1970s, and RTG power became a mainstay for that effort. The SNAP–19 technology, used successfully for the Nimbus

Table 5.2 Summary of U.S. RTG Development (after Angelo and Buden, 1985; U.S. DOE, 1987)

Power Source	Planned Application	Power (We)	Mass (kg)	Design life	Fuel	Comments
SNAP-1	Satellite (US Air Force)	500	-	60 days	Ce-144	Replaced by longer-lived SNAP-1A in 1959; Hg-Rankine design.
SNAP-1A	Satellite (US Air Force)	125	91	1 year	Ce-144	Program canceled in 1959.
SNAP-3	Thermoelectric demonstration device (AEC)	2.5	2	90 days	Po-210	Shown to President Eisenhower in 1959.
SNAP-3B	Navigation satellites (US Navy)	2.7	2	>1 year	Pu-238	Launched on Transit 4A and 4B in 1961.
SNAP-9A	Navigation satellites (US Navy)	25	12	>1 year	Pu-238	Launched on two Transit 5Bs in 1963. Launch vehicle failure of third launch in 1964.
SNAP-11	Surveyor lunar lander (NASA)	25	14	90 days	Cm-242	Requirement canceled in 1965. Five-year life test on electrically-heated unit. Fueled demonstration at ORNL in 1966.
SNAP-13	Thermoelectric demonstration device (AEC)	12.5	2	90 days	Cm-242	Fueled demonstration at ORNL in 1965. Program completed in 1966.
SNAP-17A and 17B	Communication satellite (Air Force)	30	14	>1 year	Sr-90	Design and component test completed in 1965.
SNAP-19A	Imp satellite (NASA)	20	8	>1 year	Pu-238	Design and component test completed in 1963. Not used on Imp due to radiation interference with payload instrumentation.
SNAP-19A	Various satellites (AEC)	250	-	>1 year	Sr-90	Six different studies in 1964.
SNAP-19A	Surveyor lunar rover (NASA)	40	10	1 year	Pu-238	Design and integration study completed in 1964.
SNAP-19A	Extended Apollo missions (AEC)	1500	-	30-90 days	Po-210	Design and feasibility studies completed in 1964.
SNAP-19B	Meteorological satellite (NASA)	30	14	>1 year	Pu-238	First Nimbus III launch failure in 1968 and fuel recovered from off-shore waters. Replacement satellite launched in 1969.
SNAP-25	Various satellites (AEC)	75	16	>1 year	Pu-238	Program canceled.
SNAP-27	Apollo Lunar Surface Experiment Packages (NASA)	63.5	31 (+11 kg cask)	>1 year	Pu-238	First ALSEP deployed by Apollo 12 astronaut in 1969. Second unit landed in Pacific Ocean on Apollo 13 failure. Additional deployments on Apollos 14 to 17 in 1971 and 1972.
SNAP-29	Various satellite and lunar missions (US DoD and NASA)	400-500	180-225	90 days	Po-210	Program canceled in 1969 with completion of component tests.

NUCLEAR SYSTEMS

Table 5.2 Summary of U.S. RTG Development (continued)

Power Source	Planned Application	Power (We)	Mass (kg)	Design life	Fuel	Comments
SNAP-29	Various missions (US DoD and NASA)	250	-	5 years	Pu-238 Sr-90 Cm-244	Three design studies completed in 1967.
	Thermionic module development (AEC)	100	9	-		Program suspended after completion of preliminary design in 1967 and component tests in 1970.
	Isotope Brayton ground test (AEC and NASA)	3000-15,000	-	>1 year		Preliminary heat source designs completed in 1966; fuel capsule development 1967-1970; NASA completed 2500 hour life test on electrically heated system in 1970.
Transit-RTG	Navigation satellite (Navy) improved Transit	30	14	5 years	Pu-238	Launched in 1972; RTG still operating. (Triad-01-1X)
SNAP-19	Pioneer F and G Jupiter flyby (NASA)	30	14	3 years	Pu-238	RTGs still operating: Pioneer 10 launched in 1972; spacecraft is beyond solar system, 6.2 billion miles from Earth; NASA officially ended mission on 31 March 1997; Pioneer 11 launched in 1973; spacecraft successfully operated to Jupiter, Saturn, and beyond.
SNAP-19	Viking Mars Landers (NASA)	35	14	>2 years	Pu-238	Launched in 1975; Viking 1 RTGs operated for over 6 years (until lander was shut down); Viking 2 RTGs operated for over 4 years until relay link was lost.
MHW-RTG	LES 8 & 9 (US DoD) Voyager 1 & 2 (NASA)	100-200	-	5-10 years	Pu-238	RTGs still operating. LES 8&9 launched in 1976; Voyager 1&2 launched in 1977 and have left solar system.
DIPS	Various missions (NASA and US DoD)	500-2000	215	7 years	Pu-238	Program ran from 1975-1980; demonstrated 11,000 hour Rankine cycle including 2,000 hour endurance test; mission studies and preliminary design of Brayton version in 1987-88 for BSTS satellite.
GPHS-RTG	Planetary exploration (Galileo, Ulysses, Cassini)	300	56	5-10 years	Pu-238	Galileo launched in 1989, now orbiting Jupiter; Ulysses launched in 1990, now in highly elliptical solar polar orbit; Cassini launched in 1997, mission to Saturn. All RTGs still operating.

Table 5.3 Design Characteristics of U.S. RTGs
(after Angelo and Buden, 1985; Bennett, 1989; U.S. DOE, 1987)

	SNAP-3B	SNAP-9A	SNAP-19 Nimbus	SNAP-19 Pioneer	SNAP-19 Viking	SNAP-27	Transit RTG	MHW-RTG Voyager	GPHS-RTG
BOL power (W)	2.7	26.8 (25)	30	40.3 (30)	(35)	73.4 (63.5)	35.6 36.8 (30)	158	290
Design life	>1 year	>1 year	>1 year	>1 year	3 years	>2 years	1 year	5 years	5-10 years	40,000 hours
Converter efficiency	5.1%	5.1%		6.2%		5.0%	4.2%	6.6%	6.8%
Mass (kg)	2.1	12.2	13.6	13.6	13.6	31 +11 cask	13.5	38.5	54.4
Specific power (W/kg)	1.29	2.2		3.0		2.3	2.6	4.2	5.3
Thermoelectric material	PbTe 2N/2P	PbTe 2N/2P		PbTe 2N/ TAGS-85		PbTe 3N/3P (PbSnTe)	PbTe 2N/3P	SiGe	SiGe
BOL fuel inventory (W_{th})	52	565	645	645	645	1480	850	2400	4400
Fuel quantity (curies)	1800	17,000	34,400 - 80,000			44,500	25,500	77,000	130,000
Pu-238 fuel form	Metal	Metal		PMC		Oxide microspheres	PMC	Pressed oxide	Pressed oxide
Compatibility liner	304 stainless	Tantalum		Ta-10W	Mo-Re foil + Ta-10W liner		Mo-Re foil + Ta-10W liner	Ir-W alloy	Ir-W alloy
Strength	Haynes-25	Haynes-25		T-111	T-111		T-111	Wound graphite impact shell	Wound graphite impact shell
Cladding	Not required	Not required		Pt-20Rh	Pt-20Rh		Pt-Rh	None	None
Reentry heat shield	None - high altitude dispersion	None - high altitude dispersion		Polycrystalline graphite	Polycrystalline graphite	Polycrystalline graphite + Be	Two pyrolytic graphite layers on clad + AXF-50 graphite in Inconel-718 can	Two-layer graphite (POCO+ Pyrocarb)	CBCF sleeve + graphite aeroshell

weather satellites, was modified in order to extend lifetime and was used for Pioneer 10 and 11, NASA's first missions to study the outer planets. Each satellite required 120 watts provided by four Pioneer/SNAP-19 RTGs. These units and both spacecraft have

NUCLEAR SYSTEMS 249

far exceeded their mission requirements and exited the solar system. NASA shut down Pioneer 10 on March 31, 1997, 25 years after launch and 6.2 billion miles from Earth, while Pioneer 11 continues to operate. SNAP–19 technology also powered the Viking 1 and 2 Mars Landers, operating through the severe temperature extremes, winds, and dust environments of the Mars surface. In addition, the Landers (and RTGs) had to undergo sterilization before launch to prevent contamination by Earth-born organisms of the Martian environment and on-board instruments. Although the original requirement was only 90 days of operation, the Viking RTGs performed for several years until Viking 1 was shut down and the Viking 2 signal was lost.

On all Apollo missions after the initial landing in 1969, lunar exploration included the deployment of ALSEP experiments on the lunar surface. SNAP–27 RTGs powered all of the ALSEPs. Interestingly, each SNAP–27 was launched as "separate shipment" meaning that the radioisotope heat source was contained in a separate protective cask. As part of the lunar deployment, an astronaut removed the heat source from the cask and inserted it into the generator assembly. The RTG was then carried to the ALSEP site and electrical connections were made. Because all of the other RTG missions were unmanned, the RTGs were fully assembled prior to launch. As listed in Table 5.1, Apollo missions 12 through 17 carried SNAP–27s, including the ill-fated Apollo 13 mission. As designed, the SNAP–27 heat source on that flight survived reentry through the Earth's atmosphere and plunged into the South Pacific where it lies at the bottom of the Tonga Trench at a depth in excess of 7000 feet. All five SNAP–27s deployed on the moon performed as planned until each of the ALSEP stations was purposely shut down by ground command.

The Apollo, Pioneer, and Viking missions launched in the first half of the 1970s all used RTG technology developed under the SNAP program. In addition, a new development effort was completed, the Multi-Hundred Watt RTG (MHW–RTG). That design employed three significant technological advances: an oxide ($^{238}PuO_2$) form of the plutonium–238 fuel, an alloy of iridium as the metal cladding containment material for the fuel, and SiGe unicouples for the thermoelectrics. Two different versions of the MHW–RTG were developed and launched. Two units of one version powered each of the Lincoln Experimental Satellites (LES 8 and 9) launched in 1976. Three units of the other version were launched on each of the two Voyager planetary exploration satellites in 1977; all are still operational, although both spacecraft have exited the solar system. The MHW–RTGs on the LES 8 and 9 satellites continue to perform as well. During the Gulf War, the Air Force repositioned LES 9 over the war zone to augment communications. Afterwards, it was restored to its former GEO slot (Ward, 1994).

Following the MHW–RTG program, the United States developed a modularized radioisotope heat source to reduce the cost of future RTGs. The MHW–RTGs cost about $25,000/watt, and the DOE goal was to reduce that to $7,000/watt (U.S. DOE, 1987). DOE developed the General Purpose Heat Source (GPHS) module for use as a basic building block. Various numbers of GPHS modules can be stacked together to create the total power levels required by different missions. In addition, most of the GPHS hardware safety testing for blast/overpressure, fragments, shrapnel, surface impacts, solid and liquid propellant fires, and out-of-orbit atmospheric reentry is generically applicable to all space missions; thus the safety analysis and review for each mission can be accomplished at a reduced cost. A stack of eighteen GPHS modules, coupled with an appropriate number of SiGe thermocouples, constitutes a standard GPHS–RTG; two were used in the Galileo spacecraft, one was used for the Ulysses spacecraft, and three were used for Cassini.

RTGs are generally applicable to missions requiring less than 1 kilowatt. To provide higher amounts of electrical power without incurring the cost and safety issues associated with producing and launching larger amounts of ^{238}Pu, the 2–3x higher efficiency of dynamic conversion offers significant benefits. The dynamic options that have received the most attention involve either rotating (Brayton or Rankine cycle turboalternators) or oscillating (free-piston Stirling engines with linear alternators) machinery. In addition to nuclear systems, NASA has evaluated dynamic systems for use with solar concentrators (see Chapter 3). Ironically, the first SNAP design, SNAP–1, was based on dynamic Mercury–Rankine conversion, but thermoelectrics matured to the point that they have been the only nuclear converters used to power U.S. spacecraft (see Table 5.2).

2.2 Space reactor power development

A nuclear reactor for spacecraft offers the potential advantage of a very compact power supply at relatively high power levels. Table 5.4 lists the characteristics of reactors designed for electrical power production under the SNAP and other programs during the 1960s, plus the various concepts evaluated under programs in the 1980s. In the late 1950s and early 1960s, projected power requirements varied widely. As a result, the SNAP reactor designs covered a broad range. In most cases, the mass (and cost) trades led to the selection of thermal neutron spectrum cores using zirconium hydride (ZrH) as the moderator. The higher power designs typically utilized fast neutron spectrum cores to minimize total mass.

As mentioned earlier, the U.S. has conducted only one reactor test in space. SNAP–10A was launched by an Atlas–Agena rocket on April 3, 1965, into a nearly circular polar orbit at an altitude of about 1300 km. It performed well for 43 days. At that time, the spacecraft telemetry signal was lost, and when re-established it was determined that reactor power had been lost. Investigators attributed the problem to a sequence of failures in the satellite electronics resulting in a signal that shut down the reactor (Corliss, 1966). SNAP–10A was not designed to be restarted. A second SNAP–10A operated in a ground test for 10,000 hours before the test was discontinued.

NUCLEAR SYSTEMS

Table 5.4 Design Characteristics of U.S. Space Reactor Power Systems
(after Angelo and Buden, 1985; Corliss, 1966; Dix and Voss, 1985; Buden, 1994)

System	Power (kWe)	Mass (kg)	Operating temp (K)	Program dates	Core type	Fuel	Core coolant	Conversion technology	Development level
SNAP-2	3-5	668	920	1955-1964	Thermal	UZrH	Hg	Hg-Rankine	Two reactors tested; 10,500 hrs max accumulation
SNAP-8	30-60	4460 @ 35 kWe	975	1960-1970	Thermal	UZrH	NaK	Hg-Rankine	Two reactors tested; 1 yr demonstrated
SNAP-10	0.3			1958-1960	Thermal	UZrH	None	SiGe thermoelectric	Early design for conductive cooling
SNAP-10A	0.5	427	810	1960-1966	Thermal	UZrH	NaK	SiGe thermoelectric	43 day flight test; 417 day ground test
SNAP-50	300-1000	2700-9000	1365	1958-1968	Fast	UN, UC	Li	K-Rankine	Fuels tested to 6000 hrs
Improved SNAPs 2, 8 and 10 technology	0.5 & 150	680			Thermal	UZrH	NaK	Turbogenerator and thermoelectric	
Advanced hydride reactor	5	91	920	1970-1973	Thermal	UZrH		Brayton and thermoelectric	PbTe thermoelectrics tested to 42,000 hrs
Advanced metal-cooled reactor	300		1480	1965-1973	Fast	UN		Turbogenerator and K-Rankine	Non-nuclear K-Rankine components tested to 10,000 hrs
Medium Power Reactor Experiment	150			1959-1966			K	K-Rankine (direct)	
In-core thermionics	5-250		2000	1959-1973	Fast or thermal driver	UO_2, UC-ZrC	NaK	Thermionic fuel elements	>1 yr TFE operation
Out-of-core thermionics	400		1675	1974-1981	Fast	UO_2	Na heat pipes	Thermionic	Limited testing of diodes
710 gas reactor	200		1445	1962-1968	Fast	Cermet	Na heat pipes	Brayton (direct)	Fuel element test to 7000 hr
SPAR	100			1979-1981	Fast	UN	Li	SiGe thermoelectric	Limited tests on core heat pipes
SP-100	10-100+			1982-1993	Fast	UN	He-Xe	SiGe thermoelectric	Fuel element tests to 7 year burnup; multiple component tests
Particle Bed Reactor				1983-1990	Fast	UC particle	NaK	Brayton	Limited fuel element tests
TFE Verification				1984-1993	Fast	UO_2		Thermionic	
Multimegawatt concepts	2 MWe			1985-1987				Turbogenerator, MHD and thermionic	Conceptual designs only
Gas core reactor				1986-1989		UF_6	-	MHD and turbogenerator	Design & analysis only

Except for the Apollo program, the ambitious space programs envisioned in the late 1950s and early 1960s diminished toward the end of that decade, and requirements for space reactor power faded. Early in 1973, development of space reactor systems was severely curtailed in the U.S. primarily due to a lack of clearly identified missions. All development projects ceased except for a few limited-scope technology efforts, primarily by Los Alamos National Laboratory (LANL). In addition, the U.S. terminated a major parallel program encompassing several projects on nuclear propulsion, commonly referred to as Nuclear Energy Reactor for Vehicular Applications (NERVA). RTG work previously done under the SNAP program continued at a reduced level, although more closely tied to use on missions planned by NASA and the Department of Defense (DoD).

Interest in space reactors resurfaced in the early 1980s, once again driven primarily by perceived future requirements for space exploration and military space activities. LANL conducted studies on a heat-pipe-cooled reactor design known as the Space Power Advanced Reactor (SPAR). That work provided the early basis for the much larger SP–100 program formally begun in 1983 as a joint effort by the Defense Advanced Research Projects Agency (DARPA), NASA, and DOE. At the beginning, the program assessed a wide range of reactor designs and power conversion options. In 1984, DoD transferred the DARPA responsibility for the joint program to the newly-formed Strategic Defense Initiative Organization (SDIO).

At that time, SDIO was evaluating many ground and space-based architecture options for missile defense. All architectures included requirements for about 100 kWe or more of continuous long-term satellite power, and the tri-agency SP–100 program continued uninterrupted. In addition, short-term burst power levels in excess of 100 MWe were necessary for electrically-driven lasers, particle beams, and electromagnetic launchers. SDIO and DOE initiated the joint Multimegawatt Space Nuclear Power Program in 1985. The study assessed a variety of closed-cycle and open-cycle nuclear options, most of which used hydrogen cooled reactors. One significant concern was possible contamination of the spacecraft by the exhaust from an open-cycle power system. The relatively clean hydrogen exhaust from a reactor system was generally considered manageable. However, parallel studies revealed that hydrogen-oxygen combustion systems had comparable mass even if water vapor had to be removed from the exhaust. In addition, the continuing SDIO architectures placed less emphasis on the high-power space systems, and the SDIO mission eventually evolved to ground-based theater missile defense. Consequently, the agencies terminated the Multimegawatt Program.

By 1984, when SDIO took over the DoD role for SP–100, the program had narrowed the system design options to fast spectrum reactors coupled to one of three conversion options: in-core thermionic, out-of-core and conductively-coupled thermoelectric, and out-of-core Stirling. In 1985, the thermoelectric option was selected for continuation into the next phase which was to build and test a prototype. Work on the other conversion options did not stop, however. SDIO, DOE, and the Air Force established

NUCLEAR SYSTEMS

the Thermionic Fuel Element Verification Program (TFEVP) to continue in-core testing of TFEs to address the technical issues that remained unanswered from Phase 1 of the SP–100 program. NASA continued work on Stirling engines that could be used with either solar dynamic systems or the SP–100 reactor to triple the power output.

In 1986, SP–100 moved into an aggressive Phase 2 prototype development. In the next few years, the program made significant technical advances in reactor technology, but progress on the conductively-coupled thermoelectric cell was limited until later in the program. Furthermore, DoD support abated as SDIO requirements diminished and access to the Russian work on thermionics increased. In 1990, SDIO contracted with the U.S./Russian joint venture INERTEK to evaluate Russian Enisy (TOPAZ–II) technology and to test unfueled units in the U.S. Plans included a space test of one unit in order to demonstrate nuclear electric propulsion.

As occurred in the late 1960s, the mission requirements for space reactors did not materialize. SDIO became the Ballistic Missile Defense Organization (BMDO), reflecting its new emphasis on theater missile defense. Other DoD applications such as space-based radar and submarine laser communications did not become active satellite programs. Because of higher priority programs, NASA plans for manned exploration of Mars and the Moon dissolved. Consequently, the early 1990s once again saw the end of U.S. space reactor development. DOE terminated the SP–100 and TFEVP programs in 1993; DoD transferred the BMDO TOPAZ work to the Defense Special Weapons Agency (DSWA) in 1995; and Congress eliminated TOPAZ funding for the 1997 budget year.

2.3 The future

At present (1999), no mission requirements are firm enough to support renewed development of space reactors. However, NASA's plans for unmanned space exploration indicate a clear need for radioisotope systems. Standard RTGs are being reevaluated in response to NASA's desire to do "better, cheaper, faster" missions. The new and smaller spacecraft planned for future planetary exploration require about 150 Watts or less and have stringent cost, mass, and size limits for the power systems. Figure 5.1 illustrates the relative size of a conceptual Pluto Express spacecraft with four of the large planetary spacecraft from previous missions. As depicted, the largest envelope dimension is reduced by a factor of five.

NASA and DOE have initiated the Advanced Radioisotope Power System (ARPS) program to develop high-efficiency devices for future smaller planetary probes. Most ARPS designs are based on the GPHS module, and candidate conversion technologies include advanced thermoelectrics, alkali metal thermal to electrical conversion (AMTEC), thermophotovoltaic, and Stirling engines. The agencies tentatively selected AMTEC as the leading candidate, and DOE initiated system development work in 1997.

254　SPACECRAFT POWER TECHNOLOGIES

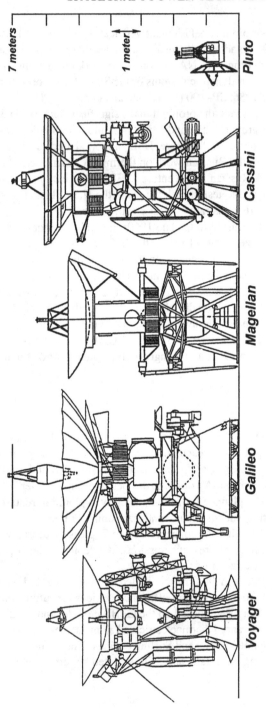

Figure 5.1 NASA Planetary Spacecraft Size Comparison (Courtesy of NASA)

3. History of the Russian space nuclear program

The FSU pursued a very aggressive development program in space nuclear power. While most of the development work occurred in Russia, organizations in Kazakhstan, Ukraine, Georgia, and Estonia participated as well. As in the U.S., the Russian program included both RTGs and reactors. However, Russia emphasized reactors more than RTGs and flew many more space nuclear systems than did the U.S. One of the reasons so many nuclear systems were launched is the basic approach used by Russia in much of its space program. While the U.S. emphasized sophisticated spacecraft designs with long operating lifetimes, Russia accomplished its missions with less complicated hardware but launched many more spacecraft to maintain operational capability. Typical lifetime for many of these missions was measured in weeks. Therefore, long life was not a pertinent requirement for the power systems.

Russian space reactor development consisted primarily of four systems: Bouk (Beechtree), Romashka, TOPAZ (Thermal Emission in the Active Zone), and Enisy (Daisy). Bouk powered the Radar Ocean Reconnaissance Satellite (RORSAT) series of ocean surveillance spacecraft. Two of the newer TOPAZ units were flight-tested in 1987 but have not been used on operational missions. Enisy and Romashka were ground tested, but neither flew in space. Because of the military nature of many of the missions, much of the information on the satellites and the nuclear power sources remained secret, and significant details did not emerge until the late 1980s. In addition, some information surfaced on reactors for propulsion and on the partial development of higher power systems, including a fast-spectrum thermionic reactor designed to produce 100 kWe for several years (Rasor, 1997). The most far-ranging interaction between the U.S. and Russia involved the SDIO and the Kurchatov Institute of Atomic Energy in Moscow. Kurchatov had been instrumental in the development of Romashka and Enisy, whereas Bouk and TOPAZ had been fielded under the auspices of Krasnaya Zvezda (Red Star). The long history of military use of Bouk hindered the release of information about it and TOPAZ. In addition, some confusion arose within the U.S. about the Russian systems. In initial discussions, it was unclear that Enisy and TOPAZ were different systems. When the existence of two Russian thermionic systems became apparent, U.S. personnel began referring to TOPAZ as TOPAZ–I and Enisy as TOPAZ–II. In this chapter, the proper Russian designations are maintained.

Table 5.5 lists the Russian space missions that were powered by nuclear sources. Bouk was launched on 32 successful RORSAT missions plus one launch that failed to achieve orbit. The RORSAT satellites used a side-looking radar to track naval vessels and orbited at altitudes of less than 300 km. Due to the atmospheric drag at this low altitude, nuclear power was selected because of its much lower projected area in the direction of flight relative to solar arrays. At the end of these missions, the RORSAT power systems were moved to a disposal orbit of about 1000 km to preclude reentry of a "hot" reactor. Unfortunately, the reboost to the higher orbit failed on three occasions,

Table 5.5 Russian Missions Employing Nuclear Power
(after Booz·Allen & Hamilton, 1995; Bennett, 1989; Makhorin et al., 1996; Purdum, 1996)

Launch Date	Spacecraft	Power Source[1]	Mean Altitude[2]	Status/Lifetime	Notes
3 Sep 65	Cosmos 84	Orion 1 RTG	1,500 km	In-orbit	Navigation?
18 Sep 65	Cosmos 90	Orion 1 RTG	1,500 km	In-orbit	Navigation?
27 Dec 67	Cosmos 198	Reactor	920 km	1 Day	---
22 Mar 68	Cosmos 209	Reactor	905 km	1 Day	---
25 Jan 69	RORSAT Launch	---	---	---	Failure ?
23 Sep 69	Cosmos 300	^{210}Po heater	Reentered	---	Lunar Probe
22 Oct 69	Cosmos 305	^{210}Po heater	Reentered	---	Lunar Probe
3 Oct 70	Cosmos 367	Reactor	970 km	1 Day	---
1 Apr 71	Cosmos 402	Reactor	990 km	1 Day	---
25 Dec 71	Cosmos 469	Reactor	980 km	9 Days	---
21 Aug 72	Cosmos 516	Reactor	975 km	32 Days	---
25 Apr 73	RORSAT Launch	---	---	---	Launch Failure
27 Dec 73	Cosmos 626	Reactor	945 km	45 Days	---
15 May 74	Cosmos 651	Reactor	920 km	71 Days	---
17 May 74	Cosmos 654	Reactor	965 km	74 Days	---
2 Apr 75	Cosmos 723	Reactor	930 km	43 Days	---
7 Apr 75	Cosmos 724	Reactor	900 km	65 Days	---
12 Dec 75	Cosmos 785	Reactor	955 km	1 Day	---
17 Oct 76	Cosmos 860	Reactor	960 km	24 Days	---
21 Oct 76	Cosmos 861	Reactor	960 km	60 Days	---
16 Sep 77	Cosmos 952	Reactor	950 km	21 Days	---
18 Sep 77	Cosmos 954	Reactor	Reentered	~43 Days	Canada Impact
29 Apr 80	Cosmos 1176	Reactor	920 km	134 Days	---
5 Mar 81	Cosmos 1249	Reactor	940 km	105 Days	---
21 Apr 81	Cosmos 1266	Reactor	930 km	8 Days	---
24 Aug 81	Cosmos 1299	Reactor	945 km	12 Days	---
14 May 82	Cosmos 1365	Reactor	930 km	135 Days	---
1 Jun 82	Cosmos 1372	Reactor	945 km	70 Days	---
30 Aug 82	Cosmos 1402	Reactor	Reentered	120 Days	South Atlantic
2 Oct 82	Cosmos 1412	Reactor	945 km	39 Days	---
29 Jun 84	Cosmos 1579	Reactor	945 km	39 Days	---
31 Oct 84	Cosmos 1607	Reactor	950 km	93 Days	---
1 Aug 85	Cosmos 1670	Reactor	950 km	83 Days	---
23 Aug 85	Cosmos 1677	Reactor	940 km	60 Days	---
21 Mar 86	Cosmos 1736	Reactor	950 km	92 Days	---
20 Aug 86	Cosmos 1771	Reactor	950 km	56 Days	---
1 Feb 87	Cosmos 1818	Reactor	800 km	~6 Months	TOPAZ
18 Jun 87	Cosmos 1860	Reactor	950 km	40 Days	---
10 Jul 87	Cosmos 1867	Reactor	800 km	~1 Year	TOPAZ
12 Dec 87	Cosmos 1900	Reactor	720 km	~124 Days	Malfunction
14 Mar 88	Cosmos 1932	Reactor	965 km	66 Days	Last RORSAT
16 Nov 96	Mars 96	Angel RTG	Reentered		Booster failure

[1] All reactor flights were RORSATS except cosmos 1818 and 1867
[2] For RORSATS, altitude shown is the disposal orbit for the reactor core. Operational altitudes were 240x280 km.

NUCLEAR SYSTEMS 257

most notably the Cosmos 954 reentry that spread radioactive debris over a remote region of Canada in 1978. After that failure, Russia redesigned the safety system to extract the nuclear core from the Bouk reactor vessel after reboost to the higher orbit. If the reboost failed (as in Cosmos 1402 in 1983), the fuel elements were also extracted so they would burn up in the atmosphere instead of impacting on Earth (IPPE, 1992). For the two TOPAZ flights, the operational orbits were about 800 km, so along with its higher ballistic coefficient, orbital lifetime was similar to the RORSAT cores.

Beginning in 1970, Russia built and tested four developmental TOPAZ thermionic reactors, plus ten thermal and structural mockups, leading to the final TOPAZ design. The first version operated successfully to a design life of 1000 hours, and the second operated about 6000 hours. Krasnaya Zvezda conducted two flight tests of TOPAZ: Cosmos 1818 launched on February 2, 1987, to an orbit of 810 x 970 km; and Cosmos 1867 launched on July 10, 1987, to an orbit of 797 x 813 km. Both TOPAZ units powered experimental electric propulsion thrusters. The altitude of approximately 800 km corresponds to an orbital lifetime of roughly 350 years, enough time for the reactor fission products to decay to an activity level of the actinides before any reentry would occur. When reentry does take place, TOPAZ is designed to burn up in the upper atmosphere so no fuel or radioactive debris should impact Earth. In addition to the two flight tests, two prototypic units were ground tested in parallel for 4500 and 7000 hours, respectively (Andreev et al., 1993; Baksht et al., 1978; Bogush et al., 1990; Gryaznov and Pupko, 1991).

In 1989, a sizable Russian contingent attended the 6th Symposium on Space Nuclear Power in Albuquerque, NM, their first participation in this annual international conference. About that same time, representatives of the U.S. SDIO and the Russian Kurchatov Institute of Atomic Energy began discussions that eventually led to the import and non-nuclear testing of Enisy units in the U.S. The first two, V–71 and Ya–21u, came to the U.S. in 1992 along with a large Russian vacuum test chamber, a Thermionic Fuel Element (TFE) test rig, and other equipment. A test facility was established at the University of New Mexico Engineering Research Institute (NMERI), and electrically-heated TFE and system tests were conducted. These units were returned to Russia in 1995 (Booz·Allen & Hamilton, 1995). Subsequently, SDIO planned to conduct a flight test using one of the Enisy reactors. In 1995, the U.S. imported four additional Enisy systems (mockups Eh–40 and Eh–41 and potential flight units Eh–43 and Eh–44) to support the flight test program. However, the flight program was discontinued due to budget reductions, and DoD transferred the program to DSWA. With the denial of funding by Congress, DSWA terminated the program and returned the four units to Russia in 1997.

4. Radioisotope systems

Figure 5.2 illustrates the current U.S. radioisotope system design, the GPHS–RTG. Envelope dimensions of each unit are 1.14 m in length and 0.41 m from fin tip to fin tip. Unit mass is about 56 kg. Radioisotope decay heat (approximately 4.4 kWt) is supplied by the pressed ^{238}PuO$_2$ fuel encapsulated within 18 GPHS modules stacked in a column along the central axis of the RTG. The thermopile consists of 572 SiGe unicouples surrounding the heat source. Beginning of life (BOL) power output is nominally 300 watts per RTG at 28–30 volts DC. For reliability, the unicouples are connected in two series-parallel circuits so that a failure of one unicouple (in either open or short circuit modes) will result in a power loss of just one unicouple. The other elements comprising the unit are the thermal insulation, heat source support, gas management system, and outer shell/radiator.

Figure 5.3 shows the current-voltage relationship typical of the GPHS–RTG. The current is nearly linear between open and short circuit. Nominal operating temperatures are 573 K (300°C) at the thermoelectric cold junction and 1273 K (1000°C) at the hot junction. For stability, the nominal operating point is slightly to the right of the peak

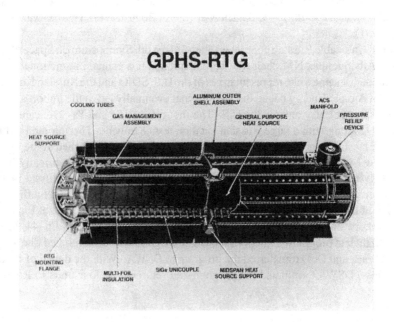

Figure 5.2 The General Purpose Heat Source Radioisotope Thermoelectric Generator (GPHS–RTG) (Courtesy U.S. Department of Energy)

NUCLEAR SYSTEMS

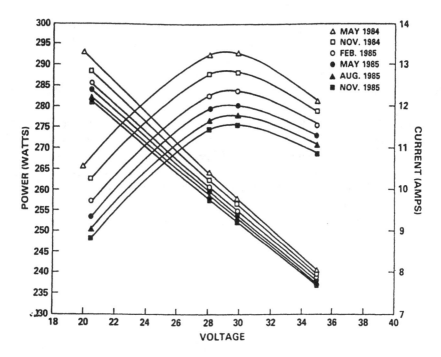

Figure 5.3 Current–voltage Curve for the GPHS–RTG (Courtesy U.S. Department of Energy)

power point. Power output is typically regulated by load matching through shunt regulation with excess power dissipated through a resistance bank. Chapter 6 provides details of the thermoelectric design.

The finned aluminum outer shell serves as the radiator for waste heat rejection and the structural support for the entire assembly. The unicouples are mounted individually by screws through the outer shell. Multifoil thermal insulation consisting of 60 alternating layers of molybdenum foil and woven astroquartz blankets is used between the unicouples and on the ends of the RTG to minimize direct heat loss. One concern with metal multifoil is the possibility of electrical shorts between unicouples. Although small leakage currents are measured, their magnitude is insignificant. Mechanical support for the heat source is provided by spring loading at the midspan and at each end. Since a radioisotope source always produces heat, an auxiliary cooling system (ACS) provides for thermal control on the launch pad. The external cooling tubes connect with an active launch vehicle cooling loop (the Space Shuttle for the Galileo and Ulysses launches) since only a limited heat load can be radiated directly to the inside of the payload bay or fairing. Prior to launch, the inside of the RTG is filled with an inert gas, usually argon or a mixture of argon and helium, which greatly increases the thermal conductivity of the multifoil. This "thermal shorting" of the insulation reduces the surface temperature of heat source to about 1100 K. During a Shuttle launch, the ACS maintains the RTG temperatures within acceptable levels. When launched on an expendable vehicle, the ACS is no longer cooling the outer shell during the ascent phase. The RTG heats up slowly due to its thermal mass, and the outer shell is exposed to space when the vehicle fairing is jettisoned. After spacecraft deployment, the gas is vented, and the GPHS–RTG temperatures increase to their operational level.

Figure 5.4 depicts a GPHS module. Each module weighs 1.43 kg including 0.6 kg of $^{238}PuO_2$ in four pressed fuel pellets generating about 250 Wt initially. The length and diameter are both 27.5 mm for the cylindrical pellet. Each is encapsulated within a 0.5 mm cladding of iridium alloyed with a small amount of tungsten. A selective frit vent is incorporated into the cladding which allows the helium atoms generated during the radioactive decay of the fuel to escape, but not the fuel particles. Iridium is a high-ductility metal and, subject to certain temperature limits, its toughness enables it to

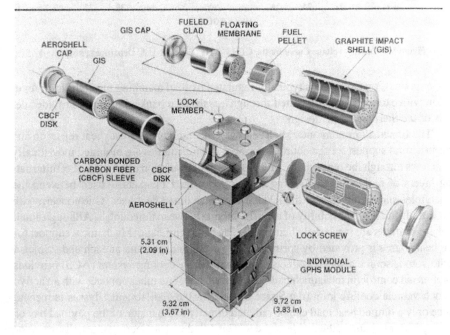

Figure 5.4 The General Purpose Heat Source (GPHS) (Courtesy U.S. Department of Energy)

deform greatly without rupturing in the event of high-velocity Earth impact or other load conditions. Below 1150 K, the material loses ductility, and above 1600 K, grain growth can negate its toughness. The RTG is designed so that the cladding temperature stays between these limits through all phases of the mission, including on-pad and ascent failures and inadvertent reentry and impact. Iridium is compatible with adjacent GPHS materials and resists oxidation after an Earth impact.

The cladded fuel is surrounded by several layers of graphitic materials that provide protection under accident conditions. Two cladded fuel pellets are assembled within each of two graphite impact shells (GISs) that provide the primary impact protection from either shrapnel/fragment impacts due to a launch vehicle failure or Earth impact due to an early launch accident or an out-of-orbit inadvertent reentry. The GIS is made of Fine Weave Pierced Fabric (FWPF® a registered trademark of AVCO) carbon-carbon composite. Each GIS is encased in a Carbon Bonded Carbon Fiber (CBCF) graphite thermal insulation sleeve, and the two assemblies are inserted into a FWPF® aeroshell. CBCF has a relatively low thermal conductivity that prevents the iridium clad from overheating during the aerodynamic heating pulse experienced during reentry, and it also maintains the iridium clad above the ductile-to-brittle transition temperature during the subsequent cooling phase of reentry to surface impact. Although the aeroshell function is primarily to withstand the aerodynamic heating and pressure loads during an out-of-orbit reentry, it also provides additional blast/overpressure and impact protection. The FWPF graphite material can tolerate very high temperatures and some of the aeroshell also ablates away to block the incident energy. Its structural properties are such that it can withstand the thermal stresses induced by the temperature gradients from the heat pulse. Lastly, graphite has a high thermal emissivity (> 0.85) so that the temperature difference is only about 100 K between the heat source and the SiMo hot shoes on the unicouples.

As seen from this discussion of the GPHS module, safety requirements have a major influence on the design of radioisotope heat sources for RTGs. These requirements are discussed later in this chapter, but the fundamental objective for modern RTGs is containment of the fuel under all credible accident conditions. Accident environments include:

- Blast/overpressure, shrapnel, fragments, heat/fires due to solid and liquid propellants, and exposure to other chemical elements from on-pad or near-pad failure of the launch vehicle
- Similar environments from ascent failures including range safety destructs
- Earth impact resulting from failures at any time during the launch, ascent, or reentry from orbit
- Aerodynamic heating and structural loads during reentry

All of the U.S. flight RTGs have used some form of Plutonium–238 (^{238}Pu) as the radioisotope source. Initially, several radioisotope fuels were considered, and some of the fuel characteristics of interest are shown in Table 5.6. The first unit, SNAP–1, used Cerium–144 (^{144}Ce) because it was readily available from reprocessing of defense reactor fuel (Sholtis, et al., 1994). At that early point in the U.S. space program, anticipated mission durations were a few months, so the 290–day half life of ^{144}Ce was acceptable. However, its beta/gamma radiation characteristics made it unattractive.

The SNAP program considered other radioisotopes that could be obtained from defense nuclear facilities. These were Polonium–210 (^{210}Po), Strontium–90 (^{90}Sr), and ^{238}Pu. Although ^{210}Po is an alpha emitter with high power density, its short half-life (128 days) limited its potential applications. The longer half lives of ^{90}Sr and Curium–244 (^{244}Cm), 18.1 and 28.6 years, respectively, were better but still limited. Radiation environments were another consideration, particularly for ^{90}Sr. Some terrestrial RTGs use a stable and insoluble form of ^{90}Sr, strontium titanate, since the weight of the radiation shield is far less important than in space systems. ^{244}Cm was a candidate, largely because the breeder reactor was the anticipated source. Since the U.S. terminated development of the breeder, ^{244}Cm was given less consideration (Lange and Mastal, 1994).

Safety considerations and evolution to longer-duration missions led to the dominant use of the long-lived radioisotope ^{238}Pu, not to be confused with Plutonium–239 used in nuclear weapons. When the safety philosophy known as "intact reentry/intact impact" was adopted for RTGs, stable and high temperature fuel forms were developed. For SNAP–19, a plutonia–molybdenum–cermet (PMC) fuel was used. The pressed plutonium oxide (PPO) form, ^{238}PuO$_2$, became more attractive because of its stability and is used almost exclusively today. For RTGs under development in the late 1960s, ^{238}PuO$_2$ was used in the form of microspheres encapsulated in metal cylinders. This form was used for the SNAP–27 RTGs. For the MHW–RTGs, pressed spheres of ^{238}PuO$_2$ were developed, and rounded cylindrical shapes are now used in the GPHS–RTG.

In all RTGs, the radioisotope fuel is encapsulated and protected by several layers of materials that provide the following individual or multiple functions:
- Chemical compatibility with the fuel
- Strength, particularly in accident situations like rocket explosions and impacts
- Cladding for chemical compatibility or protection of the strength element
- Thermal and thermal stress protection during reentry
- Venting of helium generated by the radioactive decay of the plutonium fuel

Table 5.6 includes identification of the materials used to accomplish these functions in the various flight RTGs. Prior to development of the Ir–W alloy for the MHW–RTG, tantalum alloys were used extensively for fuel compatibility and strength liners. However, as fuel inventories grew and safety criteria became more stringent, impact shells constructed from woven and sintered graphite fibers were developed for the MHW–RTG and the GPHS–RTG. In addition, the higher operating temperatures of these newer RTGs required the incorporation of the newer materials.

Subsequent to SNAP–9A, the safety design of all RTG heat sources has been based on "intact reentry/intact impact." Graphite materials are the mainstay of reentry protec-

NUCLEAR SYSTEMS

Table 5.6 Characteristics of Radioisotope Fuels for RTGs

Emitter class	Gamma	Beta						Alpha			
Isotope	^{60}Co	^{90}Sr	^{106}Ru	^{137}Cs	^{144}Ce	^{147}Pm	^{170}Tm	^{210}Po	^{238}Pu	^{242}Cm	^{244}Cm
Half-life (yrs)	5.3	27.7	1.0	30	0.78	2.6	0.35	0.38	87.8	0.45	18
Principal decay (Mev)											
Alpha								5.3	5.5	6.1	5.8
Beta	0.31	2.26	3.35	1.17	2.98	0.23	0.97				
Gamma	1.33	1.73		0.67	2.18	0.12	0.08	0.8	0.04	0.04	0.04
Fuel form	Metal	$SrTiO_3$	Metal	CsCl	CeO_2	Pm_2O_3	Tm_2O_3	Metal	PuO_2	Cm_2O_3	Cm_2O_3
Density (g/cm³ of form)	8.7	3.7	12.2	3.6	6.6	6.6	8.5	9.3	10	9	9
Isotopic purity (%)	10	50	3.3	35	18	95	10	95	80	90	95
Power density (w/cm³ of form)	15.2	0.94	13.4	0.42	25.3	1.8	9.1	1210	3.9	882	20.4
Curies/watt	65	148	102	207	126	2788	500	32	30	28	29
Lead shielding req'd for 10 mr/hr @ 1 m from 1 kWth source (in.)	9.5	6	9	4.6	10.2	1	2.5	1	0.1	0.4	2
Melting point °C	1480	1900	2450	646	2680	2270	2300	254	2280	1950	1950
Biological hazard MPC in air (μc/cm³)	$3x10^{-9}$	10^{-10}	$2x10^{-9}$	$5x10^{-9}$	$2x10^{-9}$	$2x10^{-8}$	10^{-8}	$7x10^{-11}$	$7x10^{-13}$	$4x10^{-11}$	$3x10^{-12}$

tion. They can withstand very high temperatures, and strength properties generally increase as the temperature increases. For normal operation, surface emissivity is usually high which serves to augment the radiation heat transfer to the hot side of the thermoelectrics.

The aeroshells must withstand the high heating rates and aerodynamic pressures during reentry and the thermal stresses induced by the resulting temperature gradients. The relatively dense polycrystalline graphites are excellent heat sinks which perform well for shallow angle reentries that cause high ablation of the heat shield. For steep angle reentries characterized by short but high heating rates, carbon–carbon composites are excellent at withstanding the thermal stresses that result.

5. Reactors

The basic subsystems of a space reactor power plant are the reactor core, reactivity control, radiation shield, heat transport, power conversion, and waste heat radiator. Heat is generated by nuclear fission in the reactor core. High-power, compact cores require the use of highly enriched uranium (HEU), and space reactor designs typically use enrichments of >90 percent ^{235}U. As shown in Table 5.4, space reactor designs have included thermal, epithermal, and fast spectrum cores. Moderated systems require less fuel and are lower mass at power levels of a few kWe to a few 10s of kWe. At higher powers, fast spectrum reactors are usually more mass-effective.

Reactivity is controlled either by in-core sliding control rods or by external reflector control. In the external control scheme, two approaches have been used. One is to surround the core with rotating beryllium or beryllium oxide cylinders with boron or boron carbide covering a portion of the circumference. In the shutdown mode, the cylinders are rotated completely inward so that the B_4C sections absorb the neutrons to preclude criticality. For startup and operation, the cylinders are rotated to allow the appropriate degree of neutron reflection by the beryllium in order to maintain controlled criticality at the desired thermal output power level. The second approach uses a sliding beryllium or BeO reflector surrounding the core to control the leakage of neutrons. When fully open, neutrons are not reflected back into the core so that criticality is precluded. For operation, the reflector is partially closed to create the required proportion of neutron reflection and leakage.

The spacecraft and some power system components must be protected from the harsh neutron and gamma radiation produced by an operating reactor. To minimize mass, the power system envelope configuration usually takes the form of a cone so that a shadow shield can be used on one side of the reactor to mitigate the radiation field. In this way, the heavy shielding material does not have to surround the entire reactor. Instead, the power system is mounted such that the spacecraft is in the shadow of the radiation shield, as are the thermal radiator and other components of the power system itself. Lithium hydride has been the material commonly used for space reactor neutron shielding. Gamma shielding is best accomplished using high-density materials such as tungsten, depleted uranium, and stainless steel.

In most designs, an active cooling loop transfers the fission heat to the energy conversion system. Exceptions are the Russian Romashka design wherein heat radiates directly to thermoelectric converters, and the in-core thermionic systems like TOPAZ and Enisy where the thermionic converter is an integral part of the fuel pin. The reactor cooling medium can be either gas or liquid. For two reasons, many designs use liquid metals, such as eutectic NaK or Li. First, their thermal properties create high heat transfer coefficients. Second, their magnetic properties allow the option of using electromagnetic pumps that have no moving parts and thereby offer high reliability for long-life systems.

Almost any energy conversion option can be coupled to a nuclear reactor to produce electrical power. The static options used to date are thermoelectric and thermionic. Dynamic options have been considered for many designs, particularly for the higher power concepts. As shown in Table 5.4, mercury–Rankine and potassium–Rankine turboalternator system designs have been developed and ground-tested under the SNAP program. Closed Brayton cycle converters can be used and have been baselined for gas-cooled designs like the 710 reactor. Although the SP–100 system had thermoelectrics as the baseline converter, Brayton, potassium–Rankine, and Stirling converters were all considered as candidates for power growth to several hundred kWe.

A thermal radiator dissipates waste heat, and this component usually dominates the overall dimensions of the system. The requirements of a space reactor to be compact and lightweight also lead to the use of relatively high cold side temperatures so that the radiator can be of reasonable size. Hot side temperatures are usually high as well in order to increase converter efficiency or recover the efficiency often lost due to the high cold side temperature. As a result, superalloys or refractory metals, such as tungsten, tantalum, molybdenum and niobium, and their alloys are commonly used for hot-side components in space reactors. In many designs, a secondary cooling loop is necessary to carry the waste heat from the cold side of the energy converter to the waste heat radiator.

As shown in Table 5.1, SNAP–10A is the only space reactor flown by the U.S. Table 5.7 lists the design parameters of the flight unit, Figure 5.5 illustrates the system, and Figure 5.6 illustrates the power conversion. System mass is 436 kg. The reactor uses a self-moderated UZrH fuel with Hastelloy N as the cladding. Thermal power is nominally 43 kWt. NaK coolant is circulated by a thermoelectromechanical DC conduction pump with a permanent magnet and integral PbTe thermoelectric power supply. In the region of the converter, the cross section of the cooling loop piping is D-shaped, and the hot side of the SiGe thermoelectric converter elements are bonded directly (through an intermediate electrical insulator) to the flat side of the piping. For power production, a total of 2880 SiGe thermoelectric elements is mounted on 40 cooling

Table 5.7 SNAP-10A Design Parameters (after Angelo and Buden, 1985)

Parameter	Value
Electrical power (minimum)	533 We
Voltage	30.3 volts
System mass	436 kg
Thermal power (nominal)	43.8 kWt
Fuel form	UZrH
Reflector material	Be
Design life	1 year
Radiator area	5.8 m²
Overall length	3.5 m
Mounting base diameter	1.3 m
Coolant	NaK-78
Coolant flow rate	13.3 gpm
Reactor outlet temperature	827 K
Reactor inlet temperature	790 K
Power conversion	SiGe thermoelectrics

tubes. The beginning of life power output is 580 watts. Radiator segments are bonded to the cold end of each thermoelectric element for the heat rejection subsystem (Bennett, 1989; Angelo and Buden, 1985; Schmidt, 1988).

The last major space reactor development effort in the U.S. was the joint DoD/NASA/DOE SP-100 program that began in 1983 and terminated in 1993. Figure 5.7 illustrates the system configuration and identifies the major subsystems, and Table 5.8 lists the design parameters. The nominal design point is for 100 kWe produced by SiGe thermoelectrics, but the system can be adapted over a wide range of power levels by the modularity of the thermoelectric design. Other subsystems, e.g., heat transport and waste heat radiator, are also modularized for scalability. Furthermore, the basic reactor can be coupled with an alternative energy conversion system (Stirling engines were evaluated specifically) to produce several hundred kWe. Design life is 10 years total, with 7 years at full operating conditions.

The SP-100 reactor is a fast-spectrum, lithium-cooled design. Its high operating temperatures (1400-1450 K fuel and 1350-1375 K coolant outlet) necessitate the use of various refractory metal alloys in its construction. Enriched UN is the fuel with PWC-11 cladding. An intermediate bonded rhenium liner is required for chemical compatibility, but also adds strength and aids in immersed subcriticality and containment of nitrogen released by the fuel. B_4C in-core safety rods and ex-core segments prevent pre-operational criticality. Twelve sliding BeO reflector segments surround the core for primary control. Thermoelectric electromagnetic pumps drive the lithium primary cooling loop that delivers the heat to the thermoelectric modules. The lithium secondary cooling loop transports the waste heat from the cold side of the thermoelectrics (and cools the cold side of the TEM pumps) to the radiator. Titanium flow tubes and potassium heat pipes with carbon-carbon fins reject the waste heat to space (Buden, 1994; Buksa,

NUCLEAR SYSTEMS

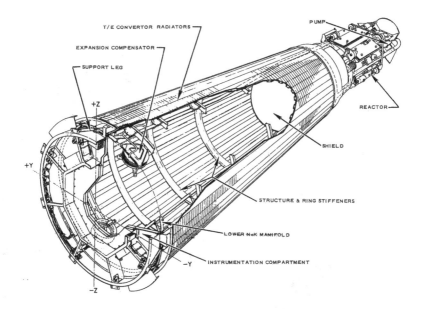

Figure 5.5 SNAP–10A Reactor Power System (Courtesy U.S. Department of Energy)

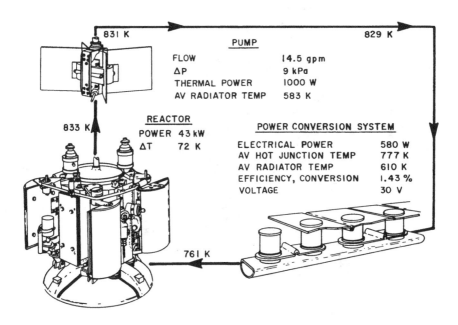

Figure 5.6 SNAP–10A Power Conversion Subsystem (Courtesy U.S. Department of Energy)

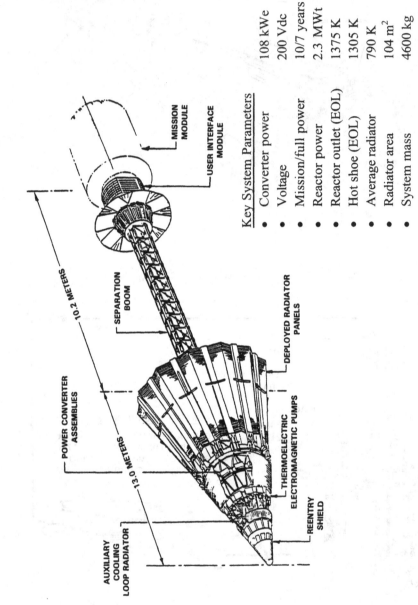

Figure 5.7 SP–100 System Reactor Power System Design (Courtesy U.S. Department of Energy)

Table 5.8 SP–100 100 kWe Design Parameters (after Buksa et. al., 1994)

Parameter	Value
Electrical power	108 kWe
Voltage	200 volts dc
System mass	4600 kg
Thermal power (nominal)	2.45 MWt
Design life	7 year ops/10 year total
Radiator area	104 m^2
Overall length	12 m w/o boom
Maximum diameter	8 m radiator cone
Reactor outlet temperature	1375 K EOL peak
Avg radiator temperature	790 K
Fuel form/cladding	UN/PWC–11/Re liner
Reflector material	BeO
Shield	LiH/depleted U
Coolant	^{7}Li
Pump	SiGe thermoelectromagnetic
Power conversion	SiGe thermoelectrics (conductively-coupled)
Radiator	Ti/K heat pipes/C–C fins

et al., 1994; Truscello and Rutger, 1992).

Although the system was never built, extensive testing of various components was accomplished, particularly irradiation tests of the fuel/clad combination. Development of the conductively-coupled thermoelectric assemblies has been less successful; by the end of the program, however, individual cells had operated for several thousand hours (Kelly and Klee, 1996). Although the basic SiGe material is similar to that used in the RTG unicouples, reactor systems producing several tens of kWe must use a thermopile that operates at much higher energy densities and heat fluxes. For conductive coupling with the lithium heat source and sink, the thermoelectric cells are bonded into a fairly complex stack of insulators and compliant pads. In the 100 kWe design, a total of 8640 cells is required. Each bonded cell stack is subsequently bonded into two 10x6 cell arrays, each producing 1.5 kWe at 34.8 volts dc. These are in turn assembled into 9 kWe modules, twelve of which produce the full power.

For compactness and low system mass, space reactors are designed to use highly-enriched (>90%) uranium fuels. Table 5.9 lists thermophysical properties and other characteristics for a few of the fuels considered for space reactors in the U.S. Uranium dioxide and uranium nitride are relatively well developed fuel forms. A variety of carbide fuels are candidates, and the Russians have undertaken the most recent R&D work on carbides. For power reactors, the nuclear fuel system must meet some very stringent requirements, including:
- Operation at high temperatures
- Stability at moderate burnup for many years

Table 5.9 Characteristics of Space Reactor Fuels (after Matthews et al., 1994)

Property	UO$_2$	UN	UC	UC$_2$	(U$_{0.2}$Zr$_{0.8}$)C$_{0.99}$
Theoretical density (gm/cc)	10.96	14.32	13.63	11.68	8.01
Theoretical U density (gm/cc)	9.66	13.52	12.97	10.60	2.88
Melting point (K)	3100	3035*	2775	2710	3350**
Thermal conductivity @ 1273 K (W/m–K)	3.5	25	23	18	30
Heat capacity @ 1500 K (J/mol–K)	71.1	65.3	66.5	98.8	57.8
Thermal expansion coeff to 1273 K (μm/m–K)	10.1	8.9	11.2	12	7.6
U vapor pressure @ 2000 K (Pa)	4x10^{-3}	2x10^{-2}	1x10^{-5}	2x10^{-5}	5x10^{-6}
Relative stability	Moderate	Low	High	High	High
Relative swelling	Low	Mid	Mid	Low	High
Relative fission gas release	High	Low	Mid	Low	Mid
Relative fabricability	Easy	Moderate	Easy	Difficult	Difficult

* Dissociation temperature **Solidus temperature

- Long-term compatibility with cladding materials
- Acceptable volumetric swelling and release of gaseous fission products
- Manufacturable in the desired form, enrichment, grain size, and stoichiometry

To varying degrees, the Russians have developed four space reactor power systems: Bouk, Romashka, TOPAZ, and Enisy. Figure 5.8 illustrates Bouk that powered 32 RORSAT ocean surveillance satellites, as listed in Table 5.5. The reactor is a fast spectrum design with uranium fuel rods containing a few percent molybdenum (Rasor, 1997). Bouk system mass is 900 kg. The RORSAT power requirement was 2.3–2.5 kWe for 2.5–3 months (IPPE, 1992), and although the longest RORSAT flight was less than 6 months, reactor life from ground testing is reported to be in excess of one year. After 6 months, however, power output was less than half of the initial 3 kWe (Rasor, 1997).

Bouk and Romashka are fast-spectrum reactors with beryllium reflector control. Both use SiGe thermoelectrics for energy conversion. However, the Bouk fuel is an enriched uranium–molybdenum alloy, while Romashka uses enriched uranium dicarbide. The Bouk reactor is cooled by a NaK loop similar to SNAP–10A, while the Romashka is designed for direct thermal radiation from the outer diameter of the reactor vessel to the thermoelectrics, somewhat similar to an RTG configuration. Six complete Bouk units are reported to exist in Russia (Rasor, 1997). It is unclear how many, if any, Romashka units Russia built or tested, but life tests of 15,000 hours are reported for the fuel and fuel elements (Kurcharkin et al., 1990).

The two Russian in-core thermionic reactor systems, TOPAZ and Enisy, are de-

NUCLEAR SYSTEMS

271

Thermal power	80-100 kWth
Electrical power	3 kWe net
Design life	~3 months
Power conversion	SiGe thermoelectrics
Reactor	Fast spectrum U-Mo alloy fuel
Coolant	NaK
Coolant temperature (Reactor outlet)	950 K
Radiator temperature	550 K
System mass	900 kg

Figure 5.8 Russian Bouk Reactor Power System

picted in Figures 5.9 and 5.10, respectively. Both are moderated cores with highly-enriched UO_2 fuel. Both produce about 5 kWe plus an additional 1 kWe to operate the electromagnetic pumps. Both weigh 1000–1200 kg. The primary difference is that TOPAZ employs a multicell TFE design, while the Enisy TFE is a single cell. The difference is illustrated is Figure 5.11. The TOPAZ multicell TFE is analogous to battery cells in a flashlight wherein individual fueled thermionic diodes are stacked in series within a cylinder. As a result, each TFE generates a voltage equal to the number of diodes in the stack times the voltage per diode. This advantage becomes more important for systems designed to produce many tens or hundreds of kilowatts. On the other hand, the Enisy single cell produces only the voltage of one diode, but the central fuel cavity enables testing and checkout of each TFE and the complete system with electric heaters substituted for the nuclear fuel. Fuel can be inserted anytime prior to launch, including on the launch pad, so logistics are simplified. Also, subsystem designs for the cesium supply and fission gas venting are somewhat simpler with the single-cell TFEs. Most of the in-core thermionic work in the U.S. has been with multicell TFEs.

Table 5.10 shows the design characteristics and performance after 90 and 180 days of operation for the two TOPAZ prototypes and two flight systems (Andreev et al., 1993). Russian analysis of these data attributes the observed degradation of all four units to reactivity changes due to hydrogen loss from the ZrH moderator and to small

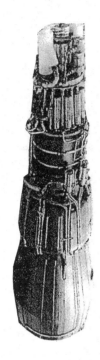

Thermal power	130-150 kWth
Electrical power	5 kWe net
Output voltage	32 volts
Design life	1 year+
Power conversion	Multicell TFEs (79 total, 5-6 cells/TFE)
Emitter temperature	1725 K
Radiator temperature	825-875 K
Radiator area	7 m²
Reactor	UO_2 fuel ZrH moderated
System mass	1200 kg

Figure 5.9 Russian TOPAZ Reactor Power System

amounts of this hydrogen entering the interelectrode gap in the thermionic fuel elements. The flight prototypes operated for 143 days and 342 days, respectively, until the flow-through cesium system exhausted its on-board supply.

Enisy is an alternative in-core thermionic design developed in parallel with TOPAZ. The power section of the Enisy core consists of 34 single-cell TFEs connected in series. Three additional TFEs provide power to the NaK loop pump. Figure 5.12 shows the Enisy single-cell TFE. Over its 20-year development period, Russia fabricated approximately 28 Enisy systems (Voss and Rodriguez, 1994). These systems are designated as either V, Ya, Eh or SM. The V units were either thermophysical or mechanical systems. Ya units were nearly prototypic and could be used for all types of ground tests, including nuclear ones. Systems developed as intended flight units carried the Eh designation. SM indicates static mockups used to check structural integrity.

Ya–81 was the most successful nuclear test, accumulating a total of 12,500 hours of operation. Power output was 4.5 kWe with a core thermal level of 105 kWt and a reactor coolant outlet of about 800 K (Voss and Rodriguez, 1994). From 1992 through 1995, U.S. researchers conducted non-nuclear performance tests of two Enisy units, V–71 and Ya–21u, at the University of New Mexico Engineering Research Institute in

NUCLEAR SYSTEMS

273

Thermal power	150 kWth
Electrical power	5 kWe net
Output voltage	29 volts
Design life	1 year+
Power conversion	Single-cell TFEs (34 + 3 for EM pump)
Emitter temperature	1900 K
Radiator temperature	850-900 K
Radiator area	7-10 m²
Reactor	UO₂ fuel ZrH moderated
System mass	1000-1200 kg

Figure 5.10 Russian Enisy Reactor Power System

Albuquerque, NM (Thome *et al.*, 1995; Luchau *et al.*, 1996). Ya–21u tests also included shock and vibration. Unfortunately, the heat input to the reactors was limited because of concerns about overheating the insulators and bellows at the ends of the TFEs (due to the flat heating profile of the electric heaters). Hence, the output power was only about 3 kWe.

6. Safety

No discussion of space nuclear power is complete without addressing the accompanying safety issues. In fact, safety has a major impact on the design of the power system, particularly the radioisotope heat source or nuclear reactor. This section focuses on the safety criteria and formal launch safety review process used in the U.S. In addition, some information is included on the Russian safety experience and on international developments on the subject.

274   SPACECRAFT POWER TECHNOLOGIES

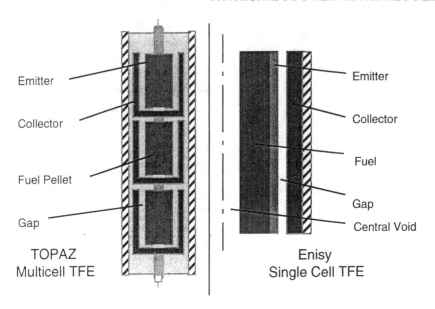

Figure 5.11 Multicell and Single Cell Thermionic Fuel Element

6.1 U.S. safety

In the U.S., the top-level safety objective is to minimize the interaction of the human population and the biosphere with the nuclear material in the system. The associated safety analysis and test program applies the principles of probabilistic risk assessment in conjunction with international guidelines for allowable radiation exposure levels. The U.S. safety record on space nuclear power has been excellent. As listed in Table 5.1, accidents have occurred on missions involving nuclear power, specifically the Transit 5BN–3 failure in 1964, the Nimbus–B1 launch abort in 1968, and the damaged Apollo–13 reentry in 1970. In all cases, the safety systems performed as designed. However, each new mission raises specific issues, either from new accident scenarios and environments or from additional information on materials and systems. Consequently, the full safety evaluation process is invoked for each mission utilizing nuclear power.

Overall safety is ultimately the responsibility of the user agency, either NASA or DoD to date. As the designer and producer of space nuclear power systems, DOE has a statutory responsibility for nuclear safety. DOE works closely with the user agencies to assure that safety requirements are met in the design of the power system, spacecraft, and mission. Although the focus of the safety program is the nuclear fuel, all hazardous materials that are part of the power system are included in the safety assessment program. Since all but one of the nuclear power systems used on U.S. missions have been

Table 5.10 Flight and Ground Test Results for Russian TOPAZ Space Reactor
(after Andreev, et. al., 1993; Bogush, et. al., 1992)

		1st Prototype[1]		2nd Prototype[1]	
		Flight[3]	Ground	Flight	Ground
Working section[2]	Design	5.7	5.7	5.6	5.75
power, kWe	90 days	5.57	5.6	5.6	5.75
	180 days	4.8	5.57	5.0	5.6
Thermal power, kWt	Design	150	150	144	147
	90 days	173	167	164	161
	180 days	177	178	176	173
NaK temperature at	Design	840	835	819	823
reactor outlet, K	90 days	868	855	848	840
	180 days	876	873	873	865
Change in reactivity,	90 days	–0.26	–0.17	–0.13	–0.04
percent	180 days	–0.46	–0.57	–0.50	–0.25
Output voltage	Design	14.8	14.0	14.5	14.0
(positive current	90 days	8.5	14.0	12.4	14.0
tap), volts	180 days	8.5	13.5	12.0	14.0
Cesium vapor	Design	597	597	585	585
generator temperature	53 days	–	609	–	–
setting, K	100 hours	–	–	575	–
	97 days	–	–	–	575

[1] The thermionic emitter in both prototypes is monocrystalline molybdenum, but the second prototype adds a monocrystalline tungsten coating.
[2] The working section is comprised of 62 multicell thermionic fuel elements. An additional 17 TFEs provide power to operate the NaK loop pump.
[3] Data listed as 180 days are actually for 142 days just prior to shut-down.

RTGs, the current safety programs, criteria, and processes are directed toward the issues related to containment of the $^{238}PuO_2$ fuel. Safety philosophy is to prevent release and minimize interaction of the fuel with the environment. One of the reasons for using the oxide form of the fuel is its stability and resistance to environmental interaction, e.g., it is non-pyrophoric and generally insoluble both in water and the human body. Allowable dose levels and health effects are based primarily on the work of the International Commission on Radiological Protection (ICRP, 1986).

The basic objective in design of the radioisotope heat source is containment of the fuel during normal operating modes and under all credible accident conditions during every phase of the mission. Although all risk situations are addressed quantitatively in the safety assessment, the primary issue related to $^{238}PuO_2$ is pulverization/atomization and airborne dispersion of any portion of the fuel at respirable particle sizes, i.e., generally smaller than about 10 microns aerodynamic diameter. Particles of this size have the potential of retention in the lungs if inhaled. Although the radiation levels are low in that event, an individual who is exposed for a long period of time has a greater potential for developing cancer.

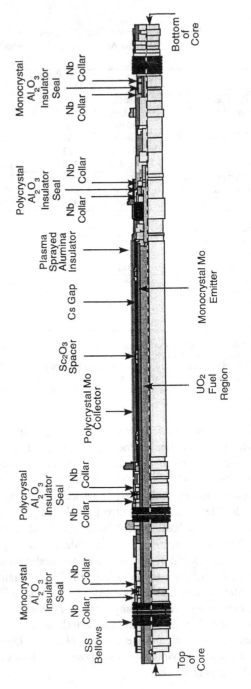

Figure 5.12 Enisy Single Cell Thermionic Fuel Element

NUCLEAR SYSTEMS

Typically, some combination of events can be postulated for which total containment of the radioisotope cannot be assured. Probabilistic risk assessment must account for the sequential likelihood of the following: 1) the incidence of a credible accident, 2) resulting amounts of fuel release, if any, 3) particle sizes contained in that release, 4) local and worldwide distribution mechanisms (including meteorological data), 5) all potential pathways to population exposure, 6) individual and collective radiation dose projections, and 7) the dose conversion factors for fatal cancer induction. The steps in the safety assessment are (NUS, 1990):
- Identification of potential accident scenarios and probabilities
- For each accident, determination of the physical environments to which the RTG will be subjected
- Evaluation of the RTG response to these environments
- Analysis of the frequency, severity, and characteristics of potential fuel releases
- Calculation of the consequences (individual and collective doses, latent cancer fatality projections, and land contamination) of any release
- Assessment of mission radiological risk (consequence and probability).

For each space mission utilizing nuclear power, a formal Safety Analysis Report is required that assesses the potential radiological risk to the world population. The space mission portion begins at the launch pad with pre-launch activities, like spacecraft integration, and extends through the entire operational phase of the system. Fuel and power system fabrication, handling, and transportation are covered by separate regulations of the DOE, Environmental Protection Agency (EPA), Nuclear Regulatory Commission (NRC), and Department of Transportation (DOT).

The Safety Analysis Report typically evolves through two or three versions. At the preliminary stage of mission planning, a Preliminary Safety Analysis Report (PSAR) is prepared that contains all relevant information that can be assembled at that early stage. For most missions, an Updated Safety Analysis Report (USAR) is also developed as detailed results become available. The key document is the Final Safety Analysis Report (FSAR) which contains the final results of the risk assessment. The FSAR must define and address all credible accidents through all phases of the mission. In addition, the probabilistic risk assessment assigns probabilities to each event and to the potential release and distribution of radioisotope fuel that may result. The accident events considered include the following (Bennett, 1981):

Pre-launch, launch and ascent phases:
- Explosion overpressure
- Projectile impact
- Land or water impact
- Liquid and solid propellant fires
- Sequential combinations of above events

Orbit and flight trajectory phases:
- Reentry
- Land or water impact
- Post-impact environment for either land or water

Although the safety assessment draws heavily on information (particularly test data) from previous work, the unique characteristics of each mission must be taken into account. For example, each version of each launch vehicle that might be used by the U.S. will have different overpressure, projectile, and propellant fire environments. Similarly, spacecraft design and RTG integration affect the launch pad abort impact conditions and post-launch reentry breakup analysis that defines the velocity, trajectory angle, and atmospheric conditions encountered by the heat source when it separates from the RTG. An issue of particular concern for the Galileo and Cassini missions is possible high-velocity reentry during an Earth flyby (used to add to the spacecraft velocity required to reach Jupiter and Saturn, respectively).

Independent safety assessment is a key element in the U.S. safety process. Although the DOE Office of Nuclear Energy typically has primary responsibility for both development of the RTG and the safety assessment for the mission, this safety assessment is subject to internal review by a separate DOE safety office. In addition, a formal independent review is conducted by the Interagency Nuclear Safety Review Panel (INSRP) by virtue of Presidential Directive NSC/25 (White House, 1996) and various agency guidelines. The panel was established specifically to assure the safe use of nuclear power in space, and its membership consists of coordinators assigned by the heads of DOE, DoD, NASA and EPA. In addition, a representative from the NRC serves as an official advisor. The White House must approve the launch of any spacecraft with reactors or radioactive sources containing more than specified amounts based on International Atomic Energy Agency transportation regulations for radioactive material (IAEA, 1990). In the case of ^{238}Pu, that amount is approximately 5 Curies.

Figure 5.13 depicts the sequence in the overall safety review process, including the INSRP role. INSRP involvement begins informally at a very early stage of mission planning, and the formal INSRP review begins when the PSAR is issued. In addition to review of the safety work done by the program, INSRP generates separate safety assessments in several areas as independent checks. INSRP subpanels perform detailed investigations in the areas of launch aborts, reentry, power systems, meteorology, and biomedical and environmental effects (Sholtis et al., 1994; Sholtis et al., 1991). The end product of the INSRP review is an independent Safety Evaluation Report (SER) for the mission wherein the coordinators' findings are reported to the project and to the Office of Science and Technology Policy (OSTP) at the White House (INSRP, 1990, for example). In conjunction with other executive entities, such as the National Security Council, the White House makes the final judgment on the benefits of the mission, justifying the risk involved.

NUCLEAR SYSTEMS

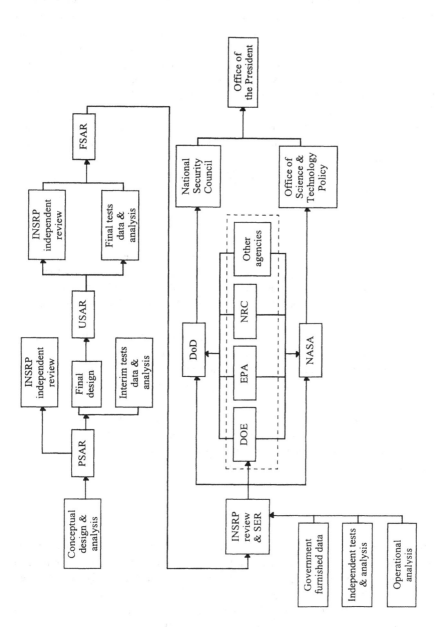

Figure 5.13 U.S. Flight Safety Review Process for Space Nuclear Power Sources

The safety approach and process for reactors are similar to RTGs, but the specific issues and requirements differ somewhat. The key safety objective for reactors is to maintain subcriticality during all pre-orbit periods and under all credible accident conditions. This objective differs from the key safety concern for terrestrial reactors for which preventing the release of irradiated fuel during and after operation is paramount. Since the U.S. has not launched a space reactor since 1965, the full safety process has not been implemented. However, specific criteria were re-established in the 1980s under the SP-100 program. These safety design criteria are:

- Remain subcritical (except for very short zero-power or low-power verification testing) through all pre-launch and launch phases until stable initial orbit is achieved
- Remain subcritical under all credible accident and post-accident conditions including:
  - On-pad explosions
  - Launch aborts
  - Reentry and Earth impact
  - Submersion and flooding in water, wet sand, or rocket propellant
- For Earth orbital missions, the operational (or final disposal) orbit shall be high enough so that the orbital lifetime is sufficient (typically about 400 years) to allow the core fission products to decay to the levels of the actinides prior to reentry
- Assure shutdown at end of operation through independent and redundant mechanisms, and post-shutdown removal of decay product heat
- In the event of inadvertent reentry, the fuel shall be either dispersed at high altitude or the core shall survive reentry and Earth impact essentially intact (to facilitate recovery)

In order to be compact and lightweight, highly-enriched uranium (HEU) is the preferred fuel for space reactors. Because of the potential use of HEU for nuclear weapons, appropriate safeguard measures must be implemented for space reactor missions. Hence, both safety and safeguards become part of the review process.

6.2 Russian space nuclear safety experience

Very little has been published on the safety process used by the Former Soviet Union. In recent years, Russia has reorganized various agencies involved in its space nuclear safety process, notably the Russian Space Agency, GosAtomNadzor RF, and the Federal Commission for Bio-Medical Issues and Emergencies. For RTGs and radioisotope heaters on the Mars 94/96 mission, Russia developed a temporary regulation while a permanent process was being established (Cook *et al.*, 1994).

NUCLEAR SYSTEMS

Much of the information released in the past has been in response to mishaps, notably the Cosmos 954 reentry over Canada in 1978. In recent years, however, several technical papers on the design of Russian space nuclear systems have been published, including information on safety criteria and experience. Since Russia routinely used reactors in the 1970s and 1980s to operate the low-orbit RORSAT ocean surveillance satellites, that experience can provide valuable insight into the adequacy of criteria and policies for future flights.

Table 5.5 includes the Russian accidents involving space nuclear systems, the latest one being the RTGs on Mars 96. However, the incidents involving reactor flights have been of greater concern since, in the event of a malfunction, the short orbit lifetimes of the RORSATs can lead to the reentry of a radiologically hot reactor. In addition to Cosmos 954, the Bouk reactor on Cosmos 1402 reentered in 1982, and the one on Cosmos 1900 nearly reentered in 1988.

In broad terms, the Russian safety policy for space reactors is similar to that outlined above for the U.S., but some design criteria differ. For example, in its launch configuration, the Enisy reactor goes critical in water (Marshall *et al.*, 1992). For the low-orbiting RORSATs, the safety approach was to separate the power system from the spacecraft at the end of mission, boost it to a high disposal orbit of > 700 km, and, for systems after Cosmos 954, separate the fuel elements from the reactor in that high orbit. If the core separation and reboost system failed, the intention was for the reactor to completely burn up in the atmosphere so that criticality after impact is precluded, and the maximum fallout would not exceed a dose of 0.5 mrem during the first year after the incident (Bennett, 1990). Unfortunately, several large pieces from the Cosmos 954 reactor and approximately 20 percent of the fuel were strewn over a large area in the vicinity of Canada's Great Slave Lake. Although it was subsequently determined that actual surface radiation levels and other impacts on the biosphere were relatively low (Bennett, 1990), the incident caused worldwide concern. Obviously, had the debris landed in a densely populated area, dealing with the problem would have been much more difficult.

Although Russia continued to fly the Bouk-powered RORSATs after Cosmos 954, some safety design changes were introduced to assure high altitude burnup and dispersion in the event of reentry. In particular, the fuel elements were extracted from the core after transfer to high orbit, or, if the transfer failed, at the low orbit or early stages of reentry. The Cosmos 1402 core apparently burned up completely during its reentry. The near reentry of Cosmos 1900 illustrated the successful operation of reactor reboost triggered by passive activation. The first line of defense on reactor reentry was ground control commands to shut down the reactor, separate it from the spacecraft, reboost it to its disposal orbit, and eject the fuel elements from the reactor vessel. Ground command was precluded when communication with the spacecraft was lost in April 1988. However, the separation and reboost of the reactor were activated passively by deviation from the stabilization angle of the spacecraft (Gryaznov *et al.*, 1992).

6.3 International developments in space nuclear safety

The Cosmos 954 reentry prompted much international concern over the use of space nuclear power. The primary forum for these discussions was the United Nations Committee on the Peaceful Uses of Outer Space (COPUOS). In November 1978, the COPUOS Scientific and Technical Subcommittee (STSC) established a Working Group on the Use of Nuclear Power Sources in Outer Space (WGNPS) to consider the relevant technical and safety issues (Bennett, 1987). The COPUOS Legal Subcommittee (LSC) participated, also.

In 1981, the WGNPS issued a consensus report that described general safety procedures and reaffirmed that nuclear power systems can be used safely (U.N., 1981). The term "consensus" means that all members were in agreement. However, that consensus was subsequently broken, and several years of negotiations within the WGNPS and STSC ensued (Bennett, 1995). The eventual result of the discussions was a set of nonbinding principles adopted by the U.N. General Assembly in 1993 (U.N., 1993). In all, eleven principles were adopted, but the key one is "Principle 3. Guidelines and criteria for safe use." Although this principle reconfirms the general safety criteria employed by the U.S., it contains several controversial details. For example, it establishes a principal dose limit of 1 mSv/year to a limited geographical region and to individuals, but allows a subsidiary limit of 5 mSv/year for some years. Consequently, the U.S. is continuing to use its current safety criteria and assessment process (Bennett, 1995). The WGNPS noted several issues related to the principles prior to their adoption by the General Assembly, and the Preamble called for reopening the WGNPS discussions in 1995.

7. References

Allied–Signal Aerospace Company brochure, *Dynamic Space Power Systems*, (Allied–Signal, Phoenix, AZ, 1988).

Andreev, P. V. et al., *The Effect of Cosmic Space on the Basic Characteristics of the "TOPAZ I" NPS and the EJES During their Joint Flight Tests*, Ref: JV–15/183, (Russian Science Center "Kurchatov Institute," Moscow, 1993).

Angelo, J. A. Jr. and Buden, D., *Space Nuclear Power*, (Orbit Book Company, Inc., Malabar, FL, 1985).

Baksht, F. G., et al., *Thermionic Converters and Low-Temperature Plasma*, Academy of Sciences of the USSR A.F. Ioffe Physico-Technical Institute, (U.S. Department of Energy DOE–TR–1, 1978).

Bennett, G. L., *Overview of the U.S. Flight Safety Process for Space Nuclear Power*, Nuclear Safety Technical Progress Review, Vol. 22–4, July–August 1981.

Bennett, G. L., *Flight Safety Review Process for Space Nuclear Power Sources*, Proceedings of the 22nd Intersociety Energy Conversion Engineering Conference (IECEC 87), Paper #879046, (American Institute of Aeronautics and Astronautics, New York, 1987), 383–391.

Bennett, G. L. and Lombardo, J. J., *Technology Development of Dynamic Isotope Power Systems for Space Applications*, Proceedings of the 22nd Intersociety Energy Conversion Engineering Conference (IECEC 87), Paper #879094, (American Institute of Aeronautics and Astronautics, New York, 1987), 366–372.

Bennett, G. L., *A Look at the Soviet Space Nuclear Power Program*, Proceedings of the 24th Intersociety Energy Conversion Engineering Conference (IECEC 89), Paper #899009, (Institute of Electrical and Electronics Engineers, 1989), 1187–1194.

Bennett, G. L., *Historical Overview of the U.S. Use of Space Nuclear Power*, Space Power, Vol. 8, No. 3, 1989, 259–284.

Bennett, G L., *Soviet Space Nuclear Power Reactor Incidents: Perception Versus Reality*, Proceedings of the Seventh Symposium on Space Nuclear Power Systems, El-Genk, M. S. and Hoover, M. D., (eds.), CONF–900109, (University of New Mexico, Albuquerque, 1990), 168–174.

Bennett, G. L., *A Technical Review of the U.N. Principles on the Use of Nuclear Power Sources in Outer Space*, Proceedings of the 46th International Astronautical Congress, October 2–6, 1995, Oslo, Norway, IAF–95–R.1.04, (International Astronautical Federation, 1995).

Bennett, G. L., Hemler, R. J., and Schock, A., *Space Nuclear Power: An Overview*, AIAA Journal of Propulsion and Power, Vol. 12, No. 5, September–October, 1996.

Bhattacharyya, S. K., *An Assessment of the Soviet TOPAZ Reactors*, (Argonne National Laboratory, February 15, 1991).

Bogush, I. P., et al., *The Main Principles of Design, Ground and Flight Tests of SNPS within "TOPAZ" Program*, Transactions of the Anniversary Specialists Conference on Nuclear Power Engineering in Space, Obninsk, May 15–19, (Ministry of Atomic Power Engineering and Industry of the USSR, 1990).

Bogush, I. P., et al., *The Main Purposes and Results of Flight Tests of SNPSs within the "TOPAZ" Program*, Space Nuclear Power Systems 1989, El-Genk, M. S. and Hoover, M. D., (eds.), (Orbit Book Company, Inc., Malabar, FL, 1992), 457–460.

Booz·Allen & Hamilton, Inc., *The TOPAZ International Program: Lessons Learned in Technology Cooperation with Russia*, 1995.

Buden, D., *Summary of Space Nuclear Reactor Power Systems (1983–1992), A Critical Review of Space Nuclear Power and Propulsion 1984–1993*, El-Genk, M. S., (ed.), (American Institute of Physics Press, New York, 1994), 21–86.

Buksa et al., *SP–100 Technical Summary Report*, JPL D–11818, Mondt, J. F., (ed.), (Jet Propulsion Laboratory, September, 1994).

Cook, B. A., Lange, R. G., and Pustovalov, A. A., *A Comparison of the Launch Approval Processes Used in the U.S. and Russia for Nuclear Power Space Exploration Missions*, Proceedings of the Eleventh Symposium on Space Nuclear Power and Propulsion, AIP Conference Proceedings 301, (American Institute of Physics Press, New York, 1994), 875–878.

Corliss, W. R., *SNAP Nuclear Space Reactors, AEC booklet*, (U.S. Atomic Energy Commission, September, 1966).

Dix, G. P. and Voss, S. S., *The Pied Piper — A Historical Overview of the U.S. Space Power Reactor Program*, Space Nuclear Power Systems 1984, El-Genk, M. S. and Hoover, M. D., (eds.), (Orbit Book Company, Inc., Malabar, FL, 1985), 23–30.

Gryaznov, G. M. and Pupko, V. Y., *TOPAZ-1*, Nature (Russian magazine), No. 10, October, 1991.

Gryaznov, G. M., Nikolaev, V. S., Serbin, V. I., and Tyugin, V. M., *Radiation Safety of the Space Nuclear Power Systems and Its Realization on the Satellite Cosmos–1900*, Space Nuclear Power Systems 1989, El-Genk, M. S. and Hoover, M. D., (eds.), (Orbit Book Company, Inc., Malabar, FL, 1992), 453–456.

IAEA, *Regulations for the Safe Transport of Radioactive Material*, 1985 Edition (as amended 1990), Safety Series No. 6, IAEA Safety Standards, (International Atomic Energy Agency, Vienna, 1990).

ICRP Publication 48, *The Metabolism of Plutonium and Related Elements*, (International Commission on Radiological Protection, 1986).

INSRP, *Safety Evaluation Report for Ulysses, Report INSRP 90–01*, Vols I, II, and III, July 1990 (predecisional), (Interagency Nuclear Safety Review Panel September, 1990).

IPPE Bulletin, *Safety, Ecology, Radiation*, Issue 2, (Institute of Physics and Power Engineering, Obninsk, Russia, 1992).

Kelly, C. E. and Klee, P, M., *Results of High Performance Conductively Coupled Thermoelectric Cell Life Tests*, Proceedings of the Space Technology and Applications International Forum, Thirteenth Symposium on Space Nuclear Power and Propulsion, AIP Conference Proceedings 361, (American Institute of Physics Press, New York, 1996), 1227–1232.

Kurcharkin, N. E., Ponomarev-Stepnoi, N. N., and Usov, V. A., *Reactor–converter 'Romashka' and the Perspectives of its Development*, Proceedings of the Seventh Symposium on Space Nuclear Power Systems, El-Genk, M. S. and Hoover, M. D., (eds.), CONF–900109, (University of New Mexico, Albuquerque, 1990).

Lange, R. G. and Mastal, E. F., *A Tutorial Review of Radioisotope Power Systems*, A Critical Review of Space Nuclear Power and Propulsion 1984–1993, El-Genk, M. S., (ed.), (American Institute of Physics Press, New York, 1994), 1–20.

Luchau, D. M., Sinkevich, V. G., Wernsman, B., and Mulder, D. M., *Final Report on Testing of TOPAZ II Unit Ya–21u: Output Power Characteristics and System Capabilities*, Proceedings of the Space Technology and Applications International

NUCLEAR SYSTEMS

Forum, Thirteenth Symposium on Space Nuclear Power and Propulsion, AIP Conference Proceedings 361, (American Institute of Physics Press, New York, 1996), 1389–1394.

Makhorin, O. I. et al., *Some Questions Concerning Safety on Emergency Landing in Dense Layers of the Atmosphere of Radionuclide Energy Sources Based on Plutonium-238 for Autonomous Station "Mars 94/96"*, Proceedings of the Space Technology and Applications International Forum, Thirteenth Symposium on Space Nuclear Power and Propulsion, AIP Conference Proceedings 361, (American Institute of Physics Press, New York, 1996), 975–979.

Marshall, A. C., Voss, S. S., Haskin, E., and Standley, V., *NEP Space Test Program Preliminary Nuclear Safety Assessment*, Second Draft, U.S. Topaz II Flight Safety Team, October, 1992.

Matthews, R. B., et al., *Fuels for Space Nuclear Power and Propulsion: 1983–1993, A Critical Review of Space Nuclear Power and Propulsion 1984–1993*, El-Genk, M. S., (ed.), (American Institute of Physics Press, New York, 1994), 179–220.

NUS Corporation, *Final Safety Analysis Report for the Ulysses Mission, Executive Summary*, ULS–FSAR–001, (NUS Corporation, March 1990).

Purdum, T. S., *Russian Mars Craft Falls Short and Crashes Back to Earth*, (New York Times, November 18, 1996).

Rasor, N., personal communication, 1997.

Schmidt, G. L., *SNAP 10A Test Program*, (Rockwell International, Canoga Park, CA, September, 1988).

Sholtis, J. A., Jr., et al., Technical Note: *The Interagency Nuclear Safety Review Panel's Evaluation of the Ulysses Space Mission*, Nuclear Safety Technical Progress Journal, Vol. 32–4, October–December, 1991.

Sholtis, J. A., Jr., et al., *U. S. Space Nuclear Safety: Past, Present, and Future, A Critical Review of Space Nuclear Power and Propulsion 1984–1993*, El-Genk, M. S., (ed.), (American Institute of Physics Press, New York, 1994), 269–304.

Thome, F. V., Wyant, F. J., McCarson, T. D., Mulder, D., and Ponomarev-Stepnoi, N. N., *A TOPAZ International Program Overview Design*, Proceedings of the Twelfth Symposium on Space Nuclear Power and Propulsion, AIP Conference Proceedings 324, (American Institute of Physics Press, New York, 1995), 877–882.

Truscello, V. C. and Rutger, L. L., *The SP–100 Power System*, Proceedings of the Ninth Symposium on Space Nuclear Power and Propulsion, AIP Conference Proceedings 246, (American Institute of Physics Press, New York, 1992), 1–23.

U.N., *Report of the Working Group on the Use of Nuclear Power Sources in Outer Space on the Work of its Third Session*, Annex II of Report of the Scientific and Technical Subcommittee on the Work of its Eighteenth Session, U.N. Document A/AC.105/287, (United Nations, February 13, 1981).

U.N., *Principles Relevant to the Use of Nuclear Power Sources in Outer Space*, General Assembly Resolution A/RES/47/68, (United Nations, February 23, 1993).

U.S. DOE, *Atomic Power in Space, A History*, DOE/NE/32117–H1, (U.S. Department of Energy, March, 1987).

Voss, S. S., *SNAP Reactor Overview*, AFWL–TN–84–14, (Air Force Weapons Laboratory, Kirtland AFB, NM, 1984).

Voss, S. S., *TOPAZ II Design Evolution*, Proceedings of the Eleventh Symposium on Space Nuclear Power and Propulsion, AIP Conference Proceedings 301, (American Institute of Physics Press, New York, 1994), 791–802.

Voss, S. S. and Rodriguez, E. A., *Russian TOPAZ II System Test Program (1970–1989)*, Proceedings of the Eleventh Symposium on Space Nuclear Power and Propulsion, AIP Conference Proceedings 301, (American Institute of Physics Press, New York, 1994), 803–812.

Ward, William, MIT Lincoln Laboratory, personal communication, 1994.

White House, *Scientific or Technological Experiments with Possible Large-Scale Adverse Environmental Effects and Launch of Nuclear Systems into Space*, Presidential Directive/National Security Memorandum 25, PD/NSC–25, issued December 14, 1977; revised May 17, 1995; revised May 8, 1996.

287

CHAPTER 6

STATIC ENERGY CONVERSION

1. Introduction

This chapter discusses the leading static energy conversion options for spacecraft power other than photovoltaics. Because photovoltaics are the mainstay of space power, Chapter 3 of this text has been devoted entirely to that topic. Four other static converter technologies are of primary interest: thermoelectric, thermionic, alkali metal thermal-to-electric conversion (AMTEC), and thermophotovoltaic (TPV, also briefly discussed in Chapter 3). For each of these converter options, this chapter explains the basic operating principles, provides a few pertinent equations, and describes the major system design features and trade-offs.

As discussed earlier in Chapter 5, thermoelectric energy conversion is the principal technology for radioisotope power sources commonly used for outer planet exploration. In addition, the Russians used thermoelectrics for the 33 Bouk nuclear reactors that powered the RORSAT satellites, as did the only reactor launched by the U.S., SNAP-10A, in 1965. The last major U.S. space nuclear reactor development program for power generation, SP-100, baselined thermoelectrics of a different configuration. Radioisotope Thermoelectric Generators (RTGs) have played a role since the beginning of the U.S. space program, first as auxiliary power for Earth satellites, then as power for experiments deployed on the lunar surface, and since the early 1970s, as power for unmanned planetary exploration. RTGs have proven to be highly reliable and have far exceeded their design lifetimes in all missions. Systems launched in the 1970s are still operating.

To varying degrees, the U.S. has investigated space power systems that use thermionic converters based on either solar, radioisotope decay, or nuclear reactors as heat sources. The bulk of the technical development occurred in the 1960s with renewed efforts in the 1980s and early 1990s. In contrast, Russia has maintained major programs in thermionic development since the 1960s; the two major system developments are the TOPAZ and Enisy space nuclear reactors using in-core thermionic fuel elements. Russia conducted two flight tests of the TOPAZ system in 1987.

No power system has flown in space using either AMTEC or TPV. Although not as well developed as thermoelectrics and thermionics, the main advantage of these two technologies is relatively high efficiency. Short-term laboratory tests on both AMTEC and TPV have shown conversion efficiencies in excess of 20%. Thermoelectric and thermionic converter efficiencies are typically less than 10%, and system efficiencies are 5% or less. U.S. interest in these technologies was renewed in the 1990s as options for the next generation of radioisotope systems for solar system exploration. In 1997, NASA and the U.S. Department of Energy (DOE) tentatively selected AMTEC as the leading candidate, and DOE initiated a system development program. However, they are continuing to work on other candidates as well.

2. Thermoelectrics

The fundamental physical process involved in thermoelectrics is the Seebeck effect, named for Thomas Seebeck who first observed the phenomenon in 1821, although he misinterpreted the effect (Angrist, 1965). The Seebeck effect is the generation of an electromotive force within two dissimilar metals when their junctions are maintained at different temperatures. A common application of this principle is the use of thermocouples to measure temperature. Two wires of different metals are joined at two points: a hot junction typically at the temperature to be measured, and a cold junction, often an ice bath, against which the thermocouple has been calibrated. The electromotive force generated by the thermocouple is counteracted by an applied voltage so that no current flows. This applied voltage is converted to a temperature measurement by comparing it with a voltage–temperature correlation for the particular type of thermocouple. The two primary differences between thermocouple temperature measurement and thermoelectric power generation are that semiconductor materials are used instead of metals, and current flows in the generator in order to produce power. Semiconductor materials have significantly higher Seebeck coefficients than metals and are thus more suited to power generation.

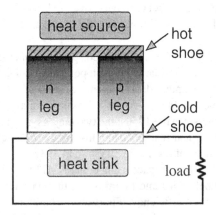

Figure 6.1 Principles of Thermoelectric Cell Operation

STATIC ENERGY CONVERSION

Figure 6.1 shows a schematic representation of a thermoelectric cell. It consists of individual legs of an n-type and a p-type semiconductor material. The hot junction between the n and p legs is formed by bonding of a hot shoe. Cold shoes are bonded to each leg, and an external load completes the circuit. In an actual converter consisting of many cells connected in series, the cold junction is formed by a current-conducting strap connecting cold shoes of adjacent cells.

The Seebeck effect in the n–type material creates a flow of excess electrons from the hot junction to the cold junction. In the p–type material, holes (missing electrons) migrate toward the cold shoe creating a net current flow from the cold junction to the hot junction. Therefore, the electromotive force in both legs is such that current flow is augmented in the same direction through the circuit. The Seebeck coefficient, α, is defined as the change in voltage per degree of temperature gradient:

$$\alpha = \frac{dV}{dT} \quad \text{volts/K (or often } \mu\text{volts/K)} \tag{6.1}$$

To generate sufficient voltage for practical use, many thermoelectric cells are connected in series, and the cell voltages are additive. Alpha typically varies significantly with temperature, so the total open-circuit voltage generated by a thermoelectric converter is given by:

$$V_{OC} = N_{Series}\left[\int_{T_c}^{T_H}|\alpha_n|dT + \int_{T_c}^{T_H}|\alpha_p|dT\right] \tag{6.2}$$

where V_{OC} is the open circuit voltage of the converter in volts, N_{Series} is the number of cells connected in series, T_C is the cold junction temperature in K, T_H is the hot junction temperature in K, $|a_n|$ is the absolute value of the Seebeck coefficient for the n leg in volts/K, and $|a_p|$ is the absolute value of the Seebeck coefficient for the p leg in volts/K.

Absolute values are indicated because Seebeck coefficients for n–type materials are often expressed as negative values, depending on convention. Eq. (6.2) assumes that the hot and cold junction temperatures are the same for all cells. Temperature variations exist in an actual converter so a more accurate analysis would be a summation of the integrals for each cell.

When the load is connected to the circuit, current will flow as a result of the voltage generated by the thermoelectric cells. The total current flow depends on two factors: the total resistance in the circuit and the number of parallel strings of series-connected cells in the converter. Parallel connections are made to improve reliability. If a cell breaks, a bond fails, or some other connection is broken in one string, power will still be

generated by the other strings. The total electrical resistance (in ohms) for the circuit is the sum of the following:
- R_{Load} = resistance of the load powered by the converter
- R_n = resistance of the n thermoelectric legs
- R_p = resistance of the p thermoelectric legs
- R_{Cnt} = contact resistance at the hot and cold shoe bonds and connections
- R_{Leads} = resistance of the leads between the thermoelectric converter and the load.

The contact resistance is usually determined empirically and a typical value is about 10–20% of the leg resistances. The lead resistance is also on the order of 10% or less, depending on the total current output of the generator. A thermoelectric converter is designed so that each parallel string will as nearly as possible produce the same voltage. Therefore, the current in each string will be nearly equal, and the total output current is the sum of these or the number of strings times the current in each string.

Unlike a thermocouple used to measure temperature, the current flow through a thermoelectric converter introduces additional physical phenomena that must be included in the analysis. The one having the most influence on the converter is the Peltier effect. Jean Peltier observed this phenomenon in 1834, but like Seebeck he failed to interpret it correctly (Angrist, 1965). Simply stated, the Peltier effect is the heating or cooling that occurs at the junction of two dissimilar conductors due to a current flowing through it. This thermal energy is in addition to the effects of Joule heating in the conductors.

In a thermoelectric cell, Peltier heat is removed from the hot junction and released at the cold junction. The Peltier coefficient is defined as the Peltier heat generated at the junction divided by the current flowing through it. The effective value of the coefficient is the difference between the values for the p and n semiconductor materials. The Peltier heat can be calculated by:

$$Q_\pi = \pi_{np} I = \left(\pi_p - \pi_n\right) I \tag{6.3}$$

where Q_π is the Peltier heat in watts, I is the current through the junction in amps (equal to the current in each string), π_{np} is the Peltier coefficient at the junction in volts, π_p is the Peltier coefficient of the p–type material in volts, and π_n is the Peltier coefficient of the n–type material in volts.

Sir William Thomson (Lord Kelvin) derived a theoretical relationship between the Peltier and Seebeck coefficients known as Kelvin's second relation:

$$\pi = \alpha T \tag{6.4}$$

The second thermodynamic effect resulting from current flow is known as the Thomson effect. In 1855, Thomson deduced that heating or cooling is induced by cur-

STATIC ENERGY CONVERSION

rent flowing along a temperature gradient (Angrist, 1965). As with the Peltier heat, the Thomson thermal energy is in addition to the Joule heating in a conductor. This thermal energy is calculated by:

$$Q_\tau = \int_{T_C}^{T_H} \tau I dT \qquad (6.5)$$

where Q_t is the Thomson heat in watts, τ is Thomson coefficient in volts/K, and I is the current through the thermoelectric leg in amps (again, equal to the current in each string).

The Thomson coefficient can be positive or negative depending on whether the effect heats or cools the leg. The magnitude of Thomson heating is relatively small, and its effect is often ignored in the analysis (Kelly, 1997).

The energy balance at the hot junction is of most interest because, along with heat losses that bypass the thermoelectrics, it defines the heat input required to operate the converter at its design conditions. Heat is removed from the hot junction by thermal conduction through both legs of the cell and by Peltier cooling. In addition, the Joule heating in each leg is dissipated at the junctions, and the usual assumption is that half goes to each. A small amount of Joule heating in the hot shoe can be neglected. The heat that must be input to each thermoelectric cell is given by:

$$Q_{Cell} = Q_{Cond} + Q_\pi - \tfrac{1}{2} Q_J \qquad (6.6)$$

where

$$Q_{Cond} = -\frac{A_p}{l_p} \int_{T_C}^{T_H} \lambda_p dT - \frac{A_n}{l_n} \int_{T_C}^{T_H} \lambda_n dT \qquad (6.7)$$

where λ_p, λ_n are the thermal conductivities in watts/cm–K for the p and n legs, and Q_π is Peltier heat in watts from Eq. 6.3.

$$Q_J = \text{Joule heat in watts} = I^2 (R_p + R_n) \qquad (6.8)$$

where I is the current through the thermoelectric leg in amps, and R_p and R_n are the resistances in ohms of the n and p legs.

The total heat required is the sum of the heat into all of the cells plus any bypass losses. In an RTG, the bypass losses are the heat flow through the thermal insulation between the cells and at the ends of the converter, and the conduction loss through the heat source supports at each end.

The preferred material properties are high Seebeck, low electrical resistivity, and low thermal conductivity. High Seebeck increases the voltage per cell. Low electrical resistivity reduces internal resistance losses, thus increasing output current. Low thermal conductivity reduces the heat losses through the thermoelectric legs. The standard measure of thermoelectric performance combines these three properties and is known as the Figure of Merit, Z, of the material. (The dimensionless product ZT is sometimes used for Figure of Merit as well.) It has units of K^{-1} and is defined as:

$$Z = \alpha^2 / \rho\lambda \tag{6.9}$$

Figure 6.2 shows the Figure of Merit for several thermoelectric materials. Although

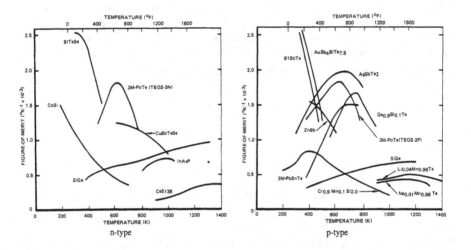

Figure 6.2 Figure of Merit for Thermoelectric Materials (after Anderson *et al.*, 1983)

the telluride compounds have much higher values of Z than SiGe, it is important to note that the temperature range of operation is significantly lower. Hot junction temperature capabilities of tellurides in increasing order are bismuth telluride (BiTe), lead telluride (PbTe), TAGS (tellurium, antimony, germanium, and silver in a solid solution of silver antimony telluride in germanium telluride), and lead tin telluride (PbSnTe). All have been used in space (except BiTe which has been used for terrestrial applications), and all SNAP RTGs used various telluride compounds for the thermoelectrics. The main

STATIC ENERGY CONVERSION

advantage that led to the introduction of silicon germanium (SiGe) is its ability to operate at higher hot side temperatures in a vacuum. Although the telluride compounds have a higher Figure of Merit than SiGe, they are limited in hot junction temperature, the maximum being around 800 K. Since both the n and p legs operate at essentially the same hot junction temperature, the design limit is set by the leg with the lowest capability. The SiGe cells used in today's RTGs typically operate at 1273 K at the hot junction and 573 K at the cold junction. Because cold side temperature has a strong influence on radiator size, optimum RTG designs with tellurides operate with a hot-to-cold temperature gradient of less than 500 K. The 100–200 K larger gradient possible with SiGe results in higher system specific powers. The upper limit is due to the sublimation rate of SiGe in a vacuum. Another significant disadvantage of the tellurides is the need for an inert cover gas to both prevent chemical reactions with oxygen and suppress sublimation. Since SiGe can operate in a vacuum, long-term sealing to retain the cover gas is not an issue. Also, heat losses are reduced because vacuum multifoil insulation can be used which has a lower effective thermal conductivity than the fibrous-type insulation (usually a form of Min–K developed by Johns Manville) used in the SNAP RTGs.

Figure 6.3 depicts the details of an individual SiGe unicouple developed for the Multi-Hundred Watt RTG (MHW–RTG) and used in today's General Purpose Heat Source RTG (GPHS–RTG). The n and p legs are the same size, 2.74 mm x 6.50 mm x 31.1 mm long. Both are segmented for higher performance and improved bonding, particularly at the cold shoe. Most of each leg is 78 atomic percent Si, with 63.5 atomic percent Si at the cold end. Boron is used as a dopant to achieve p–type semiconductor properties, and phosphorous dopant is used to create the n leg. Each leg is bonded to a SiMo hot shoe that creates the hot junction and collects the heat radiated from the GPHS. Hot

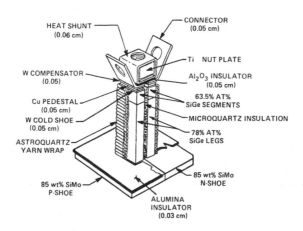

Figure 6.3 SiGe Unicouple used in U.S. RTGs (Courtesy of U.S. Department of Energy)

shoe dimensions are 22.9 mm square x 1.9 mm thick. A thin alumina insulator prevents the hot shoe from shorting to the multifoil insulation. To prevent excessive sublimation, the hot shoes and the 78 atomic percent segments of the legs are coated with silicon nitride. Microquartz thermal insulation is inserted between the legs, and astroquartz yarn is wrapped around the outside of the legs to reduce heat loss and prevent leakage currents to the molybdenum foils.

On the cold side, a stack of materials allows for bonding, compensates for thermal expansion differences, and makes the electric circuit connections. A tungsten cold shoe bonded to each SiGe leg is in turn bonded to a copper pedestal. The pedestal is attached to the copper intercell by connecting straps that are initially folded inward so the unicouple can be inserted into the multifoil stack from the hot side. Once inserted, the straps are bent out to allow a riveted connection between adjacent unicouples. To prevent electrical shorting, an alumina insulator is bonded under the straps and is thin enough so as not to add a large temperature drop between the SiGe legs and the radiator. A copper shunt is used to conduct this heat around the titanium nut that accepts the mounting screw inserted though the outer shell.

The key properties of SiGe change with time of operation due to precipitation of the phosphorus and boron dopants at the grain boundaries in the SiGe material. Although the Seebeck coefficient increases and the thermal conductivity decreases, the increase in electrical resistivity is enough to cause a net reduction in the Figure of Merit. Hence, converter performance degrades with time. Figure 6.4 shows the data for the flight RTGs on the Galileo spacecraft (Kelly and Klee, 1997). The power loss includes the loss of input heat due to radioactive decay of the $^{238}PuO_2$ radioisotope in the heat source (87.8 year half life). These loss mechanisms have been well defined and modeled, and as can be seen, RTG power exceeded predictions and requirements. During the SP–100 program, NASA's Jet Propulsion Laboratory achieved modest improvements in SiGe performance, but the manufacturing processes for these advanced materials have not been fully developed (Vinning and Fleurial, 1994).

3. Thermionics

As discussed in Chapter 5, thermionic conversion has been a technology of major interest for space nuclear reactor systems. Russia in particular has done a great deal of development, including two flight tests of the TOPAZ system, known as TOPAZ I in the U.S. The U.S. undertook considerable work in the 1960s and again in the 1980s and early 1990s, but only a few small efforts continue. Thermionic converters were developed and tested for nuclear reactor, radioisotope, and solar systems; however, the U.S. has conducted no flight tests. Figure 6.5 depicts a recent U.S. thermionic converter which can be used with either a solar concentrator or an out-of-core nuclear reactor. Similar converters were built and tested in the 1960s. In addition to space systems,

STATIC ENERGY CONVERSION

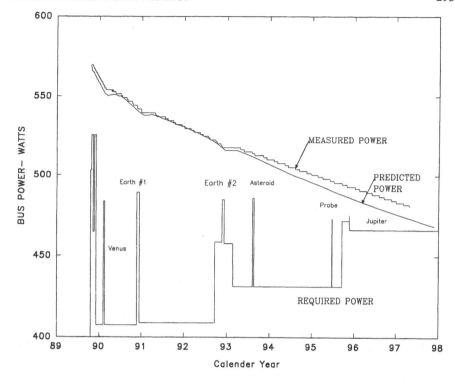

Figure 6.4 Flight Data for RTGs on the Galileo Spacecraft (Courtesy of Lockheed Martin Corporation)

thermionic technology has been a candidate for Naval propulsion and as topping systems for commercial power plants.

This R&D work covered many types of thermionic systems designed for several operating regimes. This section will identify those regimes and briefly review the ideal thermionic converter. However, virtually all of the practical designs for power production fall into the broad category of vapor thermionic converters, i.e., those that incorporate ionized gas between the electrodes. Cesium is by far the gas most commonly used for this purpose. For these reasons, the discussion will focus on cesium vapor thermionic processes in operating regimes commonly used in space power system design. More specifically, emphasis is on high-pressure diodes meaning that conditions in the interelectrode gap are dominated by collisional processes among the electrons, ions, and atoms. A diode is treated as high pressure when the product of the cesium pressure and the electron mean free path exceeds 2.5 mil–torr (Rasor, 1997). The reader is cautioned that the following discussion is a top-level, primarily qualitative, summary of thermionic diode processes and characteristics within a somewhat limited set of operating parameters and conditions. Complex local thermodynamic processes occur in the

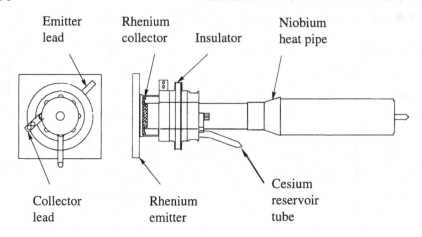

Figure 6.5 Planar 30–Watt Thermionic Converter (Courtesy of U.S. Air Force)

plasma space and electrode sheaths that are beyond the scope of this discussion. For further details and other conditions, consult the reference list, particularly Rasor, 1991; McVey and Rasor, 1992; Baksht, *et al.*, 1978; and Hatsopoulos and Gyftopoulis, 1973.

Figure 6.6 depicts the basic operating principles of a cesium vapor thermionic diode. In theory, the process is fairly simple, but reducing it to practical devices has proven challenging. The process is simply to heat an emitter material (cathode) to a

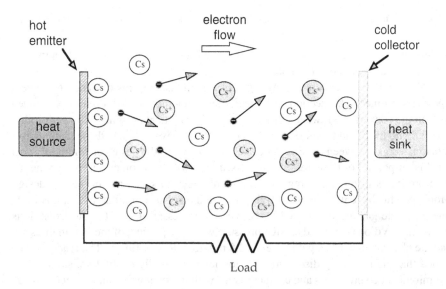

Figure 6.6 Thermionic Operating Principles

STATIC ENERGY CONVERSION

high temperature to drive electrons off the surface. These electrons then traverse a small gap between the emitter and a cooler collector (anode). An external load closes the circuit, and the electromotive potential created by the temperature difference between the electrodes drives the current. For space power systems, emitter temperatures in the range of 1700–2100 K and collector temperatures in the range of 700–1000 K are typical. Interelectrode gap widths usually fall in the range of 0.1 to 0.5 mm.

In reality, many processes occur simultaneously in the plasma space and near the electrodes. To quote Baksht, *et al.*, 1978, the plasma processes include "elastic scattering, excitation and de-excitation of the energy levels in the atoms, ionization and recombination, excitation of vibrational and rotational energy levels in molecules, charge exchange, formation of chemical compounds and radicals, etc."

Viewed more simply, four fundamental processes occur simultaneously and interact to establish a thermodynamic balance for equilibrium operation in a cesium vapor diode. The processes are:
- Electron emission from the hot emitter
- Creation and maintenance of positive cesium ions in the interelectrode gap
- Particle kinetics in the interelectrode space
- Electron absorption at the collector surface

Emission of electrons from the emitter is the key consideration in the design of a thermionic converter. Electrons near the surface of the material can escape from the surface if their energy level is raised to the point that it will overcome the inter-atomic forces holding it in place. This barrier is known as the work function, ϕ, of the material and is expressed in electron–volts (eV). The emission process can be viewed as boiling or evaporation of electrons, and the work function is analogous to the heat of vaporization. Thermionic performance is very sensitive to ϕ, and determining the effects of other system parameters on emitter and collector work functions is fundamental to the design and comprehension of thermionic devices.

In the absence of plasma effects, electric fields, and other processes (of which there are many in a thermionic converter), the maximum electron current leaving the emitter is known as the saturation current. Ideally, it is approximated by the Richardson–Dushman equation:

$$J_{ES} = AT_E^2 \exp(-\phi_E / kT_E) \tag{6.10}$$

where J_{ES} is the emitter saturation current density in amps/cm², A is a constant equal to 120 amps/cm²–K², T_E is the emitter temperature in K, ϕ_E is the emitter work function in eV, k is Boltzmann's constant = 8.62 x 10^{-5} eV/K. The value of the coefficient A is actually dependent on the material. However, for the refractory metals typically used for thermionic electrodes, Eq. (6.10) is generally accepted. Furthermore, the experimental determination of ϕ_E often assumes $A = 120$.

In an ideal thermionic converter, the net current flow in the diode can be defined as the current flow from the emitter to collector, minus the current flow from the collector to the emitter, minus the ion flow from the emitter. Typically, the ion flow is negligible, and the current flow from collector to emitter is relatively small.

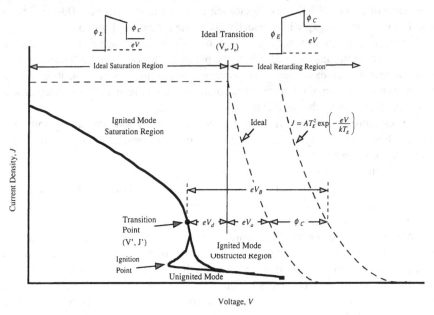

Figure 6.7 Current–Voltage for Ideal and Cesiated Thermionic Diodes

Figure 6.7 qualitatively compares the current–voltage relationship for an ideal diode to that of a cesium one. Considering the ideal case first (dashed line), operation occurs in one of two distinct regimes: saturation and retarding. These regimes are also depicted at the top of the figure as motive diagrams that show the relative energy levels at the emitter and collector. In the saturation region, the barrier to current flow is the energy needed to escape the emitter (i.e., ϕ_E). In other words, the energy imparted to the escaping electrons is more than sufficient to overcome the barrier to entering the collector (the sum of collector work function and output voltage). In the saturation region, the net current is given by the Richardson–Dushman equation, given here as Eq. (6.10) (less the small back emission from the collector). As the voltage increases further, the sum of the output voltage and the collector work function exceeds the emitter work function. This regime is known as the retarding region, and the current and power densities decrease as the voltage increases. The net current density in the retarding region is also based on the Richardson–Dushman equation, but ϕ_E in the exponent is

STATIC ENERGY CONVERSION   299

replaced by $(\phi_C + eV)$ where eV is the diode voltage in electron–volts. In this ideal approximation, output is maximum at the transition point (V_o, J_o) where $eV_o = \phi_E - \phi_C$.

The ideal case, however, does not account for several other processes that occur in a thermionic diode. When the negatively–charged electrons leave the surface of the emitter and move into the interelectrode gap, the buildup of electrons in the gap creates a negative space charge. This negative charge tends to repel additional electrons leaving the emitter and prevents them from reaching the collector. A cesiated diode enhances movement of electrons to the collector by placing positive cesium ions in the gap, producing a plasma and neutralizing the space charge.

Equally important, the cesium affects the work functions of the electrode surfaces, particularly the emitter. The emitter surface adsorbs some of the cesium atoms forming a partial monolayer. This effect reduces the emitter work function below its bare or vacuum value and increases the saturation current per Eq. (6.10). The new work function is called the cesiated work function, and to lower it is a primary design goal. The bare work function of the electrode is affected by crystal grain structure and surface finish, so it can be non-uniform for some metals and geometries. Interestingly, the ability to adsorb cesium ions increases as bare work function increases in refractory metals. The extent of this increase is such that a minimum cesiated work function is usually achieved with materials having high bare work functions. Hence, the preferred grain orientation, surface treatments, and coatings for a given emitter material are based on increasing its bare work function as a way of attaining a lower cesiated work function.

Cesium atoms are introduced into the interelectrode gap as a gas from a source usually external to the working space, typically from a reservoir containing a pool of liquid cesium. Some designs use graphite that forms intercalation compounds with cesium. Cesium gas is generated by heating the reservoir, and cesium pressure is set by the reservoir temperature, T_R, a critical parameter affecting the performance of the diode. In design studies and testing that examine cesium pressure effects, the results are often expressed in terms using T_R as a normalizing factor, e.g., T_E/T_R or T_C/T_R.

Referring again to Figure 6.7, the cesiated diode can also operate in a number of modes. In operational steady state for some designs or during startup of an ignited diode, the primary source of cesium ions is emission from the emitter surface. This situation constitutes operation in the unignited mode. During startup, the diode operates in the unignited mode until temperatures are sufficient to create the arc between the electrodes. Until ignition, the external load is usually kept low or even at short circuit. At ignition, the diode parameters change rapidly to adapt to the sudden change in ionization and the arc drop (defined below). The load is adjusted to account for these variables, and equilibrium is established at the design point.

Output is much higher in the ignited mode. Here, the primary source of energy to create the cesium plasma is a low-energy arc between the electrodes. This mode results in the highest power output in practical diodes, even though additional energy is required to support the arc. This energy loss is referred to as the arc drop, V_d. The difference in voltage between actual and ideal is known as the barrier index or back voltage, V_B, given by:

$$eV_B = \phi_C + eV_d + eV_a \qquad (6.11)$$

It is composed of three components that reduce the current flow from that released from the emitter (Rasor, 1991). The collector work function ϕ_C is the material barrier at the collector. The arc drop V_d is the voltage needed to sustain the ignited plasma of cesium ions necessary to neutralize the space charge. The third voltage loss, V_a, is due to the current attenuation by the plasma or by non-ideal electrode surfaces and is given by (Rasor, 1991):

$$V_a = kT_E \ln(J_E / J) \qquad (6.12)$$

Neglecting back emission from the collector, the general expression for current density in the ignited mode is:

$$J = AT_E^2 \exp[-(eV + eV_B)/kT_E] \qquad (6.13)$$

The internal effects for an ignited thermionic diode are very complex. Figure 6.8 depicts motive diagrams between the emitter and collector for the various operating regimes. One important phenomenon is the existence of sheath barriers, eV_E and eV_C, near the emitter and collector, respectively. The sheaths are defined as the very thin areas near the electrodes wherein collisions among the particles do not occur, unlike the plasma arc region that relies on collisions to form the cesium ions. However, local

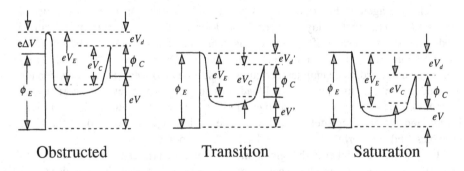

Figure 6.8 Motive Diagrams for Ignited Mode Operation

conditions within the sheath dictate the emission characteristics of the electrode because of its close proximity. (Hatsopoulos and Gyftopoulis, 1973, analyze the high-pressure diode with both a sheath and a transition region between the electrode surface and the plasma region.)

Analogous to the ideal case, a transition point exists where the emitter work function equals the sum of the output voltage, emitter work function, and arc drop. At this point, the positive ions generated in the plasma are just equivalent to that required to neutralize the space charge. Under these conditions, the electric field at the emitter is zero. The back voltage that determines current in Eq. (6.13) is given by:

$$eV_B = \phi_C + eV_d' + eV_a' \qquad (6.14)$$

The output voltage at the transition point is:

$$eV' = \phi_E - \phi_C - eV_d' \qquad (6.15)$$

The higher output voltages occur in the obstructed region. However, here the voltage drop between the electrodes would be insufficient to maintain the ignited plasma were it not for the existence of a negative space charge barrier of height ΔV that forms at the emitter. The barrier limits the electron emission from the emitter to a level that can be neutralized by the arc (Rasor, 1991). This condition can be viewed as a "virtual" emitter with an equivalent work function of $\phi_E + e\Delta V$, and the obstructed mode output voltage is given by:

$$eV_{Obs} = \phi_E + e\Delta V - \phi_C - eV_d \qquad (6.16)$$

The final operational state is the saturation region where voltages are less than the transition point voltage. Here the arc drop contains energy in excess of that needed to neutralize the plasma. Again, the local thermodynamic processes are complex, but it is believed that nearly all of this excess energy ΔV produces positive ions within the emitter sheath, and these ions enter the emitter instead of the plasma. The large positive ion current gives rise to the Schottky effect, especially near the transition point. This effect is a slight augmentation of electron emission due to the presence of the electric field.

Several processes contribute to the thermal and electrical energy distribution in a cesium diode. The total heat energy into the emitter is comprised of two components: the heat input from the solar, nuclear, or other heat source, and the Joule heating from the current flow in the emitter. This heat input to the emitter is dissipated by the following:

- Cooling by the electron flux leaving the surface
- Conduction through the cesium vapor (small and usually neglected)

- Conduction loss through the emitter lead and structural members
- Thermal radiation to the collector.

Ideally, the electron cooling is calculated by:

$$q_{EC} = (J/e)(\phi_E + 2kT_E) \tag{6.17}$$

Conduction losses are calculated by the standard conductive heat transfer equation. Because of the close spacing between the emitter and collector, radiation heat loss can be approximated by the standard radiation heat transfer equation assuming equal area for the electrodes and a view factor of 1.0. In the obstructed mode, however, resonant and non-resonant radiation loss from the plasma to the electrodes can be significant. In addition, effective or virtual work functions should be used for Eq. (6.13), and ion heating is also present. Over the years, numerous correlations and approximations have been developed to better calculate the performance of the plasma diode. In the U.S., most are embodied in a computer program named TECMDL that has become an industry standard for thermionic analysis. TECMDL correlates well with data around the operating point of typical diodes, but is less accurate at other parts of the J–V space, particularly in the obstructed mode (McVey and Rasor, 1992).

Performance optimization and selection of the design point for a thermionic converter depend on the priority of output parameters. The J–V characteristics and design operating point are not the same for maximum efficiency and maximum power output. At the transition point, the arc drop is at its minimum, so the diode efficiency is maximum. Peak power typically occurs at a voltage slightly less than the transition point voltage. However, operation in the saturation region near the transition point can sometimes result in a thermal runaway condition known as the Schock instability. Unless carefully controlled, an increase in emitter temperature can cause desorption of cesium from the surface which reduces the emitted current. Since electron cooling is in turn reduced, the emitter temperature can climb higher causing further desorption. In addition, higher output voltage carries some system benefits because a thermionic converter is inherently a high-current, low-voltage device. Therefore, the design operating point for an ignited mode diode is usually in the obstructed region near the transition point.

Selection of the optimum collector temperature also depends on several factors. To improve current density and output voltage (reduced barrier index), the collector work function should be low. However, measured current is generally lower than that predicted by the work function of the collector (Rasor, 1993). Figure 6.9 depicts an empirical correlation often used to relate collector temperature and work function.

Another consideration is that because the emitter operates at a very high temperature, a small amount of emitter material will evaporate over time and deposit on the cooler collector. Although the layer that builds up on the collector is usually quite thin, it may still be sufficient to affect the collector work function. Since space power system

STATIC ENERGY CONVERSION

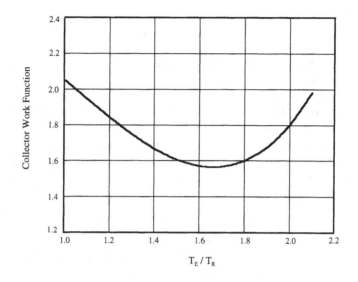

Figure 6.9 Empirical Approximation for Collector Work Function
(after Dahlberg, 1994; McVey and Rasor, 1992)

design is usually set by end-of-mission requirements, the effect on performance of any change in collector work function must be taken into account.

High collector temperature has two benefits: reducing the size of the radiator required to reject waste heat to space and reducing the radiation heat transfer from the emitter to collector. However, it also reduces Carnot efficiency and increases back emission of electrons from the collector. The net result of these several effects is that typical collector temperatures are in the range of 700–1000 K.

The key common components of all cesiated thermionic diodes are the emitter, collector, cesium supply, insulators, and seals. All of these components operate at high temperatures and for in-core nuclear systems, high radiation environments. Emitter and collector properties dominate diode performance. For emitters, the desired characteristics are:

- General
  - High bare work function (low cesiated work function)
  - Low emissivity
  - Low transport of material to collector (low vapor pressure and resistance to chemical transport mechanisms)
  - Good fabricability
  - Low brittle–ductile transition temperature
  - Low electrical resistance

- Compatibility with residual and system-generated gases in interelectrode gap (Cs, CsOH, O_2, H_2, N_2, H_2O, and possibly C, CO, CO_2, Al, Sc)
- In-core nuclear systems
  - Small cross-section to limit capture of thermal neutrons
  - Low swelling in fast neutron flux
  - Compatibility with nuclear fuel
  - Low permeability to diffusion of fuel constituents and fission products
  - High creep strength

The high temperature requirements for emitter and collector materials lead to the use of refractory metals almost exclusively. Typical emitter materials include tungsten, tantalum, molybdenum, and rhenium. Collector materials include molybdenum, niobium, tantalum, rhenium, and palladium. Rhenium has excellent emission properties, but its cost and high thermal neutron cross section have limited its consideration for in-core nuclear systems. High-strength alloys of tungsten and molybdenum are typically used in core. Molybdenum emitters are preferred for moderated reactor systems, and creep strength has been improved by alloying with niobium. A tungsten coating is typically applied to molybdenum emitters to improve surface properties. Neutron cross section is least with the ^{184}W isotope, but that material is expensive and typically used as a coating rather than a solid emitter material. Russia has developed single crystal molybdenum and tungsten alloys for emitters both for improved creep strength and to reduce migration of fuel constituents and fission products through the emitter.

Insulator materials often have the conflicting requirements of high electrical resistance and low thermal resistance. Various crystalline forms of aluminum oxide are typically used in thermionic systems. For in-core thermionic fuel elements, insulation is applied to the outer diameter of a cylindrical collector. The alumina insulator is either bonded or plasma is sprayed on the collector surface. In a configuration known as a trilayer, the collector–insulator assembly consists of a bonded stack of collector/insulator/stainless steel outer sheath. In order to limit stresses due to differential expansion, the insulator is actually several layers of graded metal-alumina compositions. The Russian Enisy system uses several scandium oxide spacers along the length of the single-cell diode for alignment and to help prevent shorting between the emitter and collector.

Alumina is also the usual choice for diode seals. One of the key challenges for seal design is providing a transition from a metal to a ceramic material at high temperatures. Stainless steel is typically used for parts external to the electrode working space, and joining methods have been developed for these attachments. As with the collector trilayer, differential thermal expansion is the primary difficulty.

Following are a series of parametric test data obtained for a variable spacing planar diode developed by Thermoelectron Technologies Corporation. This particular diode #46 has a chloride deposited tungsten emitter and a niobium collector (McVey and Rasor, 1992; Dahlberg, 1994). Figures 6.10, 6.11, and 6.12 illustrate the effects on diode performance of emitter temperature, cesium pressure (as indicated by reservoir tempera-

STATIC ENERGY CONVERSION

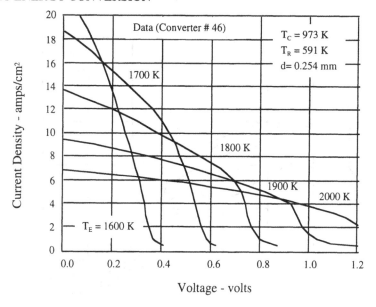

Figure 6.10 Effect of Emitter Temperature on Diode Performance
(after Dahlberg, 1994; McVey and Rasor, 1992)

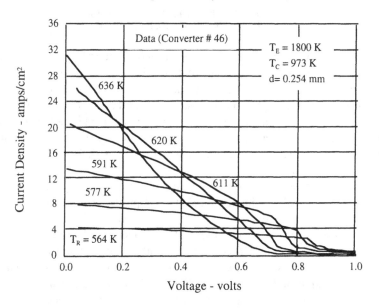

Figure 6.11 Effect of Cesium Pressure on Diode Performance
(after Dahlberg, 1994; McVey and Rasor, 1992)

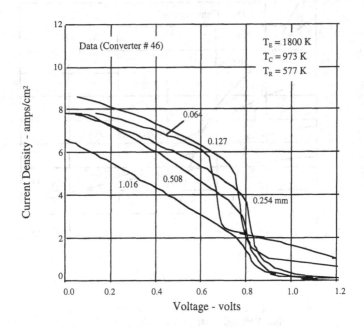

Figure 6.12 Effect of Gap Spacing on Diode Performance
(after Dahlberg, 1994; McVey and Rasor, 1992)

ture), and interelectrode gap spacing. In general, maximum emitter temperature and minimum gap spacing are desirable because output voltage increases as do usually both power density and efficiency. Emitter temperature is most often limited by structural properties (e.g., creep strength for in-core systems where the emitter must resist swelling of the nuclear fuel), and gap spacing is set by manufacturing technology or emitter deformation in an in-core system. Optimum cesium pressure represents a balance among the several processes at work in the diode, namely ionization, cesiation of the emitter, and gap kinetic and conduction losses.

Today, the cesiated plasma diode is the mainstay of thermionic systems. A variety of methods have been applied with mixed results to improve performance of these and other diode types. Among these are electrode surface treatments and additives to the system such as barium and oxygen (Tsakadze, 1995; Magera, *et al.*, 1992; Hatch, *et al.*, 1987). In addition, work has been done on diodes with spacings of only a few microns (Fitzpatrick, *et al.*, 1996; Nikolaev, *et al.*, 1993). These close-spaced converters can perform well as vacuum diodes or in the Knudsen mode (defined as collisionless current flow with the cesium only affecting the work functions of the electrodes). Adapting such small spacing to a practical system has not been perfected as yet.

4. AMTEC

The basis of AMTEC operation is the selective ionic conductivity of Beta–alumina solid electrolyte (BASE). Ford Motor Company Scientific Laboratory developed this material as the electrolyte for NaS batteries. In 1968, Ford researchers recognized that the same principle could be applied to a thermally regenerative electrochemical energy conversion system. Through 1979, Ford supported much of the development on what was then called the Sodium Heat Engine. In the 1980s, DOE supported further work on higher power cells and electrode development for terrestrial applications. To varying degrees, several other U.S. and foreign firms began work on the technology also. Interest in space applications surfaced in this time period, and NASA began to support development as well. In 1994, the Air Force initiated AMTEC technology development that could be applied to a solar-heated power system for medium Earth orbiting spacecraft (Hunt, 1997). NASA and DOE are currently investigating AMTEC as a candidate high-efficiency converter coupled with radioisotope heat sources to power future planetary spacecraft. These agencies have tentatively selected AMTEC for the next generation systems, and DOE has contracted for development of a prototype system delivering at least 100 We at the end of a 15-year mission. In the U.S., Advanced Modular Power Systems, Inc. (AMPS) is conducting much of the recent development and test of AMTEC cells for both space and terrestrial applications.

The heart of an AMTEC system is the BASE material. It is a transparent crystalline ceramic in which alumina is stabilized with either lithia or magnesia. A typical composition by weight percent is 8.85% Na_2O, 0.75% Li_2O, and 90.4% Al_2O_3 (Hunt, 1997). Its melting point is 2253 K, and it remains non-reactive with sodium at temperatures as high as 1300 K (Cole, 1983). For AMTEC, this material is generally manufactured in tubular form, and Figure 6.13 shows a schematic of a typical configuration used in current designs. A pressure differential is imposed on a high-temperature working fluid, usually sodium, between the inner and outer diameters of the BASE tube. At the BASE–sodium interface on the inside of the tube, Na^+ ions form and migrate through the BASE electrolyte. Beta"–alumina has a high conductivity for sodium ions but essentially zero conductivity for electrons and neutral sodium atoms. Thus, excess electrons are produced at the anode and are collected and shunted through an external load to produce power. When returned to the cathode on the outer diameter of the tube, the electrons recombine with the Na^+ ions. Porous refractory metal electrodes at the anode and cathode allow diffusion of the sodium between the vapor spaces and the BASE. Typical electrodes are sputtered molybdenum, titanium nitride, or tungsten alloys (Underwood, et al., 1992). For adequate diffusion, the electrodes must be very thin, so current collectors are necessary to prevent excessive axial voltage drops. Copper and molybdenum screens are typically used for current collection.

The basic building block of a system is the AMTEC cell that consists of several BASE tube assemblies connected in series. Figure 6.14 shows such a cell and depicts

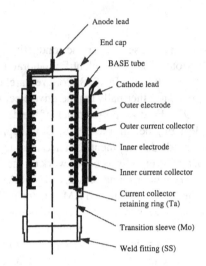

Figure 6.13 Schematic of AMTEC BASE Tube (Courtesy of Advanced Modular Power Systems, Inc.)

the key components. The BASE tubes are contained in a common pressure boundary and share the same sodium inventory. The sodium transport system operates all of the tubes. Since no tube can operate independently, a multi-tube configuration is considered a cell, instead of individual anode/cathode tube assemblies. In addition to the BASE tubes, the other key cell components are the condenser, return artery, evaporator, vapor plenum, insulation, and external cell wall.

Within the cell, sodium flows in a closed cycle as follows. Heat is rejected from the condenser at the top of the cell. This is the lowest temperature point in the cell and maintains the sodium vapor outside the BASE tubes at low pressure. Cold side conditions are in the ranges of 400 K to 800 K and 10^{-4} Pa to 500 Pa. The artery employs the surface tension principle of a heat pipe to wick the liquid sodium from the condenser to the hot end of the cell. This artery has a free surface near the hot end that allows the liquid sodium to evaporate into the high temperature and pressure region of the cell. Heat added at the hot side (bottom) of the cell evaporates the sodium in the artery evaporator section and also heats the BASE tubes. The sodium vapor accesses the inside of all of the BASE tubes through a common plenum. Typical sodium temperatures around the tubes range from 900 K to 1100 K, although recent designs are as high as 1223 K. Sodium pressure inside the tubes is typically in the range of 7 to 75 kPa. Sodium then flows out of the top end of the BASE tubes back into the common condenser area. The cell wall forms the outer hermetic boundary and is designed to minimize heat loss by conductive transfer from the hot to the cold end. Heat shields are attached to the cell wall and low emissivity surfaces or coatings are used to further reduce heat loss. Flow

STATIC ENERGY CONVERSION 309

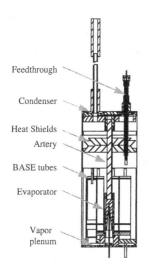
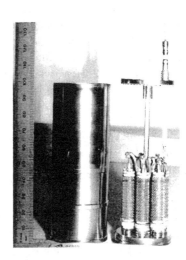

Figure 6.14 Multitube AMTEC Cell (Courtesy of Advanced Modular Power Systems, Inc.)

rates are low, typically about 1 gm/hour/amp in each tube, and the total sodium inventory can be as low as 40 cm³/kWe. Heat input to the evaporator is approximately 1 watt/tube/amp (Sievers, *et al.*, 1998; Sievers, *et al.*, 1997; Hunt, 1997).

Performance modeling of an AMTEC cell involves simultaneous solutions of interdependent thermal, fluid flow, and electrical relationships, and a complete discussion is well beyond the scope of this chapter. For space systems, three significant modeling efforts are ongoing (Hendricks, *et al.*, 1998; Schock(b), *et al.*, 1997; Tournier, *et al.*, 1997), but all are derived from the same basic physical principles and equations. However, the analysis is complicated by the fact that local conditions vary along the length of the BASE tube. Local open circuit voltage is given by the Nernst equation:

$$V_{OC} = \frac{RT_a}{F} \ln\left(\frac{p_a}{p_c}\right) \qquad (6.18)$$

where V_{OC} is the local open circuit voltage along the BASE tube, R is the universal gas constant (8.3145 J/K–mole), T_a is the local sodium vapor temperature at the anode, F is the Faraday constant (96,485 coulombs/mole), p_a is the local sodium vapor pressure at the anode, and p_c is the local sodium vapor pressure at the BASE–porous cathode-electrode interface.

When an external load is applied so that current flows, the voltage decreases from open circuit due to the ionic resistance across the BASE tube; ohmic losses in the electrodes, current collectors, intertube connections, and leads external to the cell; and the increase in the pressure required at the electrode–BASE interface to drive the vapor through the electrode and across the vapor space (Cole, 1983; Schock(b), *et al.*, 1997; Hendricks, *et al.*, 1998). The fundamental relationship for cell output is given by (Underwood *et al.*, 1992; Hendricks, *et al.*, 1998):

$$V = V_{OC} - IR_{int} - \eta_a - (-\eta_c) \qquad (6.19)$$

where V is the cell voltage, I is the cell current, R_{int} is the internal resistance across the BASE tube, and η_a and η_c are the anode and cathode overpotentials that vary with current density, temperature, and pressure.

A critical design requirement of the AMTEC cell is to prevent sodium condensation inside the BASE tubes which can result in tube-to-tube shorting in a multi-tube cell. (A single tube cell can operate with liquid sodium.) Thus, it is necessary to maintain the coldest point in the BASE tube higher than the evaporator temperature. This temperature difference is known as the temperature margin in an AMTEC cell. A typical evaporator temperature is 1000 K with a temperature margin of 50 K. Design temperature goals for the next generation cells are 1100 K, 1150 K, and 1225 K for the evaporator, minimum BASE tube, and hot end, respectively (Sievers, *et al.*, 1998).

Cell power increases as the sodium pressure inside the tubes increases, and this pressure is set by the evaporator temperature. By positioning the evaporator close to the hot end of the cell, its temperature can be raised within the required temperature margin. Establishment of the temperature margin must account for sodium pressure and evaporator temperature which vary with current flow. Hence, evaporator standoff length is an important design parameter. Of course, increasing the hot end temperature also increases thermal losses due to conduction and radiation.

Analysis tools are being developed for both multitube cells and multicell systems. For spacecraft radioisotope power, AMPS, Orbital Sciences Corporation, and Lockheed Martin have conducted most of the system design work, both independently and jointly (Carlson, *et al.*, 1998; Schock(a), *et al.*, 1997; Hendricks, *et al.*, 1997; Ivanenok and Sievers, 1996). Figure 6.15 shows one conceptual configuration (Hemler, 1997). It is designed to employ the $^{238}PuO_2$ GPHS described in Chapter 5.

In recent years, rapid progress has been made in AMTEC technology. AMPS has built several developmental cells, first with single tubes — Series I — and then with multitubes — Series II (Sievers, *et al.*, 1997). Design improvements include better thermal insulation, micromachined condensers, TiN electrodes, and Mo current collectors (Sievers, 1997; Hendricks, *et al.*, 1997; Schock(b), *et al.*, 1997; Izenson and Crowley, 1996). Figure 6.16 illustrates the projected and measured performance of the PX–5A Series II cell. One goal was to improve cell performance by increasing the hot end temperature to a level compatible with GPHS operating temperatures, in the range of 1200 K to 1300 K. Cell output voltage at this temperature level exceeds 3 volts. Future

STATIC ENERGY CONVERSION 311

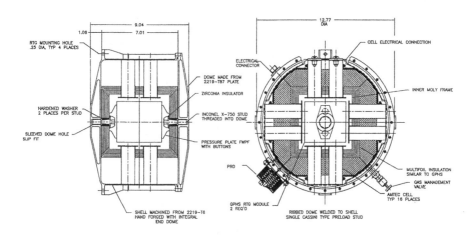

Figure 6.15 Multicell AMTEC System with GPHS Radioisotope Source
(Courtesy of Lockheed Martin Corporation)

designs must also account for shock and vibration launch loads and quantifying the long-term degradation mechanisms that affect cell and system lifetime. For example, tests with copper current collectors have degraded rapidly, although one multitube cell test was stable for nearly 3000 hours after an initial 25% degradation in power (Sievers, *et al.*, 1997; Merrill, *et al.*, 1997). Molybdenum has been substituted in the most recent cells. Uncertainties about long-term material compatibility, e.g., sodium depleting the oxygen level in the BASE tubes, must be resolved as well.

5. Thermophotovoltaics

The thermophotovoltaic (TPV) cell is a photovoltaic cell designed for maximum performance when exposed to photons at energy levels in the infrared region of the electromagnetic spectrum. Hence, TPVs are sometimes referred to as infrared photovoltaics and are usually coupled to radiant heat sources in the temperature range of 1000 K to 2000 K. TPV development began in the early 1960s and continued at various levels until the present. For space applications, the most recent work has been funded by NASA and DOE who have interest in TPVs as a candidate high-efficiency converter for advanced radioisotope systems.

Designers of TPV converters strive to match the energy of the incident photons that excite the cell to the energy bandgap of the semiconductor material that comprises the cell. In this way, more of the photon energy creates electrical current, and less of the

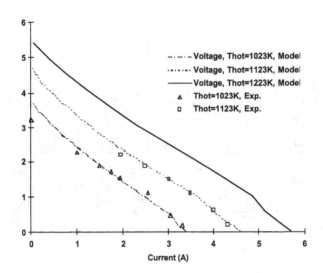

Figure 6.16 PX–5A Cell Performance, Measured and Projected (Courtesy of Advanced Modular Power Systems, Inc.)

energy is wasted in heating the material. The photon energy E_p is related to the wavelength v by:

$$E_p = hv \qquad (6.20)$$

where h is Planck's constant (6.626×10^{-34} J–sec). Figure 6.17 shows the spectral emissive power of a blackbody source at different emitter temperatures and indicates the bandgaps for several candidate TPV materials. Much of the recent cell development and testing has been done with gallium antimony (GaSb), but several investigators are working on low-bandgap tertiary (particularly $In_xGa_{1-x}As$) and quaternary compounds that can be tailored to some degree to better match the heat source wavelength near peak energy (Bhat, et al., 1996; Sundaram, et al., 1997; Uppal, et al., 1996; Wojtczuk, 1996).

Figure 6.18 depicts the general arrangement of a filtered TPV converter, GaSb in this case, along with the associated spectral energy flows. Heat emitted by the heat source first radiates a spectral filter that reflects the longer wavelength infrared photons, but transmits the high-energy, short-wavelength energy to the TPV cell. Although maximum cell output is theoretically achieved at the bandgap of the cell material, in practice, transmission of energy at a slightly shorter wavelength produces better performance.

As with any solar cell, the TPV cell consists of layered n– and p–type semiconductors on a substrate with accommodations for electrical insulation, cooling, and current collection. The filter and cell are often constructed as a single or even integral subas-

STATIC ENERGY CONVERSION

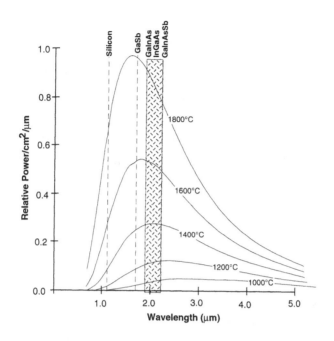

Figure 6.17 Blackbody Emission and TPV Semiconductor Bandgap Energies
Note: The bandgap for tertiary and quaternary materials varies with composition.
(Courtesy of EDTEK, Inc.)

sembly. Some combined filter/cell designs include a reflective layer behind the cell that reflects wavelengths (for which the cell is transparent) back through the cell and front-face filter to the heat source (Horne, *et al.*, 1996). Figure 6.19 illustrates one recent configuration that has been fabricated and tested.

The efficiency of the TPV cell and filter combination is given by (Horne, *et al.*, 1996):

$$\eta_{Cell} = \frac{V_{OC} FF \int\limits_{\lambda=0}^{\lambda_{sG}} I(\lambda) d\lambda}{\int\limits_{\lambda=0}^{\lambda=\alpha} [\tau_{Filt} \alpha_{Cell}(\lambda) + \alpha_{Filt}(\lambda)] d\lambda} \qquad (6.21)$$

where η_{Cell} is the cell efficiency, V_{OC} is the cell open circuit voltage in volts, FF is the cell fill factor, $I(\lambda)$ is the cell current as a function of photon wavelength in amps, τ_{Filt} is the transmittance of the filter, $\alpha_{Cell}(\lambda)$ is the cell absorptivity as a function of photon wavelength, and $\alpha_{Filt}(\lambda)$ is the filter absorptivity as a function of photon wavelength.

314 SPACECRAFT POWER TECHNOLOGIES

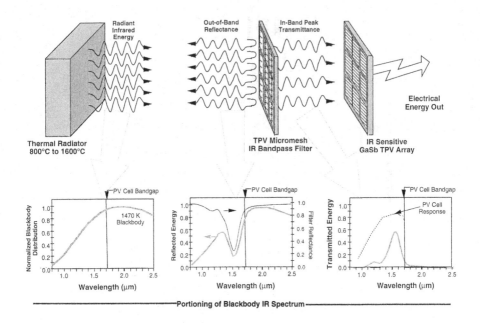

Figure 6.18 Primary Components of a Filtered TPV Converter (Courtesy of EDTEK, Inc.)

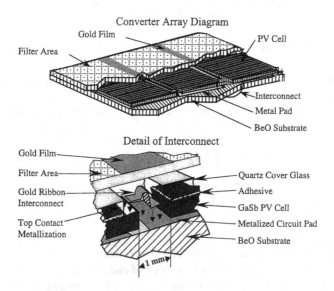

Figure 6.19 TPV Array Utilizing the EDTEK Continuous Gold Film/Filter (Courtesy of EDTEK, Inc.)

STATIC ENERGY CONVERSION 315

Because IR sources typically emit broadband radiant energy, the narrow bandpass filter is a critical component. Furthermore, irradiation occurs at varying angles of incidence, so off-normal reflectance and transmittance are factors also. Figure 6.20 illustrates the importance of filter performance on efficiency. (Packing factor is the ratio of active to inactive area.) In the desirable region of high reflectance, efficiency depends strongly on filter reflectance, and filter development has paralleled cell development since the 1960s. A typical approach is deposition of alternating layers of dielectric materials with different indices of refraction. Bandpass is determined by the optical thickness of the layers. Silicon and silicon dioxide were some of the early filter materials considered. Later, indium tin oxide and other coatings were investigated.

More recently, EDTEK Corporation has pioneered development of a new type of antenna filter utilizing e-beam and ion-beam lithography techniques developed by the microelectronics industry. Figure 6.21 illustrates this concept for a bandpass filter.

This antenna filter consists of a very thin (0.0005 mm) Au coating on a quartz substrate. Its active area has about 2 x 10^8 submicronic holes per cm^2 with mesh dimensions comparable to the wavelengths of the electromagnetic radiation to be either transmitted or reflected. Interaction of the electric and magnetic fields produces either in-

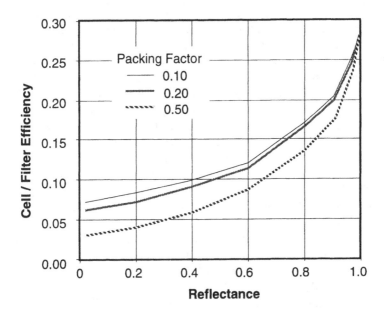

Figure 6.20 Typical Dependence of TPV Cell/Filter Efficiency on Long Wavelength Reflectance for a Blackbody Source at 1473 K (Courtesy of EDTEK, Inc.)

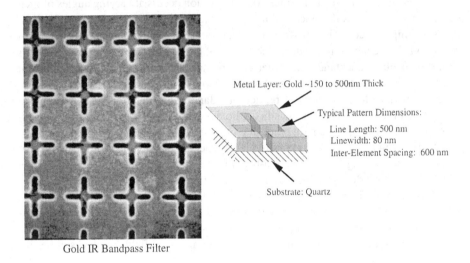

Figure 6.21 Antenna Bandpass Filter Elements (Courtesy of EDTEK, Inc.)

ductive (bandpass) of capacitive (band reject) resonance. Slots in a metal film produce a bandpass filter; dipole elements on a dielectric produce a band reject filter. Performance depends on the size and shape of the elements, resistivity of the metal film, and the dielectric and optical properties of the substrate. As with some other filter designs, the filter windows are aligned with active areas of the cell underneath the filter substrate. Test filters have been fabricated using direct-write e-beam lithography (DEBL), but masked ion beam lithography (MIBL) shows promise for much faster and cheaper production of filters. A bandpass antenna filter matched specifically to a radioisotope heat source has been developed and tested (Horne, et al., 1996).

Unfortunately, cell performance depends strongly on temperature as shown in Figure 6.22. Since both V_{OC} and FF decrease with temperature, cell efficiency is strongly affected. The temperature dependency is a significant drawback for space applications where waste heat must be rejected by radiation since it results in relatively large radiators for TPV systems.

To illustrate the system design considerations, Figure 6.23 shows two conceptual designs of radioisotope/TPV converters for potential application to a NASA outer planet probe (Schock, et al., 1996; Horne, et al., 1998). Both utilize the same 125 thermal watt $^{238}PuO_2$ heat source derived from the GPHS design described in Chapter 5. A typical operating temperature for the heat source surface is 1200 K. To prevent carbon contamination of the filter, a refractory metal housing surrounds the graphitic surface of the heat source. These concepts use molybdenum with an iridium compatibility coating on

STATIC ENERGY CONVERSION

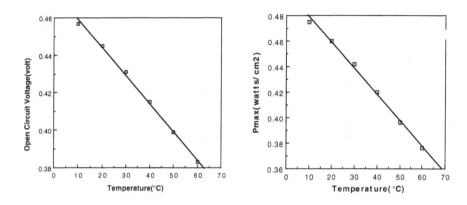

Figure 6.22 Effect of Temperature on TPV Cell Performance Parameters (Courtesy of EDTEK, Inc.)

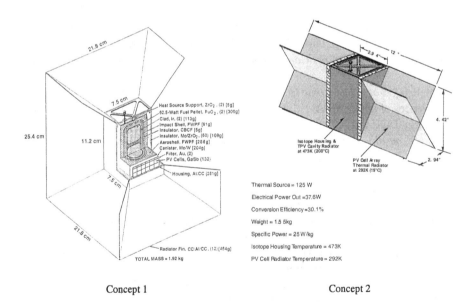

Concept 1   Concept 2

Figure 6.23 Two Conceptual Designs for Radioisotope–TPV Systems
(Courtesy of U.S. DOE and EDTEK, Inc.)

the inner surface and a tungsten coating on the outside, roughened to increase emissivity. Orbital Sciences designed concept #1 to produce 20 watts at 28 volts (Schock, et al., 1996). Using this design as a baseline, EDTEK derived concept #2 incorporating several improvements (Horne, et al., 1998). Although still conceptual, this design is projected to produce 37.6 watts in a significantly smaller package.

It is important to note that these designs are based on very low radiation sink temperatures similar to a Pluto encounter. However, for near Earth missions, solar flux and Earth albedo are significant. Hence, the sink temperature would be significantly higher, and a radiator temperature as low as 19°C would require more radiator area. Although rapid progress is being made in TPV system development, a number of issues remain, primarily related to system fabricability or lifetime. As with other photovoltaic cells, TPV cells are susceptible to degradation when exposed to radiation. Although ^{238}Pu emits alpha particles primarily, it also emits a background level neutron flux. This low-level neutron flux over a long enough period can degrade TPV cells. However, GaSb cell tests showed only about 12% power degradation for an accumulated dose equivalent to about 10 years in proximity to a GPHS heat source (Horne, et al., 1998).

6. References

Anderson, R. V., et al., *Space Reactor Electric Systems*, ESG-DOE-13398, (Rockwell International, Canoga Park, CA, March 29, 1983), IV 362-363.

Angrist, S. W., *Direct Energy Conversion*, (Allyn and Bacon, Inc., Boston, 1965), 116-166.

Baksht, F. G., et al., *Thermionic Converters and Low-Temperature Plasma*, Academy of Sciences of the USSR A. F. Ioffe Physico-Technical Institute, (U.S. Department of Energy DOE-tr-1, 1978).

Bhat, I. B., et al., *TPV Energy Conversion: A Review of Material and Cell Related Issues*, Proceedings of the 31st Intersociety Energy Conversion Engineering Conference (IECEC 96), Paper #96330, (Institute of Electrical and Electronics Engineers, Piscataway, NJ, 1996), 968-973.

Carlson, M. E., Giglio, J. C., and Sievers, R. K., *Design and Fabrication of Multi-Cell AMTEC Power Systems for Space Applications*, Proceedings of the Space Technology and Applications International Forum, Fifteenth Symposium on Space Nuclear Power and Propulsion, AIP Conference Proceedings 420 (American Institute of Physics Press, New York, 1998), 1486-1490.

Cole, T., *Thermoelectric Energy Conversion with Solid Electrolytes*, Journal of Science, Vol. 221, No. 4614, September, 1983, 915-920.

Dahlberg, R. C., et al., *Review of Thermionic Technology: 1983 to 1992, A Critical Review of Space Nuclear Power and Propulsion 1984-1993*, Mohamed S. El-Genk, (ed.), (American Institute of Physics Press, New York, 1994) 121-165.

Fitzpatrick, G. O., et al., *Close-Spaced Thermionic Converters with Active Spacing Control and Heat-Pipe Isothermal Emitters*, Proceedings of the 31st Intersociety Energy Conversion Engineering Conference (IECEC 96), Paper #96145, (Institute of Electrical and Electronics Engineers, Piscataway, NJ, 1996), 920-927.

Hatch, G. L., Korringa, M., and Sahines, T. P., *Thermionic Characteristics of Planar and Cylindrical Oxygen Additive Tungsten-Niobium Diodes*, Proceedings of the 22nd Intersociety Energy Conversion Engineering Conference (IECEC 87), Paper #879125, (American Institute of Aeronautics and Astronautics, New York, 1987), 2007-2010.

Hatsopoulos, G. N. and Gyftopoulis E. P., *Thermionic Energy Conversion*, (MIT Press, 1973).

Hemler, R. J., Lockheed Martin Corporation, personal communication, 1997.

Hendricks, T. J., Huang, C., and Sievers, R. K., *AMTEC Radioisotope Power System Design & Analysis for Pluto Express Fly-by*, Proceedings of the 32nd Intersociety Energy Conversion Engineering Conference (IECEC 97), Paper #97382, (American Institute of Chemical Engineers, 1997), 501-508.

Hendricks, T. J., Borkowski, C. A., and Huang, C., *Development & Experimental Validation of a SINDA/FLUINT Thermal/Fluid/Electrical Model of a Multi-Tube AMTEC Cell*, Proceedings of the Space Technology and Applications International Forum, Fifteenth Symposium on Space Nuclear Power and Propulsion, AIP Conference Proceedings 420 (American Institute of Physics Press, New York, 1998), 1491-1501.

Horne, W. E., Morgan, M. D., and Sundaram, V. S., *IR Filters for TPV Converter Modules*, Proceedings of the Second NREL Conference on Thermophotovoltaic Generation of Electricity, AIP Conference Proceedings 358 (American Institute of Physics Press, New York, 1996), 35-51.

Horne, W. E., Morgan, M. D., and Saban, S. B., *Performance Tuned Radioisotope Thermophotovoltaic Space Power System*, Proceedings of the Space Technology and Applications International Forum, Fifteenth Symposium on Space Nuclear Power and Propulsion, AIP Conference Proceedings 420 (American Institute of Physics Press, New York, 1998) 1385-1393.

Hunt, T. K., Advanced Modular Power Systems, Inc., personal communication, 1997.

Ivanenok, J. F., III and Sievers, R. K., *20-500 Watt AMTEC Auxiliary Electric Power System*, Proceedings of the 31st Intersociety Energy Conversion Engineering Conference (IECEC 96), Paper #96187, (Institute of Electrical and Electronics Engineers, Piscataway, NJ, 1996), 2232-2237.

Izenson, M. G. and Crowley, C. J., *Micromachined Evaporators for AMTEC Cells*, Proceedings of the 31st Intersociety Energy Conversion Engineering Conference (IECEC 96), Paper #96262, (Institute of Electrical and Electronics Engineers, Piscataway, NJ, 1996), 2226-2231.

Kelly, C. E., Lockheed Martin Corporation, personal communication, 1997.

Kelly, C. E., and Klee, P. M., *Cassini RTG Acceptance Test Results and RTG Performance on Galileo and Ulysses*, Proceedings of the 32nd Intersociety Energy Conversion Engineering Conference (IECEC 97), Paper #97435, (American Institute of Chemical Engineers, 1997), 2211-2216.

Magera, G. G., Davis, P. R., and Lamp, T. R., *Barium Interaction with Partially Oxygen-Covered Nb(110) Surfaces*, Proceedings of the Ninth Symposium on Space Nuclear Power Systems, AIP Conference Proceedings 246, (American Institute of Physics Press, New York, 1992), 623-642.

McVey, J. B. and Rasor, N. S., *The TECMDL Thermionic Converter Computer Model*, Proceeding of the 27th Intersociety Energy Conversion Engineering Conference (IECEC 92), Paper #929426, (Society of Automotive Engineers, 1992), 3.505-3.512.

Merrill, J. M., et al., *Vacuum Testing of High Efficiency Multi-BASE Tube AMTEC Cells*, Proceedings of the 32nd Intersociety Energy Conversion Engineering Conference (IECEC 97), Paper #97379, (American Institute of Chemical Engineers, 1997), 1184-1195.

Morgan, M. D., EDTEK, Inc., personal communication, 1997.

Nikolaev, Y. V., et al., *Close-Spaced Thermionic Converters for Space Systems*, Proceedings of the Thermionic Energy Conversion Specialists Conference, (Chalmers University of Technology and University of Göteborg, Sweden, 1993).

Rasor, N. S., *Thermionic Energy Conversion Plasmas*, IEEE Transactions on Plasma Science, Vol. 19, No. 6, (Institute of Electrical and Electronics Engineers, 1991).

Rasor, N. S., *The Collector Enigma*, Proceedings of the Thermionic Energy Conversion 1993, Specialists Conference, May 5-7, 1993, Göteborg, Sweden.

Rasor, N. S., personal communication, 1997.

Schock, A., Or, T. C., and Kumar, V. *Design of RTPV Generators Integrated with New Millennium Spacecraft for Outer Solar System*, Proceedings of the Space Technology and Applications International Forum, Thirteenth Symposium on Space Nuclear Power and Propulsion, AIP Conference Proceedings 361, (American Institute of Physics Press, New York, 1996), 1003-1022.

Schock(a), A., Noravian, H., and Or, C., *Coupled Thermal, Electrical, and Fluid Flow Analyses of AMTEC Converters, with Illustrative Application to OSC's Cell Design*, Proceedings of the 32nd Intersociety Energy Conversion Engineering Conference (IECEC 97), Paper #97182, (American Institute of Chemical Engineers, 1997), 1156-1164.

Schock(b), A., Noravian, H., Kumar, V., and Or, C., *Design and Performance of Radioisotope Space Power Systems Based on OSC Multitube AMTEC Converter Designs*, Proceedings of the 32nd Intersociety Energy Conversion Engineering Conference (IECEC 97), Paper #97530, (American Institute of Chemical Engineers, 1997), 489-500.

Sievers, R. K., Advanced Modular Power Systems, Inc., personal communication, 1997.

Sievers, R. K., et al., *PX-5 AMTEC Cell Development*, Proceedings of the Space Technology and Applications International Forum, Fifteenth Symposium on Space Nuclear Power and Propulsion, AIP Conference Proceedings 420 (American Institute of Physics Press, New York, 1998), 1479-1485.

Sievers, R. K., et al., *Series II AMTEC Cell Design and Development*, Proceedings of the 32nd Intersociety Energy Conversion Engineering Conference (IECEC 97), Paper #97335, (American Institute of Chemical Engineers, 1997), 1125-1129.

Sundaram, V. S., et al., *GaSb Based Ternary and Quaternary Diffused Junction Devices for TPV Applications*, Proceedings of the Fourth NREL Conference on Thermophotovoltaic Generation of Electricity, (American Institute of Physics Press, New York, 1997).

Tournier, J. M., et al., *Performance Analysis of a Multitube Vapor-Anode AMTEC Cell*, Proceedings of the 32nd Intersociety Energy Conversion Engineering Conference (IECEC 97), Paper #97378, (American Institute of Chemical Engineers, 1997), 1172-1179.

Tsakadze, L. M., *Investigation of (110)Mo, (110)W Monocrystals and Nb Polycrystal Implanted by Oxygen Ions used as TEC Electrodes*, Proceedings of the 30th Intersociety Energy Conversion Engineering Conference (IECEC 95), Paper #CT-224, (American Society of Mechanical Engineers, Fairfield, NJ, 1995), 121-124.

Underwood, M. L, et al., *An AMTEC Vapor-Vapor, Series Connected Cell*, Proceedings of the Ninth Symposium on Space Nuclear Power Systems, AIP Conference Proceedings 246, (American Institute of Physics Press, New York, 1992), 1331-1337.

Uppal, P., et al., *Development of High Efficiency Thermophotovoltaics for Space Power Applications*, Proceedings of the Space Technology and Applications International Forum, Thirteenth Symposium on Space Nuclear Power and Propulsion, AIP Conference Proceedings 361, (American Institute of Physics Press, New York, 1996), 987-991.

Vining, C. B. and Fleurial, J. P., *Silicon-Germanium: An Overview of Recent Developments*, A Critical Review of Space Nuclear Power and Propulsion 1984-1993, El-Genk, M. S., (ed.), (American Institute of Physics Press, New York, 1994), 87-120.

Wojtczuk, S., *Low Bandgap InGaAs Thermophotovoltaic Cells*, Proceedings of the 31st Intersociety Energy Conversion Engineering Conference (IECEC 96), Paper #96327, (Institute of Electrical and Electronics Engineers, Piscataway, NJ, 1996), 974-978.

CHAPTER 7

DYNAMIC ENERGY CONVERSION

1. Introduction

A wide variety of thermodynamic cycles are candidates for spacecraft power and surface powerplants for various bodies in the solar system. Compared to photovoltaics, dynamic systems hold the promise of higher efficiency and better scaling to high power levels. However, questions persist about the reliability of dynamic machinery, and spacecraft designers prefer power systems that will degrade gradually if failures do occur. Consequently, added redundancy and other reliability enhancements have negated some of the potential advantages of dynamic conversion. Despite long-life demonstrations of several critical components, no complete dynamic power conversion system has yet been flown in space.

By far, the most intensive assessment and development work has been focused on three basic cycles: Stirling, Brayton and Rankine. With a few exceptions, the required lifetime for the system dictates the use of closed cycles. The remaining sections of this chapter describe the operation, development, and key engineering issues for these three closed-cycle systems. The basic equations for ideal cycle work, heat, and efficiency are the same for all three and can be found in any basic text on thermodynamics. They simply relate the change in enthalpy of the working fluid as it undergoes the various processes in the cycle. These pertinent equations are listed below, all per unit mass flow of the working fluid:

The heat added, q_{in}, is defined as the difference of specific enthalpies (in watt–hrs/kg) between the heater outlet and inlet:

$$q_{in} = h_{htro} - h_{htri} \tag{7.1}$$

The thermodynamic or indicated work produced during the expansion process, w_{exp}, is the difference of specific enthalpies between the expansion (rotary turbine or linear piston stroke) outlet and inlet:

$$w_{exp} = h_{expo} - h_{expi} \tag{7.2}$$

The thermodynamic work required to compress the fluid, w_{comp}, is the difference of specific enthalpies between the compression (pump or compressor) outlet and inlet:

$$w_{comp} = h_{compo} - h_{compi} \qquad (7.3)$$

The thermodynamic efficiency, η_{th}, is defined as the net work divided by the heat input:

$$\eta_{th} = \frac{w_{exp} - w_{comp}}{q_{in}} \qquad (7.4)$$

2. Stirling cycle

Early in the 19th century, a Scottish minister named Robert Stirling invented an external combustion engine using air as the working fluid. Known as the hot air engine, it was a serious competitor to the steam engine. However, interest faded with the introduction of the internal combustion engine and the electric motor. In the 1930s, the N.V. Philips Company of the Netherlands developed much improved versions of the Stirling engine by using hydrogen and helium as working fluids and increasing the operating pressure. Philips investigated a wide variety of potential applications and also developed cryocoolers based on the Stirling cycle (Dudenhoefer, et al., 1994). In the 1970s, concerns over air pollution and the cost of petroleum resulted in accelerated development of Stirling engines for automotive and solar power generation in several countries, notably the U.S., the Netherlands, and Sweden. Its consideration for space power began in the 1970s because of the potential for relatively high conversion efficiency at moderate temperatures with either solar or nuclear energy sources.

The Stirling engine employs two pistons, either in separate cylinders or in a single cylinder. A low-mass displacer piston shuttles the working fluid between the hot and cold spaces of the engine through a regenerator. The high-mass power piston delivers the mechanical work to the load. Over the years, designers have developed two general categories of Stirling engines. The most common type is the kinematic engine that employs a mechanical drive linkage to both the displacer and the power piston. Although special mechanisms like the rhombic drive are often used, the kinematic Stirling is analogous to other engines in that power take-off is through a mechanical linkage. However, the mass of these engines is relatively high. For space power, the preferred type is the free piston Stirling engine (FPSE). In addition to lower mass, the FPSE can be configured within a hermetically sealed vessel so that lubricants and high-pressure seals are not required. For electric power generation, the power piston is connected directly to the armature of a linear alternator.

DYNAMIC ENERGY CONVERSION

Figure 7.1 illustrates the ideal operation of the FPSE. The general Stirling cycle consists of two constant-temperature and two constant-volume processes, but in the FPSE, the relative pressures between the pistons control the motion of the piston and displacer. The common version of the FPSE has no external mechanical connections to either the displacer or power piston. Instead, both oscillate freely between gas springs, i.e., gas spaces wherein the pressure varies as the pistons traverse their respective strokes. A relatively large volume bounce space stays at a nearly constant pressure. A displacer rod passes through the power piston so that both pistons respond to the bounce space pressure. The relative motion of the two pistons is set by the relative masses and the exposed areas of the piston, displacer, and displacer rod. Typically, two hydrostatic pressurized helium bearings support the assembly: the power piston itself is one bearing, and the other is located at the inside diameter of the alternator armature. Both bearings are located on the cold end of the engine. No piston rings are required, but the clearances between the piston and displacer rod and between both pistons and the cylinder wall must be very small with tight tolerances to minimize bypass flow.

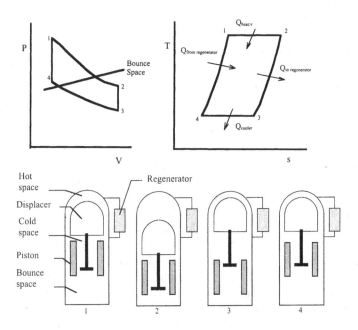

Figure 7.1 The Ideal Stirling Cycle

The power stroke from points 1 to 2 is ideally an isothermal expansion as heat is added from the external source, causing both the displacer and power piston to move downward. At point 2, the pressure in the bounce space has exceeded the pressure in the expansion space. The low-mass displacer moves upward from 2 to 3, and, in an ideally constant-volume process, pushes the expanded gas in the expansion space through the regenerator and into the cold space. At point 3, the bounce space pressure has decelerated the larger mass piston. It now begins its upward stroke from 3 to 4, compressing the cold gas. Ideally, this process is isothermal with waste heat removed by the cooling system. At point 4, the gas is fully compressed, the displacer moves downward from 4 to 1, and, in an ideally constant-volume process, forces the cold gas through the regenerator into the hot space. Gas flow into and out of the compression space is controlled by ports in the cylinder walls that are exposed or covered as the pistons move past them; thus, no valves are necessary.

Of course, ideal efficiency cannot be achieved because of the various energy losses in the engine. These include:
- Heat conduction through the walls of the pressure cylinder, regenerator, displacer, and insulation
- Pumping power for the displacer piston
- Oscillating flow losses in the regenerator
- Bypass flow and pressure losses around the displacer, displacer rod, and power piston
- Friction losses in the bearings
- Thermal, magnetic, and electrical losses in the linear alternator, including eddy currents and hysteresis

Since the power piston is coupled to the linear alternator, the electrical load will also affect the motion of the piston. In addition, the moving pistons exert vibration forces on the cylinder that must be damped out. One common method is to match two identical engines in a dual-opposed piston configuration so that the phased vibrations cancel each other. Helium is the gas usually employed in Stirling engines. It serves not only as the working fluid, but also pressurizes the gas springs and hydrostatic bearings. Performance would be improved with hydrogen, but sealing would be very difficult since high-pressure hydrogen will diffuse through hot metal walls.

A major advantage of the Stirling cycle is relatively high efficiency for designs that cover a wide range of power levels, and for a wide range of load conditions for a specific design. Indeed, the T–S diagram in Figure 7.1 can be viewed as the rectangular shape of the Carnot cycle skewed into a parallelogram of similar area. In addition, Stirling cycle efficiency remains competitive even with relatively high cold side temperatures, an advantage for heat rejection. However, approaching this ideal efficiency presents several technical challenges. Since the engine is heated from external sources, several components must operate at high temperature, including the heat delivery system, pressure cylinder, regenerator, and displacer. For space Stirling engines, superalloys are typically used for these components, and refractory alloys have been considered in some high-performance conceptual designs.

Highly-efficient regeneration is also a key requirement. Regenerators typically consist of wire meshes, metal foils, sintered metal fibers, or graphite fibers. Effective heat transfer to the working fluid must be traded against the pressure drop in the oscillating flow. In addition, heat loss due to thermal conduction along the length of the regenerator must be minimized. Furthermore, the maximum amount of working fluid must be shuttled through the regenerator. Any "dead space" in the hot side heater or expansion space, or in the cooler heat exchanger, contains fluid that does not add to the production of power. As a result, heat must be delivered to the gas by a very compact heat exchanger on the hot side of the engine. The same is true for the cooler, but that heat exchanger is less of a design challenge.

In the U.S., most of the Stirling engine development for space applications has been done by or for NASA's Lewis Research Center (LeRC). System concepts have been advanced for both solar and nuclear heat sources. Much of the key technical development began in the 1980s when LeRC investigated coupling of a high-power, long-life FPSE with the SP–100 nuclear reactor. The overall objective of the work was to achieve scale-up of the FPSE from its previously-demonstrated level of 3 kWe to a level of 25 kWe needed for space exploration applications. The key technical challenges were related to increasing the FPSE hot side temperature to the SP–100 coolant temperature of 1300 K and the cold side temperature to 650 K to enhance heat rejection. The mass-optimum temperature ratio is about 2.0 for unmanned applications (minimum shielding). For solar dynamic systems, concentrator and receiver masses are strongly affected by converter efficiency so the optimum temperature ratio is 2.5–2.7. Radioisotope systems optimize at about 2.2–2.5 (Dudenhoefer, *et al.*, 1994).

LeRC proceeded to develop the advanced FPSE in a series of steps, illustrated in Figure 7.2. Under contract to LeRC, Mechanical Technology, Inc. (MTI) conducted much of the development and testing. To verify that such scale-up was possible while maintaining high efficiency, the first-generation Space Power Demonstrator Engine (SPDE) was designed, built, and put on test in about 15 months (Dhar, *et al.*, 1987). From that hardware was derived the Space Power Research Engine (SPRE) that was used to further investigate FPSE parameters. That information fed the design and development of the Stirling Space Power Converter (SSPC) and its subsystem, the Component Test Power Converter (CTPC), to demonstrate significant increases in both hot and cold side temperatures (Dochat and Dudenhoefer, 1994). Figure 7.3 shows a schematic of the SSPC with major components identified.

The objective of the SPDE program was to demonstrate operation of a dynamically-balanced FPSE producing 25 kWe at 25% system efficiency and engine mass of 8 kg/kWe with a heat exchanger temperature ratio of 2.0 (630 K/315 K). In order to achieve the mass goal, the design frequency and a mean helium pressure were 105 Hz and 150 bar, respectively. These values were double the frequency and two-and-one-half times the pressure of past FPSE parameters. For balance, an opposed-piston configuration was employed so that each half of the engine was designed for 12.5 kWe. Hot

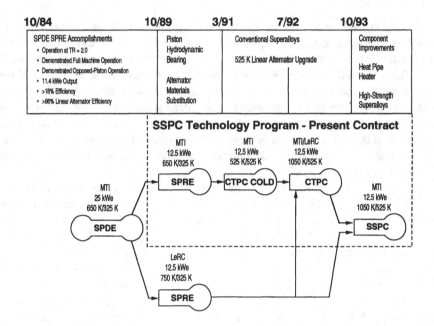

Figure 7.2 Evolution of the NASA Space Free Piston Stirling Engine (Courtesy of NASA)

molten salt (trade name HITEC) supplied heat, and a water-glycol loop cooled the engine (Dhar, et al., 1987).

Parametric tests were conducted with the SPDE for temperature ratios of 1.6–2.0, mean pressures of 75–150 bar, and piston strokes of 10–20 mm. By the end of the program, all of the design point goals were nearly met, the major deficiency being that high eddy current losses in the linear alternators reduced measured electrical power to 17 kWe. Indicated power and efficiency from engine measurements were 25 kWe (vs. a goal of 28.8 kWe) and 22% (vs. a goal of 28%) (Dhar, et al., 1987).

Following completion of the SPDE tests, LeRC conducted further development using the SPRE, formed by taking one-half of the SPDE and mating it to a dynamic vibration absorber. The SPRE served as a test bed for developments in hydrodynamic gas bearings, high efficiency linear alternators, centering port optimization, displacer clearance seals, and cooler design (Dudenhoefer, et al., 1994). Several design improvements incorporated into the SPRE resulted in alternator output of 11.2 kWe and system efficiency of about 19% (Dudenhoefer and Winter, 1991). Pressure–volume indicated efficiencies of 23–27% were demonstrated over a wide range of power levels and operating pressures (Dudenhoefer, et al., 1994).

The SSPC incorporated major advances in space FPSE development through improvements in heater head and cooler design and fabrication, regenerator design, re-

DYNAMIC ENERGY CONVERSION

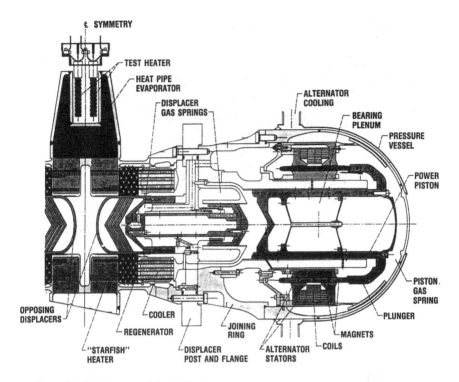

Figure 7.3 Configuration of the Stirling Space Power Converter (SSPC) (Courtesy of NASA)

duced bearing losses, and control (Dochat and Dhar, 1991). Most importantly, hot side temperature increased to 1050 K, considered the limit for superalloys, and the cold side operated at 525 K. Like the SPDE/SPRE, the opposed-piston configuration was used, and tests were conducted on one-half of the engine. Heat was supplied by a single, large annular sodium heat pipe that feeds multiple fins. This "Starfish" heater head is illustrated in Figure 7.4. To achieve the design lifetime of 60,000 hours, the SSPC would use Udimet 720 for its creep strength. However, the CTPC heater head was fabricated from Inconel 718, a superalloy that is much better characterized and more readily fabricated.

The CTPC design and results constitute the state of the art for high-temperature, high-power FPSE technology. If lifetime can be achieved with the Udimet 720, a 1050 K/525 K Stirling system could be employed with either a reactor or solar heat source. Initial testing was done with the cold end of the CTPC. This test hardware is shown in Figure 7.5. Except for the lack of lifetime testing, this hardware successfully demonstrated all of the primary cold side performance goals: 12.5 kWe alternator at 20% efficiency, operation from ambient to 525 K, full displacer and piston stroke of 28

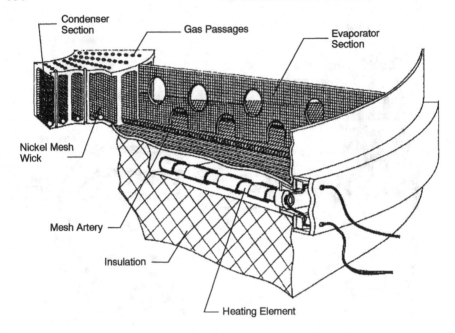

Figure 7.4 Starfish Heater Head/Heat Pipe Assembly (Courtesy of NASA)

mm, 150 bar mean helium pressure, and 70 Hz frequency. In the linear alternator, the stator coils and insulation operated at 573 K, and the samarium cobalt (Sm_2Co_{17}) permanent magnets operated at 548 K (Dudenhoefer, *et al.*, 1994). Full design operation was also achieved with the Starfish heat pipe heater head installed. Figures 7.6 and 7.7 depict these results.

LeRC analysis of the SP–100/Stirling system indicates that power density could be nearly doubled if a 1300 K/650 K system could be developed (Dudenhoefer and Winter, 1991). However, major material substitutions would be required on both the hot and cold sides of the FPSE. As shown in Table 7.1, LeRC has rated the suitability of several refractory alloys considered candidates for FPSE application. In general, complex refractory parts are difficult to manufacture and must operate in very high vacuum environments unless protective coatings can be developed. Compatibility with liquid metals, including low levels of impurities, is a significant issue, and one that is not yet fully resolved even with superalloys. On the cold side, 650 K operation necessitates new insulating and magnetic materials. Hence, the transition from 1050 K/525 K to 1300 K/650 K constitutes a major development step.

In the 1990s, interest in high efficiency converters for radioisotope systems has prompted work on small FPSEs producing a few watts. Small FPSE engines using radioisotope decay as the heat source have been developed and tested. As of July 1998,

DYNAMIC ENERGY CONVERSION

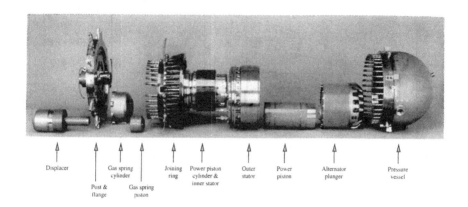

Figure 7.5 Component Test Power Converter (CTPC) Hardware (Courtesy of NASA)

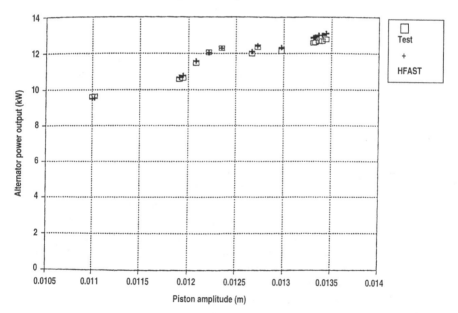

Figure 7.6 Measured CTPC Power (Courtesy of NASA)

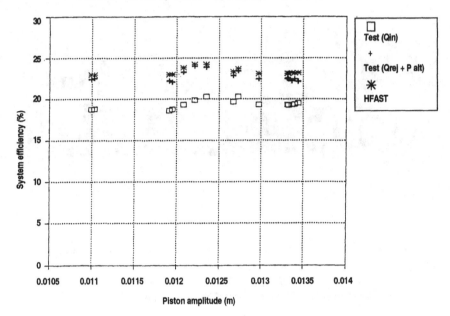

Figure 7.7 Measured CTPC Efficiency (Courtesy of NASA)

one 11 watt test engine had been endurance-tested for 44,000 hours (5 years) (White, 1998). However, the intended application was remote site power rather than space, so the cold side temperatures were low (293–368 K) (Montgomery et al., 1996).

3. Closed Brayton cycle

Investigation of Brayton cycle systems for space applications began in the early 1960s based on technology developed by the aircraft industry. NASA and the U.S. Department of Energy (DOE) sponsored development of closed Brayton cycle (CBC) components and systems for both nuclear and solar heat sources. Designs were developed for a wide range of space applications from 500 watts to 100 kWe. Among the most significant of these developments was the Brayton Rotating Unit (BRU) that accumulated 41,000 hours of operation on one of the units at a nominal power of 10.7 kWe. The BRU was later modified with foil bearings and an upgraded alternator to 15 kWe. Designated BRU–F, it was used by NASA in the 1980s as a test bed for development of the Solar Dynamic Power Module, planned at that time for the Space Station (Overholt, 1994). Another unit, the Mini–BRU was developed as a part of the Brayton Isotope

Table 7.1 Refractory Material Candidates for 1300 K Stirling Engine Components (after Dudenhoefer and Winter, 1991)

Base metal	Melting point (K)	Density (gm/cc)	Alloy name	Composition (wt %)	Join-ability	Fabric-ability	Alloy availability	Data availability	Vacuum (torr)
W	3680	19.3	W-25Re-HfC	24-26% Re 1% HfC	5	4	4	3	10^{-6}
Ta	3270	16.6	ASTAR-8 11C	8% W 1% Re 1% HfC	8	8	10	5	10^{-8}
Mo	2880	10.2	TZM	0.08% Zr 0.5% Ti	2	8	10	4	10^{-6}
			TZC	1.25% Ti 0.1% Zr 0.15% C	2	6	10	4	10^{-6}
MoRe	2780	15.5	Mo–47.5 Re	47.5% Re bal Mo	8	6	8	3	10^{-6}
Nb	2740	8.6	FS-85	11% W 28% Ta 1% Zr	8	8	5	4	10^{-8}
			B-88	27% W 2% HfC	7	7	4	2	10^{-8}
			C-103	10% Hf 1% Ti 0.7% Zr	10	10	10	7	10^{-8}
			PWC-11	1% Zr 0.1% C	10	10	10	7	10^{-8}
			Nb-1Zr	1% Zr	10	10	10	8	10^{-8}

Power System (BIPS) and produced 1.3 kWe during a 1000 hour endurance test at an overall efficiency of 28%. In 1995, NASA tested that same Mini–BRU unit with upgrades at 2.0 kWe as part of the Solar Dynamic Ground Test Demonstration (SDGTD) project. The most recent nuclear-based CBC system was the Dynamic Isotope Power System (DIPS) designed in the 1980s, also largely based on the Mini–BRU technology. The objective of the DIPS program was to extend the power range of Radioisotope Thermoelectric Generators (RTGs) into the 1–10 kWe range by replacing the thermoelectrics with a high-efficiency CBC system. A system design was developed for 6 kWe end-of-life operation.

Figure 7.8 illustrates the ideal temperature–entropy diagram. In the CBC, gas is pressurized and heated, then expanded through a turbine to produce work. The ideal CBC consists of isentropic compression, isobaric heat addition, isentropic expansion, and isobaric cooling. Ideal performance is reduced by heat and flow losses, bearing losses, and irreversibilities in the compression and expansion processes. Cycle efficiency is improved significantly by adding an intermediate heat exchanger known as a recuperator. The recuperator recovers a large portion of the heat from the turbine exhaust and transfers it to the compressed gas flowing into the heat source. Hence, less heat is required from the heat source.

Space CBC systems are normally optimized to minimize mass, although limiting the area of the waste heat radiator or increasing efficiency to reduce radioisotope inventory sometimes takes precedence. The key independent cycle parameters are (Overholt, 1994):

- Output power
- Compressor inlet temperature
- Turbine inlet temperature

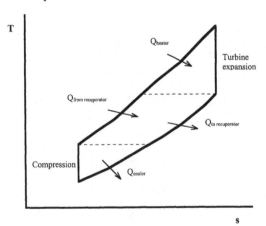

Figure 7.8 The Ideal CBC Temperature–Entropy Diagram

- Recuperator effectiveness
- Compressor pressure ratio
- Pressure loss parameter
- Shaft speed
- Working fluid and molecular weight

The design trades among these parameters are many. Output power is determined by the application, including mission duration and power system degradation over that time period. For high efficiency, low compressor inlet temperature is desired, but compressor in*let al*so sets the temperature of the waste heat radiator which in turn sets radiator size and mass. Although a recuperator increases efficiency, it adds mass and usually lowers the cold side temperature, thus adding to the size and mass of the radiator. Increasing recuperator effectiveness also increases its size and mass, but system mass generally optimizes at very high recuperator effectiveness. Recuperated CBCs optimize at relatively high turbine temperatures, which in turn reduces the pressure ratio required across the turbine. Lower pressure ratio simplifies the design of the turbine. Turbine inlet temperature is typically limited by the strength and compatibility of the materials used. Design speed is chosen to avoid resonance points, limit bearing loads, and achieve high alternator efficiency. The working fluid must be compatible with CBC materials and have desirable heat transfer and flow properties.

The state of the art for CBC systems for space applications is embodied in the NASA solar dynamic designs and the DOE DIPS. Both CBC systems draw heavily on development work in the 1970s, particularly the BRU and Mini–BRU technology. Chapter 3 discusses the solar dynamic systems in more detail. The DIPS system diagram and state points are shown in Figure 7.9 and Table 7.2, respectively (Rockwell, 1988).

The materials used for the various components are listed in Table 7.3 (Rockwell, 1988). The BRU and Mini–BRU constitute the state of the art for turboalternator–compressor (TAC) technology for space applications. Figure 7.10 illustrates the Mini-BRU TAC.

For power levels less than about 100 kWe, radial turbomachinery can be used instead of more complex multistage axial schemes. Figure 7.11 shows several radial compressor and power turbines designed for space systems. The power conversion portion of the CBC is typically packaged as a single TAC unit with all three components mounted on a single shaft. The major advantage of this configuration is reliability because the alternator is housed within the gas system, so no external mechanical drive and associated seals are required. The TAC shaft is the only rotating component in that type of CBC design. This configuration also enables the use of self-actuated foil gas bearings, illustrated in Figure 7.12. Working fluid bled from the compressor discharge provides both lubrication and cooling for these bearings. Foil bearings have the additional advantages of no running contact, good shock and vibration resistance, no constraint on rotor speed, increased load capacity at higher speed, and relatively high tolerance to support misalignment. Application of this bearing technology to the BRU and

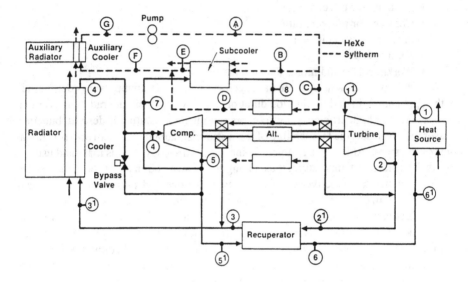

Figure 7.9 The DIPS Block Diagram (after Rockwell, 1988)

Table 7.2 State Points for DIPS (after Rockwell, 1988)

State point	Location	Temperature (K)	Pressure (psia)	Mass flow (gm/sec)
1	Heat source outlet	1038	52.55	306.4
1^1	Turbine inlet	1033	52.49	306.4
2	Turbine outlet	861	31.27	306.4
2^1	Lo-P recuperator in	852	31.26	311.6
3	Lo-P recuperator out	417	30.99	311.6
3^1	Gas cooler in	419	30.95	317.9
4	Gas cooler out	290	30.34	317.9
4^1	Compressor inlet	290	30.31	317.9
5	Compressor outlet	384	53.86	317.9
5^1	Hi-P recuperator in	384	53.79	306.4
6	Hi-P recuperator out	826	53.10	306.4
6^1	Heat source inlet	826	53.04	306.4
7	Bleed flow out	384	53.86	15.86
8	Bleed flow to TAC	337	53.84	15.86
A	Pump outlet	337	–	12.68
B	Bleed cooler inlet	337	–	3.05
C	TAC inlet	337	–	9.63
D	TAC outlet	365	–	9.63
E	Bleed cooler out	367	–	3.05
F	Cooler inlet	366	–	12.68
G	Cooler outlet	336	–	12.68

Net power = 6 kWe; TAC speed = 49,632 RPM; He–Xe molecular weight = 61.7

Table 7.3 Materials for DIPS Components (after Rockwell, 1988)

Component	Material
Turbine shroud, nozzle and scroll	Inconel 617
Turbine rotor	Inconel 713 LC
Bearings	
Thrust rotor	15–5PH
Thrust carrier	CRES 347
Thrust foil	Inconel X–750
Foil journal spacer	Inconel X–750
Foil carrier	CRES 347
Compressor	
Diffuser	CRES 347
Impeller	Ti–6Al–4V (Forged)
Scroll	CRES 347
Alternator	
Shroud	Copper
Stator	AISI 4340/Inconel 718
Laminations	ARMCO ingot iron
Main housing and seals	CRES 347
Recuperator	
Plate	Hastelloy X
Fin	Hastelloy X
Structure	Hastelloy X
Cooler	High chromium steel

Mini–BRU was a major advancement in CBC reliability (Rockwell, 1988; Overholt, 1994). Magnetic bearings are another option for high-efficiency, long-life CBC systems.

Turbine inlet temperature is limited to about 1050-1150 K for the state-of-the-art nickel-based superalloys used for the turbine, shroud, and hot side ducts. Refractory materials, such as niobium alloys, can raise hot side temperatures about 300 K, but these materials are very susceptible to oxygen corrosion and some impurities. In a CBC loop that also contains nickel alloys or stainless steel, the transport of oxygen in the working fluid is an important lifetime concern. The automotive industry has developed ceramic (silicon carbide and silicon nitride) radial turbomachinery that can be used up to about 1700 K, but these materials also have compatibility questions and can fail catastrophically. Future developments in carbon-carbon composites may allow hot side temperatures above 2000 K, but require reliable coatings. In all advanced material cases, the manufacture of complex shapes is an issue (Overholt, 1994; Gilmour, et al., 1990).

Figure 7.13 shows a cross section schematic of the DIPS alternator (Rockwell, 1988). The Rice (modified Lundell) alternator has been used in all space CBC systems to date. It is a brushless, nonrotating coil, synchronous machine. The fully redundant, stationary field coils provide excitation. A conventional 3–phase winding is used for

SPACECRAFT POWER TECHNOLOGIES

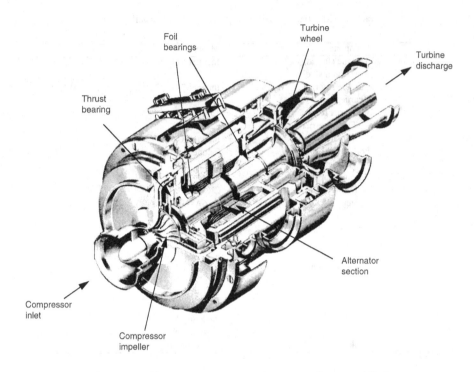

Figure 7.10 The Mini–BRU Turboalternator Compressor (Courtesy of NASA)

TYPE	DIA. IN	SPEED RPM	FLUID	PRESSURE RATIO	PERFORM. %	SPONSOR
RADIAL	4.16	45,000	ARGON	2.0	82.5	AIRESEARCH
RADIAL	3.2	64,000	ARGON	2.06	80.5	USAF
RADIAL	6.0	38,500	ARGON	2.38	80.0	NASA
RADIAL β^2	6.5	38,500	ARGON	2.38	83.0	NASA
RADIAL β^2	4.25	36,000	Xe-He	1.9	83.0	NASA
RADIAL β^2	10.75	35,000	Air	2.5	86.0	AIRESEARCH
RADIAL β^2	2.12	52,000	Xe-He	1.53	77.0	NASA

TYPE	DIA. IN	SPEED RPM	FLUID	PRESSURE RATIO	PERFORM. %	SPONSOR
RADIAL	3.5	53,000	ARGON	1.56	87	NASA
RADIAL	4.6	50,500	ARGON	1.56	88	NASA
RADIAL	6.0	38,500	ARGON	1.56	88	NASA
RADIAL	5.0	36,000	Xe-He	1.87	90	NASA
RADIAL	2.85	52,000	Xe-He	1.5	84	NASA

Figure 7.11 Radial Compressor and Power Turbines Designed for Space Systems (Courtesy of NASA)

DYNAMIC ENERGY CONVERSION

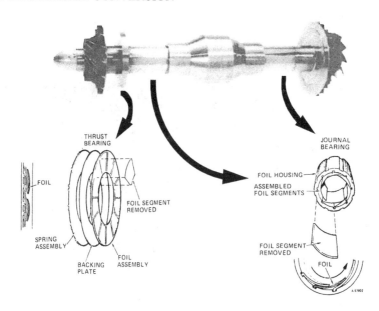

Figure 7.12 Self-Actuated Foil Gas Bearings for CBC (Courtesy of NASA)

the stator. The alternator rotor is an integral part of the TAC rotor and consists of three brazed components: two 4340 steel end pieces that carry the flux and an Inconel 718 center section. For startup, the alternator acts as an induction motor when external power is applied. For systems generating over 100 kWe, permanent magnet generators probably weigh less (Rockwell, 1988; Overholt, 1994).

For space power systems, the working fluid is typically a mixture of helium and xenon. High-purity inert gases add to the reliability of unattended, long-life systems by greatly reducing the potential for corrosion. The mixture ratio is selected to balance the favorable heat transfer characteristics of helium (molecular weight = 4) and the improvements in turbomachinery performance afforded by the higher molecular weight (131) of xenon (Overholt, 1994). NASA designed the SDGTD system for a He/Xe mixture with a molecular weight of 83. In the DIPS design, the molecular weight of the He/Xe mixture was 61.7 (Rockwell, 1988).

CBC systems employ heat exchangers for various purposes. Heat input and waste heat rejection radiators are necessary in all CBC systems, and most designs incorporate a recuperator to enhance efficiency. As shown in Figure 7.10, the DIPS design employs an auxiliary loop to cool the alternator stator, and this loop entails two heat exchangers: a subcooler and an auxiliary radiator. Other designs, like the NASA SDGTD, use an intermediate heat exchanger between the primary CBC gas loop and a liquid-based waste heat radiator. The waste heat rejection radiator typically constitutes a significant portion of the mass and size of the CBC system; hence, lightweight materials with good

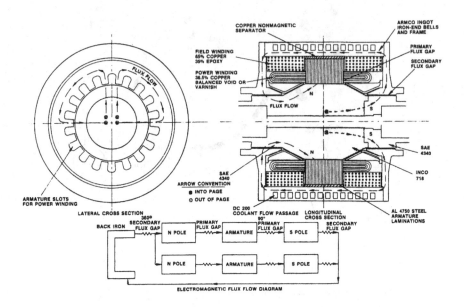

Figure 7.13 Cross-section of the DIPS Four-Pole Rice Alternator (Courtesy of NASA)

heat transfer characteristics are necessary. In addition, this radiator must fit in the launch vehicle, and deployable radiators may be necessary for larger power systems. Because of high temperature operation, the heat input heat exchanger and the recuperator must use lightweight materials with high structural and creep strength at high temperature. The recuperator technology from the BIPS program in the 1970s is state-of-the-art for CBC systems. This plate-fin design has a demonstrated effectiveness of 0.975, and has been subjected to accelerated life tests, including 200 thermal shocks, without formation of internal or external leaks (Overholt, 1994). Chapter 9 provides more detail on thermal management technology.

4. Rankine cycle

The most common example of the Rankine system is the steam thermodynamic cycle used extensively for commercial power production. However, other fluids are desirable for space applications, and operation in zero-g or microgravity requires some specially designed components. NASA, DOE (and its predecessor agencies), and the Air Force sponsored development of Rankine designs for both solar and nuclear energy sources. However, most of the Rankine system development was based on either nuclear reac-

DYNAMIC ENERGY CONVERSION

tors or radioisotope heat sources. During the SNAP program in the 1960s, the 3–5 kWe SNAP-2 and the 35–50 kWe SNAP-8 liquid metal Rankine systems were developed and tested. In addition, the larger (300 kWe nominal) potassium (K–Rankine) SNAP-50 was designed. Some SNAP-50 components were tested, but no system tests were conducted. Oak Ridge National Laboratory (ORNL) designed two versions of its Medium Power Reactor Experiment (MPRE): a moderate temperature version with stainless steel, and a high temperature version with refractory metals.

All of the aforementioned designs used liquid metals as the working fluid. Other lower temperature designs use organic fluids as the working media. A 6 kWe organic Rankine cycle (ORC) system was tested in the 1960s. The most recent ORC system development was the 1.3 kWe Kilowatt Isotope Power System (KIPS) developed by DOE in the 1970s as part of the DIPS program. (In that program, ORC was chosen over Brayton for further development, and both are often designated DIPS, not to be confused with the later DOE/Air Force DIPS Brayton program in the 1980s). In the 1980s, early designs for the Space Station used solar dynamic modules in addition to photovoltaics. Although NASA eventually selected Brayton (solar dynamic was later dropped), some additional ORC component work was performed (Anderson, 1983; Angelo and Buden, 1985; Bennett and Lombardo, 1987; Bloomfield, 1994; Chaudoir, et al., 1985; Voss, 1984).

Figure 7.14 illustrates the ideal Rankine cycle T–s diagram with superheat. The key attribute of Rankine systems is the alternate boiling and condensing of the working fluid that allows the use of very effective, lightweight, and compact heat exchangers. The cycle consists of isentropic compression of the liquid in a subcooled state by the pump, isobaric heat addition in the boiler which vaporizes and superheats the fluid, isentropic expansion in the turbine, and isothermal heat rejection by the condensing radiator. By superheating the vapor beyond its vapor state at the boiling temperature,

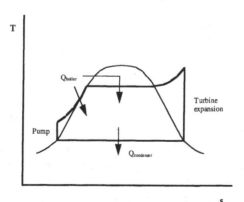

Figure 7.14 Temperature–Entropy Diagram for Ideal Rankine Cycle with Superheat

Carnot efficiency is improved and liquid droplets can be largely eliminated in the expansion process. The droplets are undesirable because they increase erosion of the turbine blades. Efficiency can be improved further by reheating the fluid between turbine stages or adding a regeneration step. Regeneration transfers heat from the fluid exiting the turbine to the liquid entering the boiler. Both reheat and regeneration add components and complexity to the system, and the efficiency gains must be traded against reliability and total system mass.

Selection of the working fluid and its operating pressures establishes the boiling and condensing temperatures of the cycle. At the turbine inlet, higher enthalpy (temperature and pressure) fluid is desirable to improve Carnot efficiency. Table 7.4 summarizes typical turbine inlet conditions. However, the turbine and ducting materials must be compatible with each other and the fluid and must have sufficient structural and creep strength at temperature to withstand the pressure, thermal stresses, and any other loads. Low fluid enthalpy in the condenser increases efficiency but also increases radiator size. Toluene and Dow Chemical Company's Dowtherm A® (a eutectic mixture of diphenyl and biphenyl oxide) are the organics that have received the most attention for space ORC designs. The Air Force has considered phosphorous halides, phosphoryl halides, and thiophosphoryl halides as possibilities for raising Rankine operating temperatures (to as high as 1000 K boiler temperature), but very little work has been done for space systems (Grzyll, *et al.*, 1988). Liquid metal options include mercury and alkali metals like potassium, cesium, and rubidium. Potassium is of particular interest because of possible vaporization temperatures of 1400 K. However, material problems are much more challenging with liquid metals than with ORCs.

Table 7.4 Typical Turbine Inlet Temperature and Pressure for Rankine Cycle Fluids
(after Bloomfield, 1994)

Fluid	Temperature		Pressure	
	K	°F	kPa	psia
Organic	650	700	2200	320
Mercury	950	1250	1830	265
Potassium	1420	2100	1120	160

Additional trades occur at the system level. Liquid metal designs are primarily intended for reactor-based systems where the reactor coolant is also a liquid metal. For example, both SNAP–2 and SNAP–8 used mercury as the working fluid, but the boiler in each was an intermediate heat exchanger to a NaK loop that cooled the reactor. SNAP–50 and the follow-on NASA work that investigated K–Rankine technology also used an intermediate heat exchanger between the converter loop and the reactor coolant, lithium in the latter case. The ORNL MPRE program investigated direct loop designs wherein the potassium working fluid also cooled the reactor. The advantage of the direct loop is

DYNAMIC ENERGY CONVERSION   343

elimination of the intermediate high temperature heat exchanger. However, the reactor must now serve as the boiler which complicates reactor design and testing. Indirect systems simplify the development and qualification of the system since the reactor loop and the Rankine loop can be tested independently. Another advantage is that any nuclear activation of the reactor coolant does not affect the Rankine components, nor do any impurities or other materials that may originate in the core.

Discussion of the state of the art of Rankine technology for space is best approached by considering the ORC and liquid metal systems independently. ORC systems provide moderate efficiency at lower temperatures than liquid metals and work best for lower power systems. The best established technology comes from the KIPS work in the 1970s, much of which derives from the 6 kWe program of the 1960s. In addition, a very large and relevant data base exists from terrestrial use of ORC systems (Niggemann and Lacey, 1985). Figure 7.15 shows the block diagram, and Table 7.5 lists the state points for the 1.3 kWe KIPS system.

In this design, only one Rankine loop is included, but any future flight system will probably incorporate a least one redundant loop for reliability. The ground demonstration system tests accumulated over 11,000 hours between December, 1977 and December, 1980. These hours include an endurance test of 2000 hours that was terminated intentionally. Table 7.6 lists some of the measured component performance parameters. The system produced 1.3 kWe at 28 Vdc. Its mass is 210 kg, and the radiator area is 10.8 m^2. For a turbine inlet temperature of 645 K, system efficiency was measured at 18.5% based on the ac output of the alternator.

The design is based on the 1970s vintage Multi-Hundred Watt radioisotope source used in the RTGs that powered the Voyager 1/2 and LES 8/9 missions, but the current General Purpose Heat Source (GPHS) can also be easily accommodated. Key ORC

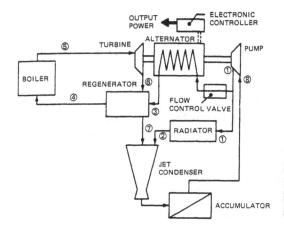

Figure 7.15 Block Diagram for 1.3 kWe Organic Rankine System (after Anderson, 1983)

Table 7.5 State Points for 1.3 kWe Organic Rankine System (after Anderson, 1983)

State point	Temperature (K)	Pressure (psia)
1	374	96
2	348	86
3	380	73
4	520	69
5	630	57
6	559	0.137
7	389	0.105
8	373	31

Table 7.6 Measured Performance Parameters for 1.3 kWe Organic Rankine System (after Anderson, 1983)

Parameter	Value
Input electrical power	7.515 kWe
AC power	1.387 kWe
AC efficiency	18.5%
DC power	1.239 kWe
DC efficiency	16.56%
Turbine flow rate	14 gm/sec
Rotational speed	34,610 RPM
Turbine efficiency	68%
Pump efficiency	71%
Regenerator effectiveness	97%
Alternator efficiency	93.5%
Bearing and seal losses	95 watts
Turbine inlet temperature	645 K

components are an axial flow impulse turbine, homopolar inductor alternator, centrifugal feed pump, working fluid lubricated film bearings, once-through boiler, jet condenser, and pumped liquid working fluid radiator. The liquid outlet from the pump is split by a flow control valve between the power loop and the cooling loop. Liquid in the power loop first cools the alternator, then passes through the regenerator to extract energy from the turbine exhaust vapor, is vaporized in the boiler surrounding the radioisotope heat source, and finally expands through the turbine. The turbine drives both the alternator and the pump, all mounted on a single shaft as a combined rotating unit (CRU). Liquid from the pump entering the cooling loop flows first through the radiator to reject waste heat, and it is then injected into the jet condenser to condense the vapor from the power loop. The combined liquid returns to the pump through the accumulator that collects noncondensibles that develop in the flow.

The Dowtherm A® working fluid is boiled in a heat exchanger that surrounds the radioisotope heat source. The turbine, pump, alternator, condenser, regenerator, and accumulator are all packaged in a single compact arrangement shown in Figure 7.16.

DYNAMIC ENERGY CONVERSION   345

For operation in zero-g, the jet condenser is a key innovation. Subcooled liquid Dowtherm A® is injected at approximately 30 m/s through a central nozzle coaxially with vapor flowing in around it from the regenerator. The cold liquid condenses the vapor. The diffuser section allows recovery of part of the momentum of the injected liquid. To function properly, the jet condenser needs a very low vapor pressure fluid like Dowtherm A®. Nonreversible floodout may occur if the jets are defocused for any reason, e.g., by noncondensible gases or shock loads. The original ground test hardware was reworked to solve this problem. Furthermore, for power levels of several kWe, the high volumetric flow rates necessary with the low pressure Dowtherm A® results in large heat exchangers and other necessary equipment to avoid excessive pressure drops. Toluene is the probable alternative, but jet condenser operation becomes more difficult. To improve reliability, other zero-g condenser options have been studied, notably the Rotary Fluid Management Device (RFMD) and the rotating jet condenser (RJC) (Bland, et al., 1987). Sundstrand operated a prototypic RFMD for over two years in the 1980s (Niggemann and Lacey, 1985), and NASA conducted microgravity tests for short intervals with 54 KC-135 flights in 1987 (Bennett and Lombardo, 1987). Although some development work has been done for these condenser options, reliable, long-life operation in zero-g has not been fully established for ORC systems.

Table 7.7 summarizes the properties of three organic fluid options: toluene, Dowtherm A®, and RC-1 (hexafluorobenzyne–pentafluorobenzyne). In the KIPS program, slight pyrolytic decomposition of the Dowtherm A® occurred during thermal cycling at high temperatures (Anderson, et al., 1983). Liquid droplets of high molecular weight formed at the boiler outlet. In the 1980s, during investigation of ORC for the

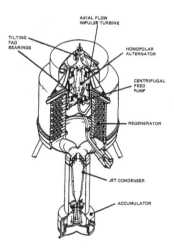

Figure 7.16 Combined Turbine, Pump, Alternator, Condenser, Regenerator and Accumulator for 1.3 kWe Organic Rankine System (after Anderson, 1983)

Space Station, NASA funded dynamic loop tests to examine toluene stability (Havens, et al., 1987). Test time was 3410 hours at 670 K with short excursions to 730 K. Fluid samples contained both noncondensible gases (hydrogen, methane, ethane, and propane) and liquids (benzene, benzaldehyde, biphenyl, bibenzyl, and a variety of related species). The total accumulation of noncondensibles was about 200 scc and about 0.3 weight percent liquids. These results are not considered excessive, but point out that the degradation of organics with time must be understood and accounted for in ORC design.

For the liquid metal Rankine technology, the data base has not changed appreciably since the major programs ended in the 1960s and early 1970s. Table 7.8 summarizes the U.S. Rankine cycle technology test experience for space applications. Since K–Rankine technology promises to deliver the best performance, and because the later programs focused on this liquid metal, emphasis here is placed on its status. Except for the boiler, the state of the art of K–Rankine derives from two programs: the two-loop,

Table 7.7 Summary of Organic Fluid Properties (after Boretz, 1986)

Items	Dowtherm A®	Toluene	RC–1
Chemical composition	0.265 $(C_6H_5)_2$+ 0.735 $(C_6H_5)_2O$	C_7H_8	0.60 C_6HF_5+ 0.40 C_6F_6
Molecular weight	166	92.12	175.26
Upper cycle temperature	615 K (650 °F)	670 K (750 °F)	755 K (900 °F)
Boiling point	–	384 K (231 °F)	351 K (172 °F)
Freezing point	285 K (54 °F)	178 K (–140 °F)	231 K (–44 °F)
Flash point	397 K (255 °F)	278 K (40 °F)	None
Specific heat	0.579 cal/gm-K	0.471 cal/gm-K	–
Critical point temperature	772 K (930 °F)	594 K (610 °F)	509 K (457 °F)
pressure	3241 kPa (470 psia)	4254 kPa (616 psia)	2828 kPa (410 psia)
System design impacts	Easier to adapt to low power system	Higher pressure levels	Higher pressure levels
	Low head drop in expansion — low tip speed	Stable at higher temperature	May be stable at higher temperature
	Low turbine exit pressure	Easier startup due to low freezing point	Non-toxic and nonflammable
	Requires pressure recovery to provide suction head at pump	Less fluid inventory	Low head drop — low tip speed — large mass flow rate
			High fluid inventory

indirect-heating system technology developed under the SNAP–50 and the subsequent NASA work on components, and the single-loop direct-heating of the ORNL MPRE.

The ORNL stainless technology is by far the most mature. However, efficiency can be greatly improved (or radiator area greatly reduced) with the higher temperature capability of refractory alloys. Figure 7.17 shows the cycle schematic for the refractory version of the MPRE. This diagram is simplified in that additional components are not shown. Nb–1Zr is the refractory alloy used for most of the high-temperature components. Under MPRE, niobium underwent 15,000 hours of thermal convection corrosion testing and 3300 hours of forced convection testing at temperatures up to 1365 K. In a K–Rankine system, it is difficult to superheat the fluid sufficiently at the inlet to avoid condensate formation in the turbine (Angelo and Buden, 1985). Interstage reheat and condensate removal are necessary to achieve long life of the turbine. If condensate does form, turbine lifetime is impacted, although some condensate can be tolerated in the colder stages.

The materials and component technology base developed by NASA for K–Rankine includes (Bloomfield, 1994):
- Compatibility and corrosion loop testing of 5000 hours for niobium alloys and 10,000 hours for tantalum alloys at temperatures up to 1500 K
- Fabrication, performance test, and endurance test of potassium condensers and once-through potassium boilers/lithium-to-potassium heat exchangers

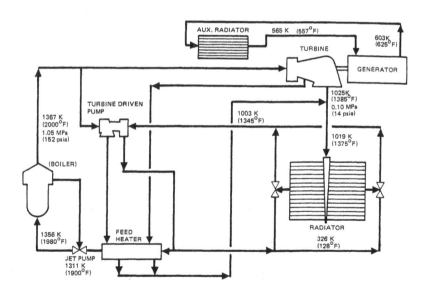

Figure 7.17 Block Diagram and State Points for the Refractory Version of the MPRE K–Rankine System (after Anderson, 1983)

Table 7.8. Summary of K–Rankine Test Experience (hours > 810 K) (after Anderson, et al., 1983)

	Aerojet General	AiResearch	Allison (GM)	General Electric	NASA JPL	NASA LeRC	ORNL	Pratt & Whitney	Rocketdyne	United Nuclear
Corrosion test systems										
Boiling systems										
Thermal convection	–	–	–	10,500	–	–	43,600	12,000	–	–
Forced convection	–	1300	–	5000	–	–	19,200	–	–	–
All liquid systems										
Thermal convection	–	–	–	–	–	–	–	–	–	100,000
Forced convection	–	–	53,000	–	–	–	–	–	–	3000
Component test systems (boiling)	100	5900	–	19,500	–	1000	2800	4900	200	–
Simulated power plants	–	–	–	–	~1000	–	10,200	–	–	–
Component–power plant										
Boilers <35 kW	–	1300	–	15,500	1000	–	71,400	16,900	–	–
>35 kW	100	5900	–	19,600	–	1000	4400	–	200	–
Turbines with K lubricated bearings										
<10 kW	–	3000*	–	–	–	–	5000*	–	100	–
>10 kW	–	50*	–	–	–	–	–	–	–	–
Turbines with oil lubricated bearings >10 kW	–	–	–	5100*	–	–	–	–	–	–
Boiler feed pumps										
Electromagnetic	100	7200	–	24,600	3600	1000	26,900	4900	200	3000
Centrifugal	–	–	–	–	–	–	5000*	–	–	–
Radiator	–	–	–	–	–	600	–	–	–	–
Condensers										
Liquid metal loop	–	–	–	2500	–	600	5000	4900	–	–
Air-cooled	–	5900	–	16,100	–	–	43,600	4900	200	–
Radiator (combined)	100	1300	–	16,500	~1000	400	27,500	12,000	–	–
Pumps (liquid syst)	–	5900	53,000	18,600	3600	1600	6100	1300	1600	3000
Potassium seals	–	3050*	–	5100*	–	–	–	–	–	–
Bearing test rig										
Instrumented	–	3000*	–	–	–	–	–	–	1600	–
Single endurance	–	300	–	–	–	–	4500	–	–	–

* Accumulated during operation of the turbine indicated in that column

- Fabrication and 10,000 hours of endurance test of the boiler feed electromagnetic (EM) induction pump
- 5000 hour performance and endurance testing of a two-stage potassium vapor turbine
- 5000 hour performance and erosion test of a three-stage potassium vapor turbine
- 10,000 endurance testing of 980 K electrical components (solenoid, transformer, and stator)

Despite this array of data on high-temperature materials and components, significant uncertainty remains about the long-term performance of the K–Rankine system. Most of the issues center on the need for high-strength refractory metal alloys and include fabricability of complex parts, system-level materials compatibility, and corrosion. Particular technology needs include (Bloomfield, 1994):
- Fabrication processes for T–111 and ASTAR–811C alloys
- Tip speed limitations and potassium erosion of large potassium vapor turbines
- Turbine bearings and seals capable of operating at higher temperatures and pressures
- Non-magnetic generator winding seals

5. References

Anderson, R. V., et al., *Space Reactor Electric Systems*, ESG–DOE–13398, (Rockwell International, Canoga Park, CA, March 29, 1983), IV 362–363.

Angelo, J. A. Jr., and Buden, D., *Space Nuclear Power*, (Orbit Book Company, Inc., Malabar, FL, 1985).

Bennett, G. L. and Lombardo, J. J., *Technology Development of Dynamic Isotope Power Systems for Space Applications*, Proceedings of the 22nd Intersociety Energy Conversion Engineering Conference (IECEC 87), Paper #879094, (American Institute of Aeronautics and Astronautics, New York, 1987), 366–372.

Bland, T. J., Sorensen, G., and Clodfelter, K., *Two-Phase Fluid Management for a Rankine Cycle Isotope Power System*, Transactions of the Fourth Symposium on Space Nuclear Power Systems, El-Genk, M. S., and Hoover, M. D., (eds.), CONF–870102–Summs., (University of New Mexico, Albuquerque, 1987), 209–214.

Bloomfield, H. S., *Dynamic Power Conversion Systems for Space Nuclear Power, Section II: Rankine Cycle*, A Critical Review of Space Nuclear Power and Propulsion 1984–1993, M. S. El-Genk, (ed.), (American Institute of Physics Press, New York, 1994), 345–357.

Boretz, J. E., *Supercritical Organic Rankine Engines (SCORE)*, Proceedings of the 21st Intersociety Energy Conversion Engineering Conference (IECEC 86), Paper #869474, (American Chemical Society, Washington, DC, 1986), 2050–2054.

Chaudoir, D. W., Niggemann, R. E., and Bland, T. J., *A Solar Dynamic ORC Power System for Space Station Application*, Proceedings of the 20th Intersociety Energy Conversion Engineering Conference (IECEC 85), Paper # 859085, SAE P-164, (Society of Automotive Engineers, Warrendale, PA, 1985), 1.58–1.65.

Dhar, M., Jones, D., Huang, S., and Rauch, J., *Design and Performance of a 25-kWe Free-Piston Stirling Space Power Demonstrator Engine*, Proceedings of the 22nd Intersociety Energy Conversion Engineering Conference (IECEC 87), Paper #879163, (American Institute of Aeronautics and Astronautics, New York, 1987), 133–137.

Dochat, G. R. and Dudenhoefer, J. E., *Performance Results of the Stirling Power Converter*, Proceedings of the Eleventh Symposium on Space Nuclear Power and Propulsion, AIP Conference Proceedings 301, (American Institute of Physics Press, New York, 1994), 457–464.

Dochat, G. and Dhar, M., *Free-Piston Stirling Component Test Power Converter*, Proceedings of the 26th Intersociety Energy Conversion Engineering Conference (IECEC 91), (Society of Automotive Engineers, Warrendale, PA, 1991), 5.239–5.244

Dudenhoefer, J. E. and Winter, J. M., *Status of NASA's Stirling Space Power Converter Program*, Proceedings of the 26th Intersociety Energy Conversion Engineering Conference (IECEC 91), (Society of Automotive Engineers, Warrendale, PA, 1991), 2.38–2.43.

Dudenhoefer, J. E., *et al.*, *Dynamic Power Conversion Systems for Space Nuclear Power, Section I: Stirling Cycle*, A Critical Review of Space Nuclear Power and Propulsion 1984–1993, M. S. El-Genk, (ed.), (American Institute of Physics Press, New York, 1994), 306–344.

Gilmour, A. S., Jr., Hyder, A. K., Jr., and Rose, M. F., *Space Power Technology*, short course, circa 1990.

Grzyll, L. R., Mahefkey, E. T., and Leland, J. E., *Identification of New Working Fluids for Use in High-Temperature Rankine Cycles*, Proceedings of the 23rd Intersociety Energy Conversion Engineering Conference (IECEC 88), Paper # 889189, (American Society of Mechanical Engineers, Fairfield, NJ, 1988), 267–272.

Havens, V. N., Ragaller, D. R., and Sibert, L., *Toluene Stability Space Station Rankine Power System*, Proceedings of the 22nd Intersociety Energy Conversion Engineering Conference (IECEC 87), Paper #879161, (American Institute of Aeronautics and Astronautics, New York, 1987), 121–126.

Montgomery, W. L., Ross, B. A., and Penswick, L. B., Third Generation Development of an 11-Watt Stirling Converter, Proceedings of the 31st Intersociety Energy Conversion Engineering Conference (IECEC 96), Paper #96371, (Institute of Electrical and Electronics Engineers, Piscataway, NJ, 1996), 1243–1248.

NASA Reference Publication 1310, *Solar Dynamic Power System Development for Space Station Freedom*, (NASA Lewis Research Center Solar Dynamic Power System Branch, Cleveland, OH, July 1993).

Niggemann, R. E. and Lacey, D., *Reactor/Organic Rankine conversion — A SOTA Solution to Near Term High Power Needs in Space*, Proceedings of the 20th Intersociety Energy Conversion Engineering Conference (IECEC 85), Paper #859338, SAE P–164, (Society of Automotive Engineers, Warrendale, PA, 1985), 1.352–1.357.

Overholt, D. M., *Dynamic Power Conversion Systems for Space Nuclear Power, Section III: Brayton Cycle*, A Critical Review of Space Nuclear Power and Propulsion 1984–1993, M. S. El-Genk, (ed.), (American Institute of Physics Press, New York, 1994), 358–379.

Rockwell International, Rocketdyne Division, *Dynamic Isotope Power Subsystem Engineering Unit Flight System Design Report (Draft)*, ER–00021, (Rockwell International, Canoga Park, CA, October 14, 1988).

Voss, S. S., *SNAP Reactor Overview*, AFWL–TN–84–14, (Air Force Weapons Laboratory, Kirtland AFB, NM, 1984).

White, M.A. presentation of status of an Advanced Radioisotope Space Power System using Free-Piston Stirling Technology, by White, M.A., *et al.*, Proceedings of the 33rd Intersociety Energy Conversion Engineering Conference (IECEC 98), Paper #98-417, (American Nuclear Society, LaGrange Park, Il., 1998).

CHAPTER 8

POWER MANAGEMENT AND DISTRIBUTION

1. Introduction

If various electronic equipment within the spacecraft could operate at the same voltage, and if the load demand by these pieces of equipment were constant and equal to the power generated by the energy source, the PMAD subsystem would simply be reduced to an interconnect harness or a bus bar. In reality, however, the electronic loads within the spacecraft need different voltages to operate, and special power demands need to be met for a variety of components within the spacecraft during its mission life. On-board computers and housekeeping equipment have very different power requirements than, for example, the solid state power amplifiers (SSPA) or traveling wave tube amplifiers (TWTA) within the communication subsystem. Solar arrays degrade with time, deliver more power and higher voltage immediately after the orbital eclipse, and do not provide power during eclipse. Batteries need to be controlled and monitored during charging and discharging and need reconditioning for long life. The output impedance of the spacecraft power bus needs to be specified, designed, and analyzed as part of an overall system. A properly fused and switched load distribution system is needed to isolate failed components from the remainder of the subsystem. The PMAD subsystem is responsible for managing these widely varying load demands and energy generation and distribution elements. As shown in the schematic for a typical PMAD (Figure 8.1), since it is also responsible for autonomous control of the spacecraft bus, it must be internally redundant for reliability and provide ground-controlled functions and telemetry.

In this chapter we will examine the various elements of a modern spacecraft PMAD subsystem, discuss advantages and disadvantages of various systems in use today, and present several prototypical examples.

1.1 The ideal power system

As spacecraft life expectancies are increased, higher power subsystem efficiency and lower mass are becoming increasingly crucial in providing spacecraft with longer life and a larger end-of-life (EOL) power margin. The ideal power system would provide this power at 100% efficiency, weigh nothing, and be infinitely small. The PMAD subsystem is an integral part of the overall power system, and its design is crucial in providing the highest efficiency solution for a given mission. The goal is to achieve the highest *system* efficiency possible and, as with any subsystem design, its optimization is secondary to the overall power system optimization.

354   SPACECRAFT POWER TECHNOLOGIES

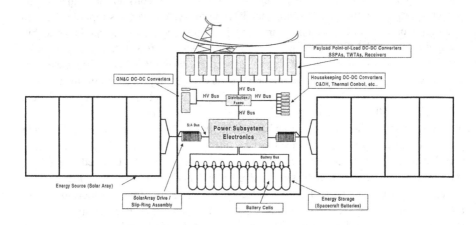

Figure 8.1 The Power Management and Distribution System

As a demonstration of the difficulty of achieving this goal, consider a typical communication satellite system whose primary function is to transmit the maximum amount of radio-frequency (RF) power in a maximum number of channels and multiple beams with the smallest, lightest possible satellite. In this case the overall spacecraft efficiency may be taken as the total equivalent isotropically radiated power (EIRP) divided by the total solar power available at the solar array (see Figure 8.2):

$$\eta = \text{EIRP} / (S_0 A)$$

where η is the communication satellite system efficiency, EIRP is the product of the total transmitter power and the antenna gain, S_0 is the solar power available from the Sun, and A is the area of the spacecraft solar array.

While much simplified, this way of looking at spacecraft performance is intended to point out the importance of the individual efficiencies in each subsystem. As we will show, the total efficiency is quite low in spite of the PMAD subsystem efficiency typically

POWER MANAGEMENT AND DISTRIBUTION

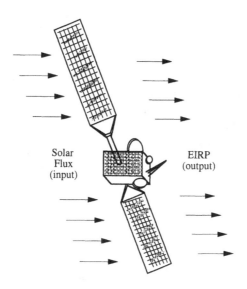

Figure 8.2 A Simplified Method of Determining Satellite Efficiency

being greater than 90%. Solar array efficiencies range between 12% and 18%, and transmitter DC to RF power efficiencies range from 35% for an SSPA to 50% for a TWTA. The overall efficiency is then described as:

$$\eta_{Spacecraft} = \eta_{Solar\ Array} \times \eta_{PMAD + TT\&C + GN\&C + Thermal} \times \eta_{Communications}$$

The ability of the power subsystem to provide the maximum amount of DC power harnessed from the energy source throughout the mission life and doing that at the maximum efficiency is of paramount importance in maximizing the available EIRP. This is particularly true in a communication satellite since this is directly related to maximizing transponder revenues. To continue the example, consider Table 8.1 which summarizes the key figures from Figure 8.3 for power allocation in a typical medium-power communications satellite.

Assume that the payload consists of 24 SSPAs, each at 20W RF output, and 24 TWTAs, each at 50W RF output. Referring again to Figure 8.3, and assuming 35% SSPA efficiency and 45% TWTA efficiency, this translates into 1372W of DC power required for the SSPAs and 2667W of DC power for the TWTAs. For a 50% plumbing

Table 8.1 Average Power Required for a Medium-Sized Communications Satellite

Component	Power (Watts)
Communications Payload	
Solid State Amplifiers	1372
Traveling Wave Amplifiers	2667
Tracking, Telemetry & Control	50
Guidance, Navigation & Control	70
Power System Ohmic Losses	50
Thermal management Subsystem	100
Total	4309

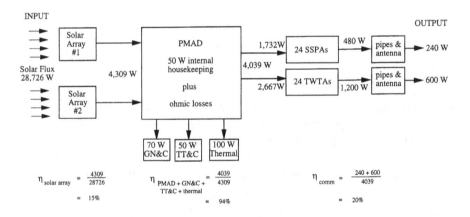

Figure 8.3 Power Efficiency Flowchart in a Communications Satellite

and antenna efficiency, the total communication subsystem efficiency is less than 20%. The typical power consumption for tracking, telemetry, and control is 50W; the guidance, navigation, and control, 70W; the power subsystem harness losses, 50W; and the thermal management subsystem, 100W. The overall housekeeping efficiency is then 94%. Assuming a solar array efficiency of 15%, the total spacecraft electrical efficiency can be calculated:

$$\eta_{spacecraft} = (0.15 \times 0.94 \times 0.2) \times 100 = 3\%!$$

POWER MANAGEMENT AND DISTRIBUTION 357

This small system efficiency illustrates the overall impact of various elements within the spacecraft subsystems. It is clear that one of the areas that could greatly benefit from improvements in efficiency is the solar array technology, followed closely by the communication transmitters. But it also points out that the efficient generation and distribution of power in a spacecraft is extremely crucial and every percent in efficiency counts.

1.2 Power subsystem overview

The design choice and complexity of a spacecraft power subsystem is primarily driven by the type of spacecraft and its mission parameters. Satellites in LEO put unique demands on a power system as compared to their GEO counterparts. Similarly, interplanetary and deep space missions require special designs that are driven by their own unique mission parameters. Once the EOL power requirements for the spacecraft are determined, a power budget and energy balance study is performed to determine the power requirements for the spacecraft at the beginning-of-life. This will guide the designers in selecting the different elements or specific technologies for the power source, storage, and distribution in a power subsystem. A typical spacecraft power subsystem, elements of which are shown in Figure 8.4, consists of the photovoltaic solar array; slip rings and solar array drive; power control electronics; battery charge, discharge, and reconditioning electronics; fuse and switched load assemblies; bus bars; and point-of-load power converters. Two critical functions of the power subsystem, the energy source and the energy storage elements, are briefly described below in order to point out their impact in the selection of the PMAD subsystem design.

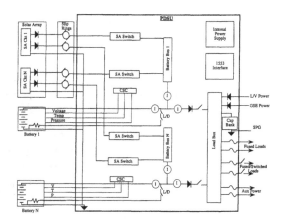

Figure 8.4 A Typical Photovoltaic Power Subsystem

Energy sources

The choice of the energy source for a spacecraft is highly dependent on the mission type. Some of the more typical spacecraft primary power sources are reviewed in this section along with comments that will be important in the context of selecting the proper PMAD design.

Solar arrays

By far the most widely used primary source of power in Earth orbiting satellites, the photovoltaic cells convert the radiant solar energy into electrical energy. Solar cells are reliable, safe, and provide sufficient power for long missions. A typical spacecraft solar array consists of a series-parallel combination of a multitude of solar cells mounted on a support structure . Slip rings are used to transfer the energy from the cells to the remainder of the power system. Two types of solar array structures are in use today: body-mounted arrays are used primarily on spin-stabilized satellites, and deployable arrays are used on the three-axis stabilized satellites (Figure 8.5). Body-mounted arrays simplify construction and thermal management; however, they have limited power capability since the available solar array area is restricted. In the case of a spinning body-mounted array, only a small portion of the array is illuminated at any given time. Deployable arrays are capable of providing much higher power (typically 5-20 kW). These arrays are wing-type structures which are folded during launch and transfer orbit, and are deployed once the spacecraft is parked on-orbit. Various deployed solar array structures have been designed and are successfully operating on orbit. The most widely

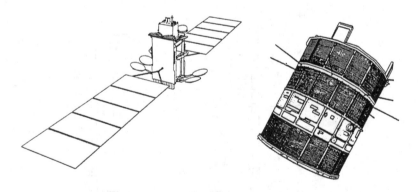

Deployable Arrays　　　　　　　　　　　Body Mounted
Figure 8.5 Two Common Array Configurations

POWER MANAGEMENT AND DISTRIBUTION

used, the rigid structure, has solar cells mounted side-by-side on a honeycomb structure, while the less-frequently seen flexible solar arrays provide lower weight and ease of storage.

Radioisotope Thermonuclear Generators (RTG)

RTGs (Figure 5.2) are used as the primary source of power in instances where use of a solar array is not practical due to reduced solar flux. These missions are usually long-duration interplanetary and outer planetary missions. The RTG uses the heat generated from the natural decay of Plutonium-238 and directly converts this heat to electrical energy using a thermoelectric couple device. Among the spacecraft that have used RTGs are the Pioneer, Voyager, Galileo, and the recently launched Cassini. The Voyager II spacecraft has now traveled beyond the solar system and has provided data continuously for over 20 years.

Primary batteries

For short duration missions, primary batteries offer a high specific energy and are often used. These batteries cannot be recharged and are used for missions such as launch vehicles or small, specialized spacecraft. The JPL Sojourner micro-rover aboard the Mars Pathfinder used a primary battery on the Martian surface for operations during the Martian night and a flat panel solar array for daytime operation. The Sojourner used three lithium-thionyl chloride cells in series with a nominal capacity of 12 A-hr. A number of primary cells have been used in space and the reader is referred to Chapter 4 for a full discussion of this technology.

Energy storage

Storage devices are needed to supply the spacecraft with power during the orbital night. These devices are charged when incident light is available and are discharged to supply the spacecraft loads in eclipse. For a GEO satellite, the eclipse seasons occur for 45 days around Vernal and Autumnal equinox with each eclipse lasting less than 70 minutes (Figure 2.5). The number of eclipses in a low-Earth orbit satellite depends on the mission and orbital altitude, but typically number about 15 per day. It is evident from this discussion that storage technologies are mission dependent and place specific demands on the rest of the power subsystem, especially the PMAD.

Secondary batteries

Secondary batteries, those that can be recharged, consist of individual electrochemical cells connected in series to meet the bus voltage requirements. In a battery dominated PMAD system, the battery provides a peak power capability and low output impedance to the spacecraft bus.

The capacity of a battery (C, in ampere-hours), for a given battery chemistry, depends mostly on the discharge current. When a battery supplies high-current loads, the battery capacity is drastically reduced. In contrast, when delivering power to low-amperage loads, the battery discharge curve reflects a constant voltage for most of the discharge period. As discussed in Chapter 4, the discharge rate is defined as the capacity divided by time. For example, C/10 is the current that is required to completely charge or discharge the battery in 10 hours. For a 100 A-hr battery, a C/10 charge rate indicates a 10A current is required. A figure-of-merit for batteries is the ampere-hour capacity, the number of ampere hours divided by the battery weight (A hr/kg). The specific energy of the battery is the ampere hour capacity multiplied by the battery voltage (W-hr/kg). The most common discriminators for spacecraft batteries are greater specific energy, cycle life, and depth of discharge. Other design drivers for a battery selection include safety, cost, predictable performance across a temperature range, voltage, shelf-life, and resistance to shock and vibration.

Nickel cadmium (NiCd) batteries have been the most widely used for spacecraft applications and have provided years of on-orbit reliable performance. In recent times, most Earth orbiting satellites have switched to nickel hydrogen (NiH_2) (Figure 4.18) batteries that can provide as much as twice the specific energy of the NiCd. They provide greater depth of discharge (up to 80% for GEO) and do not require reconditioning. Since the pressure of the hydrogen in the cell is linearly proportional to the amount of energy stored in the cell, the state of charge in a NiH_2 battery can easily be measured using a pressure monitor. Nickel hydrogen batteries cost more than NiCd but, considering the system weight savings and high launch cost, they can provide a lower system cost.

The advent of lithium-ion batteries promises to revolutionize the spacecraft battery technology. With batteries constituting 5-10% of a spacecraft's weight, another advancement in battery technology is long overdue. Once qualified for space, lithium-ion technology can provide two to four times the specific energy of NiH_2 and NiCd batteries.

PMAD

Proper conditioning and transmission of the power generated from the energy source to the batteries and the spacecraft loads are the primary functions of the PMAD system. As we have seen, various choices exist in the selection of the spacecraft energy source and

POWER MANAGEMENT AND DISTRIBUTION 361

storage elements with some devices better suited for specific missions in spite of higher cost, increased weight, or extra demands on the power management subsystem. Not surprisingly then, the choice for a PMAD system is also highly mission dependent and requires a detailed trade study before the optimum choice can be made. The electrical and mechanical design of a PMAD system requires a clear understanding of the mission requirements and consideration of such issues as energy balance over life, system stability, electromagnetic interference, corona and electrostatic discharge immunity, radiation and single event effects, reliability, redundancy, autonomous operation, fault recovery, thermal design margins, size, efficiency, weight, and cost. Choices exist between a distributed versus a centralized system, regulated versus unregulated buses, high voltage versus low voltage, etc. These options are discussed in the next sections.

1.3 Electrical power system options

Centralized versus distributed systems

A small spacecraft requiring less than a few hundred watts of power would be a good candidate for a single, centralized PMAD system. The small size of the spacecraft and low power demand from the payload do not require the distribution of high-power, low-voltage lines over long distances. For example, the Orbital Sciences Orbcom satellite, for which transmission distances are typically one meter or less, may use a centralized PMAD system due to its small size and relatively small ohmic losses. This allows the entire function of the PMAD to be performed by a single box or card. Future missions such as the JPL X2000 program are considering these systems. In a typical high-power (10-15 kW) communications satellite, the distance between the power system electronics of the PMAD and the payload components may be as long as 12 meters (roundtrip). The ohmic losses over these distances at low voltage would be prohibitive. In contrast to these small, low-power specialized missions, most commercial communication satellites are striving for more power. This has given rise to a choice between high-voltage and low-voltage buses.

High voltage versus low voltage

The power requirement for a typical Fixed Services Satellite (FSS) has tripled over the past decade. Commercial communications satellites in the mid 1980s, with a DC power level of three to four kilowatts, have now grown to power levels of greater than 15 kilowatts. At these power levels, centralized and low voltage power systems become impractical. Much like the interstate power distribution which requires high voltage transmission lines, a regulated high-voltage intermediate bus is often selected to carry the payload current over extended distances within the spacecraft. With the size of a typical FSS communication satellite bus approaching the size of a small room (a cube three meters on a side), distribution of ten to 15 kilowatts of power at low bus voltages would incur excessive ohmic loss. Secondly, high-voltage buses allow better utilization of energy density from secondary batteries. Most communication satellites now use regulated bus voltages such as 50V, 70V, 100V and even 120V. For future higher power missions, such as the space-based laser and radar systems, 270V DC regulated systems are now being studied.

AC versus DC distribution

Since the energy source in most satellites provides a DC voltage, the typical distribution system has remained DC. As demand for higher power spacecraft increases, the debate for the selection of an AC bus resurfaces. AC buses can be more economical for a spacecraft with greater than a 20 kilowatt power requirement, the International Space Station (ISS) being a current example. However, most of these AC systems would require new point-of-load AC-DC converters. Even the ISS power system was finally changed back to DC after much debate. With proper design and a larger initial investment, AC systems could provide a higher overall system efficiency at a lower mass.

Direct energy transfer (DET) versus peak-power tracker (PPT)

Power subsystems are further categorized as DET and PPT systems. The DET systems, also referred to as dissipative systems, Figure 8.6, are the most widely used today. They provide the lowest parts count and offer higher system efficiency and lower cost. In these systems, linear or switching shunt transistors are connected in parallel to the array and regulate the excess power available from the array. A PPT system, also referred to as a non-dissipative system, regulates the array power by extracting only the required load power at the array's maximum power point when the load demands it. This configuration typically involves the addition of a power converter in series with the bus and can reduce the overall system efficiency in higher power spacecraft. However, in missions that require maximum solar array power at end-of-life or in low Earth orbit applications, these systems can be advantageous.

Regulated versus unregulated bus

Most system trades are often interrelated (i.e., choice of a high-voltage bus typically also implies a distributed, regulated system). The choice of a regulated versus an

POWER MANAGEMENT AND DISTRIBUTION

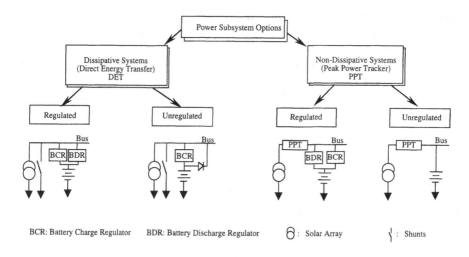

Figure 8.6 Power Subsystem Electronics (PSE)

unregulated system also implies several other factors which need to be considered. These include the type of the mission, EOL power requirements, bus voltage, and the existing heritage equipment that could require a major redesign. Regulated bus systems require a switching regulator and are much more complex than their unregulated bus counterparts. They do simplify the design of the point-of-load DC-DC converters, however, and may decrease the overall spacecraft weight.

2. Functions of PMAD

The PMAD functions can be divided into several main areas: power management and control (which includes the solar array control); battery charging, discharging and reconditioning; power distribution; system fault management and telemetry; and point-of-load DC-DC voltage conversion.

2.1 Power management and control

Most Earth-orbiting satellites use solar arrays as the primary energy source. Solar arrays do not provide power in eclipse, they degrade with time, and have a varying output voltage depending on the temperature. Batteries are used to provide power to the loads during eclipse and must be fully charged when the satellite is in sunlight. In a LEO application, the battery charging must be completed in approximately one hour, and a spacecraft battery will go through 30,000 to 50,000 cycles during its life. Batteries also have varying output voltages depending on state of charge and temperature. They cannot be overcharged and the depth of discharge must be limited to extend the battery life. Batteries also need periodic reconditioning for long life. The output characteristics of the solar array must be matched to the spacecraft bus and the batteries. Power management electronics are required to perform these functions. This power management and control function is categorized in three sub-functions: solar array control, battery charging and discharging, and, if needed, bus regulation.

Solar array control

The power generated by the solar array must be controlled to limit the maximum bus voltage, to regulate the charge and discharge of the batteries, and to maintain the energy balance in the spacecraft. As discussed earlier, two types of solar array control systems are in use today. The dissipative systems shunt the excess solar array power and maintain the bus voltage at a regulated level by dissipating the excess power in a shunt transistor. Non-dissipative systems extract only the required power from the solar array using a peak-power tracker DC-DC converter placed in series with the solar array bus or a series-switched solar array connection. Most communication satellites use the simpler dissipative systems.

Direct energy transfer systems

Typically, the two methods used in DET, or dissipative systems, are the linear and switching shunts (Figure 8.7). Further variations include partial shunts, full shunts, and sequential shunts. In the linear shunt, a power transistor, operating in its active region, is placed across the solar array string and dissipates the excess power from the array. A series resistor is sometimes used to share the dissipated power in the shunt transistor. An error amplifier compares the bus voltage with a reference voltage and sends a control signal to the shunt transistor. These systems are simple, but the heavy heatsink-mounted transistors that are required increase the subsystem mass. In a linear shunt system, a higher control bandwidth can be attained which provides a lower output impedance to the bus.

In contrast, the transistor in a switching shunt regulator is pulse width modulated (PWM) to control the solar array power. As in the linear system, the bus voltage is compared to a reference voltage and an error signal is generated. In this case, however, the error voltage is then compared with a ramp signal and the resulting output is used to modulate the shunt transistor between the saturated and cutoff states. Since the transistors are operated as switches, power dissipation is drastically reduced and lower-power transistors can be utilized.

POWER MANAGEMENT AND DISTRIBUTION

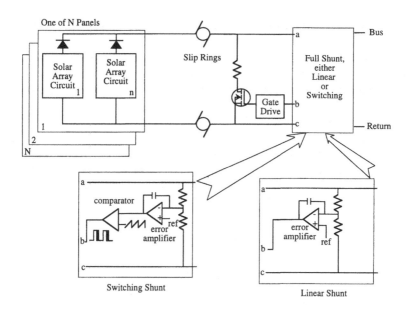

Figure 8.7 Linear versus Switching Shunts (Both Operating as Full Shunts)

In practice, however, both of these systems are used in a multistage sequential topology (Figure 8.8). In a sequential shunt configuration, several transistors are used as a switch in series with a shunt resistor. The scaled-down bus voltage is compared with a reference and the error signal is used to turn on shunt transistors one at a time. As the error voltage increases, more transistors are turned on to shunt more solar array circuits. In a PWM shunt regulator each solar array string is connected to a PWM transistor and these stages are modulated to control the solar array power. An improved version of the multistage PWM shunt circuit operates the transistors in a phase-shifted fashion so that the instantaneous number of circuits connected to the bus is reduced, resulting in lower peak current spikes.

In yet another dissipative linear shunt method, the shunt transistor is placed across a portion of the solar array string allowing a better matching of the solar array power and the load demand (Figure 8.9).

Peak power transfer systems

The typical current-voltage curve of a silicon solar cell is shown in Figure 8.10a. Note that the solar cell is capable of a producing the maximum power available from the cell at the "knee" of its I-V curve. To take advantage of this characteristic, some power systems use a peak power tracker in series with the array (Figure 8.10b). This is achieved by utilizing a DC-DC converter to control the operating point of the array. As the load demand decreases, the peak power tracker moves the operating point towards the open circuit voltage of the array. As the load demand increases, the operating point is moved to the maximum power point of the solar array. Peak power trackers decrease the power transfer efficiency in the current path, but at moderate to low power they do provide a higher overall system efficiency at lower mass.

Other non-dissipative systems have been successfully used which do not use a peak power tracker. The series-switched solar array string provides a high efficiency method of solar array control. Solar array circuits are connected to the power subsystem using a series transistor. In this system more solar array strings are connected to the bus as the load demand increases or battery charging is required.

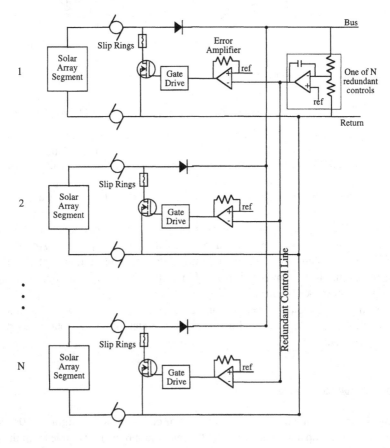

Figure 8.8 Sequential Shunts

POWER MANAGEMENT AND DISTRIBUTION

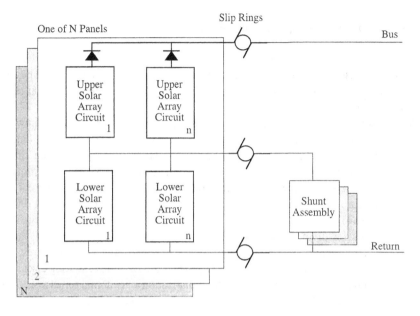

Figure 8.9 Partial Shunts

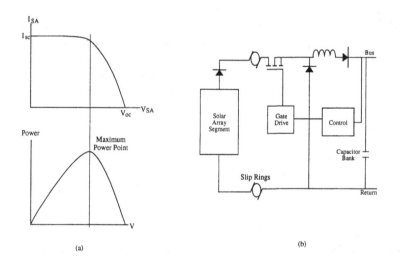

Figure 8.10 (a) Solar Array IV Characteristics
(b) A Peak Power Tracker Subsystem Using a Buck Regulator

Battery charging, discharging, and reconditioning

Batteries are sensitive devices and require special attention in every aspect of their design and use. Batteries are never used at their maximum nameplate rating. They are sensitive to overcharging and their life cycle will be reduced if repeatedly discharged to their full capacity.

Typical spacecraft battery cell charge and discharge voltage profiles are shown in Figure 8.11 for NiH_2 cells. Notice in Figure 8.11(a) the higher end-of-charge voltage at lower temperatures, similar to NiCd cells, and the linear relationship between charge input and cell internal pressure. As mentioned earlier, this linear relationship, unique to NiH_2, provides an easy method for determining the state-of-charge. Figure 8.11(b) shows the pressure and voltage of a 40-Ahr cell during charge and overcharge (200%). As Schiffer (1984) points out, that the pressure and voltage level off at full charge is an important safety factor in this system. Finally, Figure 8.11(c) shows the discharge performance of this same cell. Notice the higher capacity at lower temperatures, and again, the linear relationship between cell pressure and state-of-charge. During discharge, the voltage drops at a faster rate due to electrode polarization and internal series resistance. The cell voltage then flattens for most of the battery capacity until another rapid drop in voltage, with time, indicates the end of the battery capacity.

The most commonly used method for battery charging is the constant current approach with a change to constant voltage taper charging at full charge. To determine the proper charge rate for a battery, the battery's state of charge needs to be known. For nickel-hydrogen batteries this can be determined by measuring the cell pressure, usually using a strain gauge (Figure 8.12). This task is not as simple for nickel-cadmium batteries. In this case, battery voltage and temperature are sometimes used to approximate the state of charge. However, the most reliable method is ampere-hour integration in which the exact amount of current and duration of discharge are measured and the same number of ampere-hours is used during battery charging (Figure 8.13).

A typical battery charging circuit provides two charge modes. When commanded, a high charge-rate current can be selected from several pre-selected temperature-and voltage-compensated charge-rate curves (typically referred to as V/T curves) (Figure 8.14). When the battery voltage and temperature reach the maximum V/T limit, the charging circuit reduces the charge current into a trickle charge mode and maintains the batteries at a constant voltage.

The maximum depth of discharge allowed for a battery depends on the battery chemistry and the type of mission. In a LEO mission, the spacecraft experiences 15 eclipses daily, each lasting 30 minutes. For a five-year LEO mission this translates into more than 27,000 charge and discharge cycles imposed on the battery. To ensure that the batteries meet these requirements, the depth of discharge for a NiCd battery must be limited to less than 20%, although nickel-hydrogen batteries can tolerate a higher depth of discharge. For a similar LEO mission, nickel-hydrogen batteries can be used with a

POWER MANAGEMENT AND DISTRIBUTION

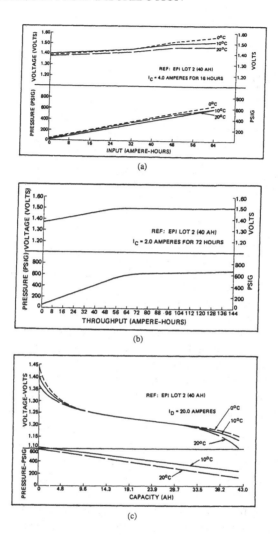

Figure 8.11 Battery Charge and Discharge Voltage Profiles (Schiffer, 1984, used with permission)

40-50% depth of discharge. For a GEO mission, the number of the charge/discharge cycles demanded from the batteries is much less (less than 1500 cycles for a 15-year mission), so a much higher depth of discharge can be tolerated.

Batteries require periodic reconditioning for long life. This is achieved by providing a simple relay/reconditioning resistor circuit to deplete the battery periodically to a full discharge state. Although this is required only for nickel-cadmium batteries, most power subsystem designs do include this feature to allow for system flexibility.

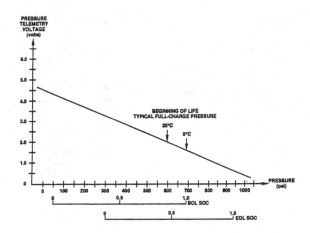

Figure 8.12 NiH_2 Battery Pressure Characteristics

Individual cell bypass diodes are required in the event of an open-cell failure. Antiparallel diodes are placed in parallel with each cell to bypass the battery in case of a failed cell. Some designs may use MOSFETs or even mechanical relays. To equalize discharge in a multiple battery spacecraft, a cell removal circuit is used to delete a cell from one of the batteries in the event of a cell failure on an opposite battery.

Thermistors are typically utilized to monitor the individual cell temperatures. This information is used for both V/T charge control and spacecraft telemetry. Strain gauge resistors are used on each cell to monitor the cell pressure and thus the battery's state of charge. The analog strain gauge output is used for battery charge and discharge control and spacecraft telemetry. Battery voltage telemetry monitors the overall battery voltage (as distinguished from individual cells) and is used for the V/T charge and discharge control, and for spacecraft telemetry. Newer battery technologies such as the lithium-ion batteries are extremely sensitive to the over and under voltage limits of the cell and require individual cell monitoring and balancing circuits.

Although NiCd and NiH_2 batteries are sensitive to overcharging, both can be slightly overcharged without damage. The overcharging can result in generation of excess heat but with proper thermal design this can easily be handled. In contrast, lithium-ion battery technology, while promising dramatic improvements in specific energy, cannot tolerate the slightest overcharge. Special charging circuits requiring individual cell voltage monitoring and charge balancing are required to prevent damage to the battery. Even small amounts of overcharge can lead to permanent cell damage or possible cell explosion. This complicates the battery charging system in a lithium-ion power subsystem

POWER MANAGEMENT AND DISTRIBUTION

design, but because of increased consumer application of this technology, great progress has been made in integrated circuits specifically designed to manage the charge monitoring of these batteries.

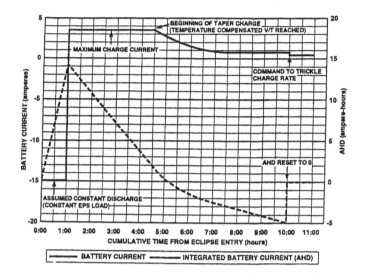

Figure 8.13 Determining the Battery State of Charge (Integration Method)

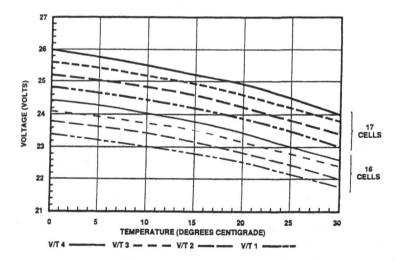

Figure 8.14 Battery Voltage/Temperature Limits

Bus regulator

In battery-dominated systems, the spacecraft bus simply follows the battery and the bus voltage is unregulated and follows the battery state of charge (e.g., from 22-36V). This is a simple PMAD subsystem and is preferred for applications with less than approximately four kilowatts in power requirements. With the majority of communication satellites in the eight to 15 kilowatt range (as in the Lockheed Martin A2100 and Hughes HS702 systems), low-voltage buses are not practical. The resulting harness mass requirement would make the system too heavy and the battery capacity would not be optimized. Regulated high-voltage buses are more practical for these systems and offer many advantages. The high-voltage bus drastically reduces the harness weight and the regulated bus simplifies the design of the point-of-load DC-DC converters. Additionally, battery capacity can be optimized at higher voltage buses.

The solar array shunt and series control provide a regulated bus in sunlight although during the orbital eclipse the bus would follow the battery discharge voltage. A series power converter is typically used to regulate this voltage. Since the regulated bus voltage is higher than the low end of the battery discharge voltage, a boost switching regulator is used. A boost converter is one of three basic building blocks in switching DC-DC converters and will be discussed in detail in the next section. In general, switching regulators are exclusively used for this purpose. Linear regulators can supply only a regulated output that is less than the input and are not very efficient; switching regulators can provide an output higher, lower, or equal to the input voltage and at very high efficiencies. Depending on the input and output voltage, power efficiencies in the 90-96% range can be realized. Some power subsystems combine the battery charge and discharge regulator into one circuit in the form of a bi-directional power converter. As we will see later, this method simplifies the regulation circuit design.

2.2 Power distribution

The power distribution includes the solar array slip rings, the bus bar, cabling, fuses, and switches. Many trades are considered during the design phase of the power distribution system and concurrent engineering is a key discriminator for this task, and the entire process is an iterative and integrated part of the initial PMAD subsystem trade study and bus voltage selection. Important considerations in the design ensure easy access during the integration and test phase of the spacecraft, proper grounding and shielding, elimination of single-point failures, and fault isolation. Particular attention must be given to implement features that ensure a fault tolerant and robust PMAD subsystem. Diode isolation techniques, proper separation of high voltage lines, double insulation of power lines, and redundancy of critical elements are a must. An example from a typical communication satellite (Figure 8.15) illustrates the process and highlights

POWER MANAGEMENT AND DISTRIBUTION

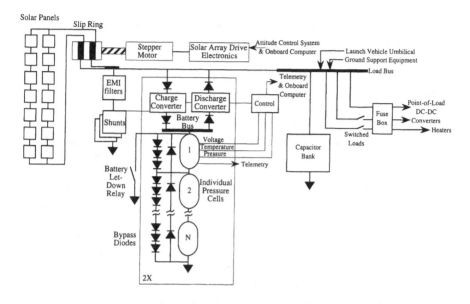

Figure 8.15 A Typical PMAD Architecture

the typical elements of the distribution, regulation, and protection circuits.

Slip rings are required to transfer the power from the solar array to the spacecraft. In three-axis stabilized satellites, the solar array is deployed in a wing. This allows more panels to be deployed for more power, and permits, using a brushless DC or stepper motor, the solar array wings to be kept pointed normal to the Sun for maximum power. This rotation requires the use of a solar array slip ring and brush assemblies to transfer the power and the telemetry and control signals to the spacecraft. A typical solar array slip ring assembly may contain over 120 individual rings for this purpose. The power-carrying capability of these rings is always derated for margin.

Power distribution also includes the individual battery bus bars (in a multi-battery system) and the main bus. From this point, multiple fused and/or switched lines carry the power throughout the spacecraft. The bus bars are double insulated to ensure reliability. Some spacecraft use a dual bus system since some of the critical payload may use previously-designed heritage hardware and it is more economical to provide them with a separate bus voltage rather than redesign each element of the payload hardware.

Proper grounding and harness routing and shielding are also extremely important. Many spacecraft have suffered from spurious commands due to poor EMI design. A single-point ground system referenced to the spacecraft structure is often used. All spacecraft loads are then fed with twisted shielded power and return lines to minimize stray magnetic fields and unwanted ground loops. Point-of-load power converters are all transformer-isolated, DC-DC converters so that primary power is kept separate from secondary power lines. This isolation also limits stray structure currents. Proper shielding and termination of these shields are also crucial.

2.3 Fault management and telemetry

The spacecraft bus supplies power to all other subsystems. It is the "life line" to the entire spacecraft and is protected against all spacecraft fault conditions. Battery cells have bypass circuits in case of a cell opening. The battery cells use double insulation to minimize the possibility of a cell short to ground. Solar array circuits are diode isolated in case of a circuit short to ground. Harnesses are double insulated and fused to preclude any fault condition from compromising the spacecraft bus, and loads are switched to isolate faulty sections.

Functional redundancy is used to prevent single-point failures. All electronic functions such as housekeeping DC-DC converters, the solar array shunts, and the control circuits are redundant. Several redundancy methods are utilized. Cold redundancy is used when one or more back up functional elements are in a cold (off) or warm (standby) mode until required to take over for a failed element. Another common redundancy method involves an N+1 system (i.e., one more than actually needed) in cases where several boards are operational but a minimum number of boards are required to complete the mission. Solar array shunt circuits are often utilized this way. Finally, some other key circuits, such as the voltage error signal for a regulated bus, may use a voting circuit to select the mid-point from three redundant error signals.

Protection features are built-in so that transients or failed components do not propagate. For example, bus over-voltage protection disables the solar array series switches or enables more solar array shunts (depending on the system) to protect the subsystem and spacecraft loads.

Data from essential telemetry points are sent to the spacecraft computer or the telemetry processor. The batteries are instrumented with thermistors to monitor the cell temperature. This information, along with the battery voltage, is sent to the spacecraft computer for proper V/T charge curve selection. Strain gauge resistors are mounted on individual battery cells for state-of-charge monitoring. This information is also sent to the telemetry processor unit for transmission to ground stations for monitoring and intervention of over-temperature or over-pressure conditions, or for V/T manual override. Within the power system, data from various other points are telemetered, including on/off status of functional elements such as the solar array shunt or series switch voltages, and temperature, battery reconditioning relay and distribution load switches, and solar array currents.

2.4 Point-of-load DC-DC converters

Not all electronic components can operate from the same voltage magnitude or polarity. The function of a DC-DC converter is to accept a single DC input voltage and to provide a single or multiple regulated DC output voltage of the same or opposite polarity and of higher, lower, or equal magnitude. (Figure 8.16) Most electronic subsystems within the satellite receive the regulated or unregulated bus and locally convert this voltage to multiple, isolated low voltages using a DC-DC converter. Such a distributed system has two advantages. First, the distribution of multiple low-voltage, high-current lines across spacecraft are impractical due to ohmic losses. The high-voltage bus can efficiently carry the payload and housekeeping power to various points within the spacecraft while the local DC-DC converter can then efficiently convert the main bus to multiple isolated outputs for a specific use. Second, isolation of the primary power bus from the secondary voltages is required at the point-of-load to reduce EMI and prevent spacecraft chassis currents. Transformer-isolated switching DC-DC converters are normally used for this purpose. Many different circuit topologies are in use today, but they are all built around a few basic arrangements of three ideal, non-dissipative components such as inductors, diodes, and transistors-operated-as-switches. A number of converter options are compared in Figure 8.17. These basic topologies will be examined, but first we begin by describing the simplest forms of a voltage regulator.

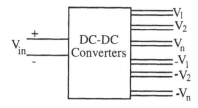

Figure 8.16 The DC-DC Converter

Linear regulators

From the early days, the basic method of regulation was the linear voltage regulator. Although the majority of regulators in use today are the switching variety, linear regulators still play an important role in the field of power electronics. When extremely quiet outputs are required (e.g., gate voltage for an SSPA), the sub-millivolt output ripple performance of a linear regulator cannot be matched. Also, linear regulators are often used as low-power secondary post regulators in point-of-load DC-DC converters.

A basic resistor divider is a simple form of a DC-DC converter, and although it is not regulated and not very practical for power conversion, it does provide a stepped down DC output from a given DC input voltage (Figure 8.18a). This simple circuit

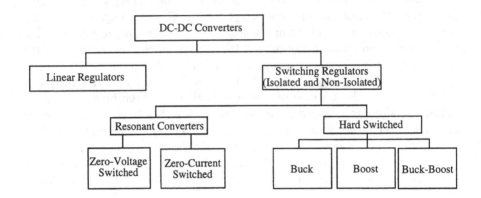

Figure 8.17 Converter Options

suffers from poor load regulation due to its high output impedance. To make this a practical voltage regulator, the input resistor can be replaced with a transistor operating in its active region (Figure 8.18b). This will serve as a variable resistor that now can be controlled to provide a regulated output voltage. The series transistor is operated as an emitter-follower and lowers the output impedance of the simple divider circuit by the transistor current gain. The output voltage is now regulated against load variations. This circuit still relies on a stable input voltage. To provide line regulation, another modification is necessary, namely the addition of a Zener diode in the base of the transistor (Figure 8.18c). This provides a stable reference to the emitter follower and the circuit is now a complete regulator that provides both an input line and output load regulation.

To further reduce the output impedance of the linear regulator and improve its load regulation, negative feedback can be employed. As shown in Figure 8.18d, an additional transistor can be added to compare a sample of the output voltage with the Zener reference voltage and produce an error signal which drives the series transistor. An operational amplifier for the error signal can also be added. The extremely high open loop gain of the op-amp drastically reduces the output impedance of the circuit (Figure 8.18e). An additional benefit of this circuit is the ease with which the output voltage can be adjusted since changing the reference voltage adjusts the output voltage. However, the output voltage regulation is only as good as the reference voltage and varying the input voltage will change the Zener current. The Zener diode breakdown characteristics indicate that there is a finite dynamic resistance in series with the Zener which affects the Zener voltage with varying current through it. Additionally, the Zener voltage varies with temperature. Further improvements are required to improve the line regulation. Replacing the input resistor to the reference Zener with a constant current source

POWER MANAGEMENT AND DISTRIBUTION

minimizes the effects of the Zener dynamic series impedance. Some circuits connect the reference circuit to the regulated output, but care must be taken to ensure that the circuit will start up. Temperature effects also deteriorate circuit performance. To avoid temperature drift problems, a diode is inserted in series with the Zener to cancel out the temperature drift of the Zener diode. The constant-current Zener circuit improves both temperature and dynamic resistance deficiencies. In practice, however, designers use commercially available laser trimmed and radiation-hardened band-gap reference circuits.

Several integrated-circuit linear regulators are available but these are typically limited in current capability and are difficult to analyze for end-of-life variations. Spacecraft

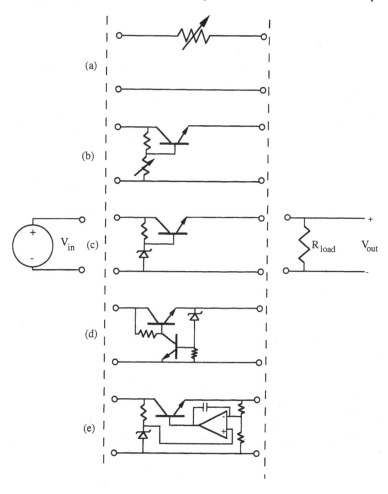

Figure 8.18 The Evolution of Linear Regulator Design

power designers prefer to use discrete devices to construct their own linear regulators. In this way, individual circuit component variations with temperature, radiation, and aging are better known and more accurate end-of-life analyses are generally possible. A complete linear series pass regulator is shown in Figure 8.19. Note that the series element bipolar transistor is replaced with a MOSFET, a change preferred by designers since it is a voltage driven device and the pass element can be directly driven from the error amplifier instead of a driver stage as with the high current bipolar transistors. MOSFETs also have a high transconductance which lowers the output impedance of the linear regulator.

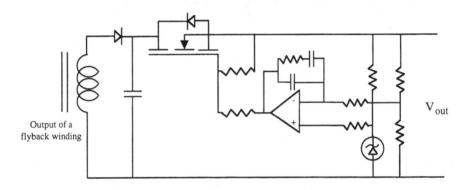

Figure 8.19 A Linear Series Pass Regulator

Switching regulators

Linear regulators are of limited use for high-power, high-efficiency power processing systems. They can provide a regulated voltage lower than the input voltage and cannot provide a voltage inversion. Switching regulators, or power converters, have dominated the field of power electronics for over 30 years. These converters provide a high-density, high-efficiency alternative to their linear regulator counterparts.

Unlike the control element in the linear regulator, the transistor in a switching regulator control element is used as an on-off switch. As it switches, it chops the input DC voltage into a variable duty cycle square wave. At a 100% duty cycle the output equals the input DC. As the duty cycle is decreased, the average voltage of the square wave drops by the duty cycle ratio. At 0% duty cycle, the switch is completely open and the average output voltage is zero. This simple variable output control element has advantages over the linear pass transistor operated in its active region and is the heart of

POWER MANAGEMENT AND DISTRIBUTION

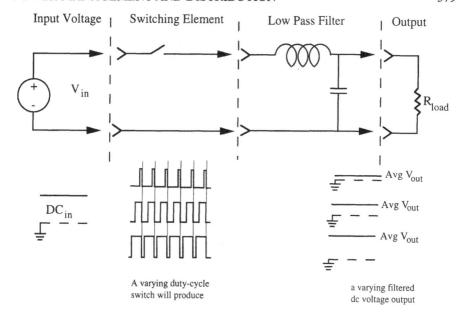

Figure 8.20 The Switching Regulator

every switching regulator. Since the transistor carries the load current when it is fully on, and carries no current when it is fully off, the ideal efficiency of this pass element can be 100% (Figure 8.20). In reality, however, the transistor is not a perfect switch and carries a finite voltage when it is fully saturated (or in the case of a MOSFET, it has a finite $R_{DS(ON)}$). Additionally, the switching time in a transistor is limited, which results in an overlapping voltage and current during the turn-on and turn-off transitions and adds to the circuit losses. Despite these shortcomings, the switching control element efficiency is much higher than its linear counterpart. In contrast to the linear regulator, however, the output of the switching control element cannot be directly connected to the load. An LC filter is typically needed to smooth this output square wave and produce a DC output.

Buck regulators

As the name implies, the buck regulator produces a regulated DC output which is always less than the input voltage. The basic theory of operation for the buck regulator was described earlier with one minor modification. A freewheeling diode is needed to provide a current path when the switch is open (Figure 8.21). Since inductor current cannot change instantaneously, this diode provides a current path for the inductor. When the

380   SPACECRAFT POWER TECHNOLOGIES

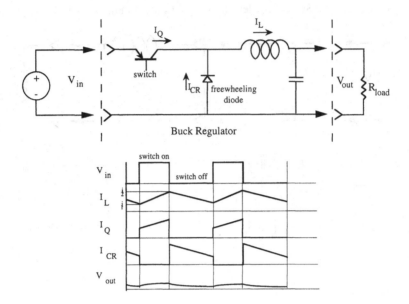

Figure 8.21 The Buck Regulator

switch is turned on, the current through the inductor ramps up. Since V=dI/dt, we can write the slope of the up ramp current as

$$\frac{dI}{dt} = \frac{V_{in} - V_{out}}{L}$$

or

$$L(i_2 - i_1) = (V_{in} - V_{out})t$$

where i_1 is the inductor current when the switch turns on. This equation also assumes the voltage drop across the pass element is zero. When the switch is turned off, the freewheeling diode will be forward biased, the current through the inductor decreases, and the slope of the down ramp current becomes

$$\frac{dI}{dt} = \frac{-V_{out}}{L}$$

or

$$L(i_1 - i_2) = (-V_{out})t$$

POWER MANAGEMENT AND DISTRIBUTION 381

where i_2 is the inductor current at the beginning of switch turn-off. This equation assumes the freewheeling diode forward voltage is negligible. If the inductor is large enough, the current in the inductor does not decrease to zero and the buck regulator is said to operate in a continuous conduction mode (CCM). When the switch is on, the current through the inductor increases by $(V_{in}-V_{out})DT$; when the switch is off, the current decreases by $(V_{out})(1-D)T$. Since the starting current in the inductor at turn-on must equal the final current during turn-off we have:

$$(V_{in}-V_{out})\,D\,T = (V_{out})\,(1-D)\,T$$

or

$$V_{out} = DV_{in}$$

where D is the ratio of the 'on' time to the total duty cycle time, T.

This indicates that with the buck regulator operating in CCM, the output voltage is independent of load current, and further indicates a very good load regulation which is inherent to CCM switching regulators. In reality, however, there is a net series resistance in the output inductor which modifies the output voltage slightly. Another method of deriving the DC relationship of the CCM buck regulator is to equate the average voltage at the left of the output inductor to the voltage to the right of it since the average voltage across an inductor during each cycle must be zero. Again, we have

$$V_o = DV_{in}$$

Boost regulators

The boost regulator produces a regulated DC output always higher than the input voltage. The operation of the boost converter is also intuitive and simple to understand (Figure 8.22). When the switch is turned on the current in the inductor begins to ramp up with a slope

$$\frac{dI}{dt} = \frac{V_{in}}{L}$$

During this time the output capacitor is supplying the load. When the switch turns off, the voltage across the inductor reverses polarity and, being in series with V_{in}, it adds to it. The diode is now forward biased and the inductor delivers its stored energy,

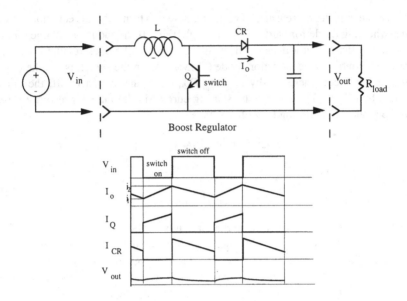

Figure 8.22 The Boost Regulator

during the turn on period, to the load and the capacitor. The current through the inductor is now discharging with a down ramp slope given by

$$\frac{dI}{dt} = \frac{V_{in} - V_{out}}{L}$$

Again, the diode forward voltage is assumed negligible. When the switch is on, the current through the inductor increases by $V_{in}D\,T/L$; when the switch is off, the current decreases by $((V_{in}-V_{out})/L)(1-D)T$. Since the starting current in the inductor at turn-on must equal the final current during turn-off, we have

$$(-1)\frac{V_{in} - V_{out}}{L}(1-D)T = \frac{V_{in}DT}{L}$$

or

$$V_{out} = V_{in}\left(\frac{1}{1-D}\right)$$

Buck-boost regulators

Once again, as the name implies, the buck-boost regulator produces a regulated DC output which can be lower (buck), higher (boost), or equal to the input voltage. However, the polarity of the output voltage will be opposite of the input voltage (Figure 8.23). The operation of the buck-boost (also called the flyback) converter is similar to the boost converter with a slight twist. As with the boost converter, when the switch is turned on, the current in the inductor begins to ramp up with a slope

$$\frac{dI}{dt} = \frac{V_{in}}{L}$$

During this time, the output capacitor is supplying the load. When the switch turns off the voltage across the inductor reverses polarity and, being in series with V_{in}, it adds to it. The diode is now forward biased and the inductor delivers its stored energy during the turn on period to the load and the capacitor. The current through the inductor is now discharging with a down ramp slope

$$\frac{dI}{dt} = -\frac{V_{in}}{L}$$

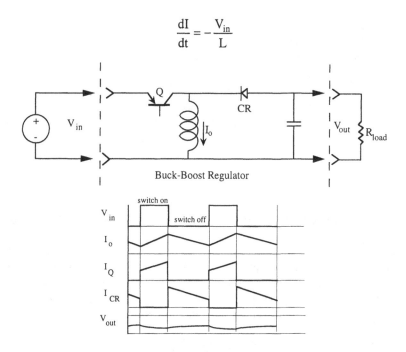

Figure 8.23 The Buck-Boost Regulator

where, again, the diode forward voltage is assumed negligible. When the switch is on, the current through the inductor increases by $V_{in}DT/L$; when the switch is off, the current decreases by $(-(V_{in})/L)(1-D)T$. Since the starting current in the inductor at turn-on must equal the final current during turn-off we have

$$-\frac{V_{in}}{L}(1-D)T = \frac{V_{in}DT}{L}$$

or

$$V_{out} = -V_{in}\left(\frac{D}{1-D}\right)$$

Note that the polarity of the voltage is reversed.

Isolated DC-DC converters

Although most pulse-width modulated DC-DC converters in use commercially are based on the three basic topologies discussed in the previous section, these converters are rarely used as shown due to the lack of isolation between the input and output grounds. Additionally, most electronic circuits require several voltages at both polarities (e.g., a solid state power amplifier may require +7V, -3V, ±10V). Although it is possible to produce multiple outputs from the basic topologies, the designs are limited. To provide both the ground isolation and multiple outputs, a transformer is needed. The most common DC-DC converters in use today are actually the same basic buck, boost, and buck-boost topologies with a transformer inserted in the circuit. The basic principles of operation remain the same, however. We will now examine several of these circuits.

Buck derived topologies: forward, voltage-fed push-pull, and bridge dc-dc converters

These circuits are listed together since they are all driven from the basic buck switching regulator topology. The metamorphosis of the buck circuit into the popular forward converter is shown in Figure 8.24. It is seen that in the forward converter the basic function of the buck switch, generating a square wave from the DC input source, is now replaced with a transistor, transformer, and a diode circuit. The function of the circuit remains unchanged, that is to provide a pulse width modulated square wave at the input of the LC filter. However, the added benefits resulting from this change satisfy both shortcomings of the basic buck circuit. Multiple ground isolated outputs with both polarities are now as easy as adding another winding to the transformer. In the forward converter, adding the isolation transformer adds certain peculiarities that need to be addressed. Considering a transformer model with a parallel magnetizing inductance and its associated current during the switch-on period, a method is needed to provide a continuous path for the demagnetizing current when the switch turns off. The common method to provide this current path is to add a tertiary winding and reset the core when the switch turns off. The forward converter is a single switch converter and is well suited for medium power DC-DC conversion. The continuous output current makes the forward converter desirable for low-voltage, high-current outputs. The drain (or collector) voltage on the switch is subjected to a voltage twice the input voltage, thus making this converter less attractive for high input voltages. However, with the advances in MOSFET technology and availability of radiation-hardened (against both total dose and single event effects), high-voltage devices, this is less of a concern. The forward converter is often used in the 50-200W output power range.

The voltage-fed push-pull (called a voltage-fed since the input to the switching converter is a low impedance voltage source) converter (Figure 8.25) is also a derivative of the basic buck converter. If the tertiary winding diode in the forward converter is replaced with a switch, a voltage fed push-pull converter results. Again it is easily seen that the basic function of the switching cell is unchanged and a pulse width modulated square wave is presented to the input of the LC filter.

The advantage is that the primary winding of the transformer is driven in both directions and a full wave rectified output is used. This allows better utilization of the

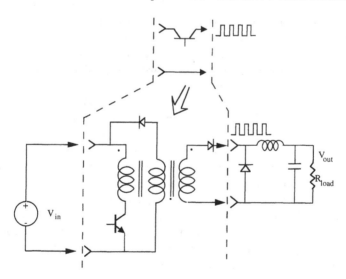

Figure 8.24 The Forward Converter

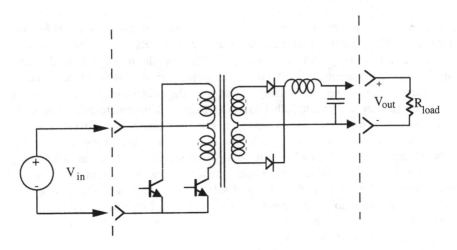

Figure 8.25 The Voltage-Fed, Push-Pull Converter

magnetic core and results in a smaller transformer. The disadvantage of the push-pull topology, as shown, is the possibility of core flux density imbalance which leads to switching transistor failures. The problem is as follows. The change in the flux density of the transformer core (ΔB) is proportional to the product of the voltage across the winding and the time this voltage is applied to the winding. This product, the volt-seconds, must be kept equal in both halves of the push-pull primary. If not, the flux density in the core will not be reset to its initial point at the beginning of the switching cycle, and the transformer core will walk-up its hysteresis curve and reach its saturation flux density (Please see the Appendix for a more detailed discussion of magnetic materials.) At this point, the transformer cannot sustain voltage and the power transistors will be destroyed due to high currents. The volt-seconds imbalance can result from any number of factors including the normal variation in the saturation voltage and the storage times of the switching transistors. With the advent of the MOSFET and current mode control techniques, this problem is less significant. MOSFETs provide an inherent negative feedback to balance the voltage across each half of the push-pull windings. The $R_{DS(ON)}$ of a MOSFET increases with temperature; therefore, if an imbalance occurs, the increased current in the transistor will increase its channel resistance resulting in an increased on-voltage and a lowering of the volt-seconds in that transformer half. Another method used to avoid the flux imbalance in the push-pull DC-DC converter is to use cycle-by-cycle current mode control for the main loop feedback. The current mode control circuit measures the individual transistor currents and limits these currents by adjusting the duty cycle of that switch. Other variations of this converter include the current-fed push-pull (in contrast to the voltage-fed push-pull, the input to the current fed topology is a high impedance current source) which moves the output inductor to the input side of the transformer primary. The advantage this affords is the complete

POWER MANAGEMENT AND DISTRIBUTION

elimination of the flux imbalance problem. More important, however, the current-fed topology is more suitable in meeting prompt-dose effects in missions that require nuclear-event hardening.

The third example of the buck derived circuit is the half bridge converter (Figure 8.26). Unlike the push-pull and the forward converters, the half bridge converter provides the benefit of lower stress on each of its two switches. One end of the single winding primary is switched between the bus voltage and ground and the other end is connected to the middle of a capacitor divider connected to the bus. A dc blocking capacitor is inserted in series with the transformer primary to prevent a flux imbalance problem seen with the voltage-fed push-pull converter. However, since this capacitor forms a resonant circuit with the reflected inductance from the secondary side, care must be taken to assure that the resonant frequency is kept well below the converter switching frequency. A good rule of thumb is between 10% and 20% of the switching frequency. With all its advantages over the push-pull and the forward converter, the half bridge converter has its drawbacks. The half bridge has two switches as compared with one in the forward. The high side switch needs a floating gate drive circuit which makes it more complicated compared to the push-pull and forward counterparts which have ground-referenced switches. Half-bridge converters are well suited for high input-voltage buses with moderately high power outputs.

Boost-derived topologies: current-fed, push-pull dc-dc converters

Having discussed several of the isolated versions of the basic buck switching regulator, we now turn to the isolated version of the boost switching topology. The modification of the basic boost circuit to its isolated counterpart is shown in Figure 8.27. The resulting converter and many other variations form a family of current-fed topologies. Current fed topologies eliminate the deficiencies that exist with the voltage-fed topologies previously discussed and are much better suited for high power and high output-voltage applications. As discussed earlier, the flux imbalance problems of the voltage-fed push-pull can be eliminated with the current fed version. The high impedance presented by the input inductor eliminates the need for extra circuits (or the blocking capacitor in the case of the half-bridge converter) in preventing the core flux imbalance. The inductors in a multi-output voltage-fed topology are replaced by a single input inductor. Several problems are thus avoided. The output inductors in voltage-fed topologies need to be sufficiently large to prevent them from operating with discontinuous current at the minimum specified load (a condition generally referred to as inductors running dry). For high output powers (>250W) and high output voltages, the size of this output inductor becomes impractical, at least for the point-of-load converter for a satellite where low mass is so critical. Having a single input inductor instead of the multiple output inductors greatly improves the cross regulation in the lower current windings of the converter.

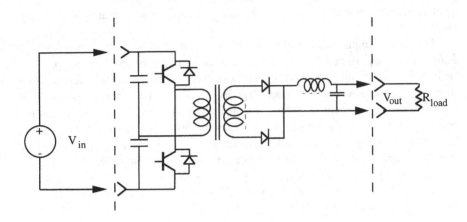

Figure 8.26 The Half-Bridge Converter

These windings are usually not directly regulated and are "slaved" to the main high-current output which is under feedback control. Since in a voltage-fed topology the point of control is at the output side of the main inductor, load variations greatly impact the slave windings' regulation. A current path needs to be provided for the current fed push-pull topology when both switches are opened. To maintain continuous current through the input inductor and prevent switch failures due to the inductive kick, a winding is coupled to the input inductor and excess energy is returned back to the primary bus or the secondaries.

Buck-boost derived topologies: flyback dc-dc converters

For low-power applications (typically less than 100W), the flyback converter is one of the most popular and simplest circuits in use today. The modification of the basic buck-boost switching regulator to its isolated flyback counterpart is shown in Figure 8.28. It is immediately observed that the flyback circuit is very different than the previously discussed isolated converters. Only one piece of magnetic material is used in this circuit, a coupled inductor, which provides both the isolation function and the filtering circuit. The circuit operates as follows. When the switch turns on, current ramps up in the primary of the coupled inductor and stores energy. No current flows in the secondary of the coupled inductor since the polarity of the secondary winding back biases the rectifier diode. When the switch turns off, the polarity of the secondary winding voltage changes, the rectifier diode is forward biased, and the stored energy is transferred to the load and the output capacitor. It is the simplicity of the circuit with one switch and one piece of

POWER MANAGEMENT AND DISTRIBUTION

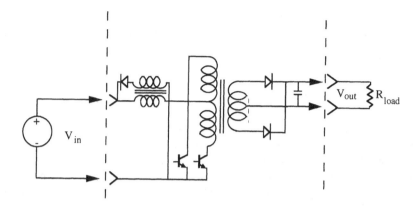

Figure 8.27 The Current-Fed, Push-Pull Converter

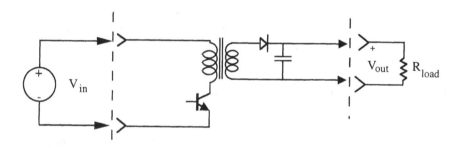

Figure 8.28 The Flyback Circuit

magnetic material that has made the flyback so popular. Another benefit is good cross regulation for a multi-output version of the converter due to lack of individual output inductors such as those found in the forward converter. However, in practice this is not easily achieved. At low power levels, with an extremely low leakage inductance coupled inductor, good cross regulation can be achieved. However, when the load variations in the outputs are far from each other, the lower power outputs suffer greatly in cross regulation due to the peak charging and current distribution effects. The peak secondary spike voltage resulting from the leakage inductance appears at the secondary winding and is rectified by the secondary diode. When the load current is at a minimum, the output capacitor is peak charged to very high levels which causes the poor cross regulation. The lack of an output inductor makes this topology suitable for high-voltage applications. Flybacks are not well suited in applications where the peak primary current is too high.

Resonant DC-DC converters

The converters that have been discussed thus far have all been of the PWM hard switched variety. They are called hard switched because the voltage and current waveforms in these converters are generally square wave. At the moment of turn-on there is full voltage across the device and at the beginning of turn-off there is full current through the device. Figure 8.29 shows the typical waveforms of a flyback PWM converter and demonstrates the switching losses at turn-on and turn-off of the switching transistor. Since there is a finite rise time and fall time for these waveforms, there is a period of overlapping voltage and current across the switch (e.g., as the drain voltage is rising the drain current is falling). This overlapping current and voltage results in the switching power dissipation in the transistor and occurs every cycle. Therefore, as the switching frequency is increased, the switching losses in the device increases. It is this loss that prevents hard switching PWM converters from operating at frequencies much above ~300KHz. Higher switching frequencies are desired since the size of a switching converter is inversely proportional to its switching frequency. This is due to the reduction in size with increasing frequency of energy storage elements such as the magnetics and capacitors.

Power supplies are typically measured by a figure-of-merit, the power-mass or power-volume density. This is usually defined as the output power of a DC-DC converter per unit mass or volume respectively (W/kg or W/m^3). While PWM topologies are the most widely used DC-DC converters in the industry, they are all limited by a maximum switching frequency limit at which point the switching losses in their switching elements would become too high, and thus, intolerable. Various methods, such as integrated magnetics and ripple current cancellation techniques (e.g., the Cuk' converter), have been employed with reasonable success to increase the power density of the converters. To drastically reduce the switching converter size and increase its density, higher switching frequencies are desired. With very high launch costs, and typically more than 100 point-of-load power converters per spacecraft, reducing the weight of each converter is important. Similarly, if some of the weight saving is exchanged for fuel, on orbit life can be extended which further reduces life cycle cost of the spacecraft and increases the revenues. Figure 8.30 illustrates these improvements over the past several decades.

Resonant converters are so called since they utilize the parasitic elements in a circuit as a tuned resonant tank. The resonant frequency of this circuit is designed close to the switching frequency of the converter. As the switching frequency of the converter is varied, the selectivity of the resonant circuit is used to control the amount of power that is transferred to the load. The benefit this method of conversion provides is the ability

POWER MANAGEMENT AND DISTRIBUTION

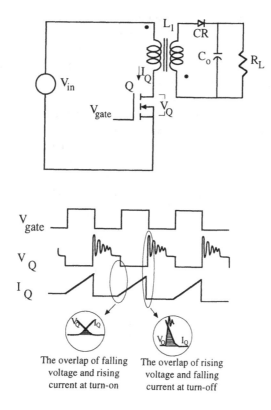

Figure 8.29 Waveforms of a Flyback PWM

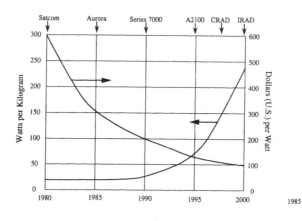

Figure 8.30 Improvements in Converter Design

to switch at frequencies an order of magnitude above hard switching converters without incurring excessive switching losses. This is due to the sinusoidal waveforms that are the natural result of the tuned resonant circuit. The main switch in a converter can now be turned on when its sinusoidal voltage naturally rings back to zero and can be turned off when the current through the device is nearly zero (Figure 8.31). Additionally, since the parasitic elements of the circuit are part of the resonant circuit, there are no uncontrolled switching spikes and high frequency ringing requiring lossy 'snubbers'. The waveforms in a resonant converter can be much closer to 'text book' and, unlike the hard-switched PWM counterparts with square waves rich in harmonics, the sinusoidal waveforms in resonant converters generate less EMI. This further simplifies the filtering requirements in the converter. Other benefits of the resonant converters include higher loop bandwidth due to their higher switching frequencies which results in better transient response.

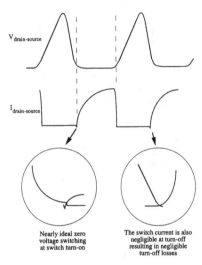

Figure 8.31 Resonant Converter Waveforms

There are some disadvantages with resonant converters but most of these can be addressed by proper design considerations. Creation of a high Q resonant tank circuit means higher transistor stresses during the switch-off time (often several times the bus voltage). The same type of ringing also exists in hard switching PWM converters (the inductive ringing and spikes) except it is an uncontrolled parasitic effect and dissipative 'snubbers' are used at a cost of lower efficiency (Figure 8.32). With the advent of higher-voltage, radiation hardened MOSFETs from several vendors (up to 500V), this is less of an issue. These devices have been successfully tested for single event effects (SEE) including single event burnout and single event gate rupture up to 80% of the

POWER MANAGEMENT AND DISTRIBUTION

device rating. While the $R_{DS\,(ON)}$ for these devices is higher, the added ohmic losses are traded against the losses that are incurred due to dissipative 'snubbers' in hard switching converters. Another disadvantage noted is the increased gate drive losses at higher frequencies. Although the gate drive losses could increase proportionally with frequency, the total gate charge for a MOSFET used in a zero voltage switching resonant converter is much less. The Miller capacitance in the MOSFET is used as part of the resonant tank and the energy stored in this capacitor is resonantly re-circulated in the resonant tank. Since the switch is turned on at zero voltage, the gate charge that is used for gate drive power calculation uses the total gate charge number without the Miller effect capacitance (Figure 8.33). Another disadvantage that is often noted is a faster drop-off in efficiency at lighter loads compared to PWM converters; however all converters drop in efficiency with lighter loads. As the load is decreased, the housekeeping power becomes a larger part of the delivered power to the load. This drop-off in efficiency tends to be flatter for PWM converters (Figure 8.34). This is because high frequency resonant converters regulate the transfer of power based on selectivity of their high Q tank circuit at either above resonance or below resonance mode of operation.

Moving away from resonance to deliver less power translates to a lower circuit power factor and higher circulating current in the tank that do not contribute to the delivered power but are rather part of the circuit losses. Secondly, in a zero voltage switching converter, a lighter load means a higher frequency of operation which lowers the efficiency further since frequency dependant losses such as gate drives and core losses are increased. This is not as bad as it may seem at first. In a typical zero voltage switching converter the loss of efficiency may be at most 2-3% from full load to 10% of

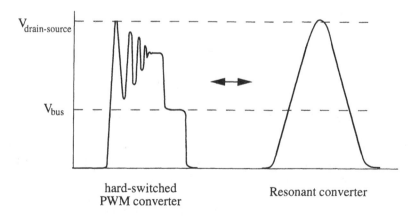

Figure 8.32 Comparison of Waveforms of a Hard-Switched PWM and Resonant Converter

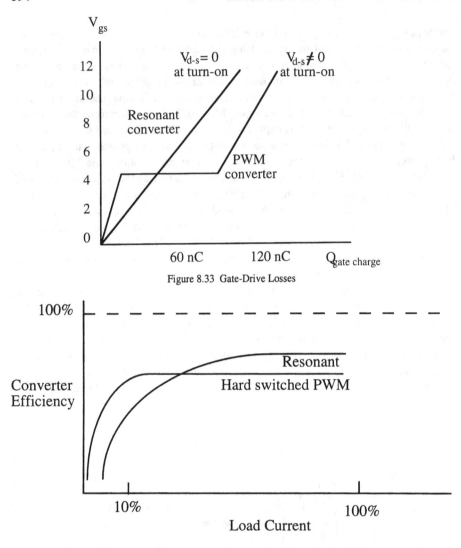

Figure 8.33 Gate-Drive Losses

Figure 8.34 Efficiency Comparisons for Hard-Switched PWM and Resonant Converters

load. Considering the 5-10% efficiency improvement, a resonant converter can provide at full load, the minimum load efficiency can still be better than its hard switching counterpart. Most resonant converters operate with variable switching frequency to regulate their outputs. This is not desirable for systems that require specified switching frequencies that are synchronized with other converters. This is to avoid generation of beat notes and interaction with sensitive instruments in the spacecraft payload. Although

POWER MANAGEMENT AND DISTRIBUTION

variable frequency resonant converters cannot be synchronized, their range of operation frequency can be specified and designed away from sensitive frequencies. The inherent unit-to-unit variation in switching frequency (since various components are operating at different load levels) has not proven to be problematic. Hundreds of variable frequency resonant converters are successfully operating on orbit aboard various communication satellites.

Operation of resonant converters is quite intuitive and easy to understand. Most engineers are very familiar with design and properties of LC filters. If a sinusoidal waveform is presented to, for example, a bandpass filter, the full amplitude waveform is seen at the output of the filter (neglecting any insertion loss) in the pass band of the filter. As the frequency of this signal is varied in the rejection bands, the sinusoidal amplitude is reduced. When the frequency is increased on the low pass side (above resonance), the peak-peak output voltage decreases and as the frequency moves closer to the corner frequency the peak-peak amplitude increases again. The opposite is true in the high pass side of this filter (Figure 8.35). The same principal holds for a simple LC tank circuit. The filter selectivity is used to control the power transfer above and below the resonant point with varying frequency. If the output of this LC circuit is presented to a transformer and is subsequently rectified, a DC voltage is produced. Therefore, varying the switching frequency can now increase or decrease this DC output. For example, in above-resonance mode, as frequency increases the DC voltage would drop (for a given load) and the output would increase again as the frequency moved back towards resonance. The opposite is true for below-resonance. It is immediately observed then, that if a negative feedback loop is closed around this circuit, a regulated

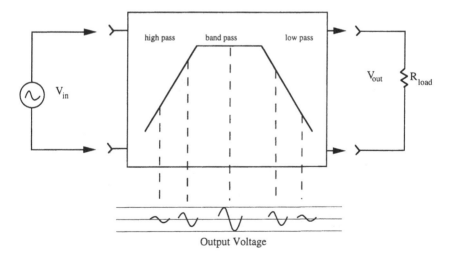

Figure 8.35 The Operation of a Simple Band-Pass Filter

DC-DC converter would result. The output DC voltage can be sensed and compared to a reference with an error amplifier. The error signal would be fed to a voltage-controlled-oscillator (VCO) which is used to drive the switching device. Therefore, for a given mode of operation, say above resonance, a decreasing output voltage would force the VCO to decrease the switching frequency and move the operation closer to resonance, thus bringing the voltage back to the correct level (Figure 8.36).

All resonant converters utilize a resonant tank circuit in one form or another in their circuits. There are two basic modes of operation. The series resonant converter places the load in series with the resonant tank and the parallel resonant converter places the load in parallel with the resonant tank capacitor (Figure 8.37). In the series resonant converter, a square wave voltage is generated from a voltage source and is presented to a resonant tank resulting in a sinusoidal current. In a parallel resonant converter, a square wave current is generated from a current source and is presented to the LC network resulting in a sinusoidal voltage. The series resonant converters lend themselves to zero current switching (i.e., the switch is turned off when there is zero current through it) and parallel resonant converters lend themselves to zero voltage switching (i.e., the switch is turned on with zero voltage across it). Any of the hard switched PWM converters can be transformed into its resonant converter by simply replacing the switch network of the PWM converter with its resonant counterpart. A zero current switching resonant buck converter is shown in Figure 8.38.

Other resonant DC-DC converters have been developed which find their roots in the RF power amplifier circuits such as the class D or class E amplifiers. Again, the basic principals of operation remain the same and they can be operated in series, parallel, or both.

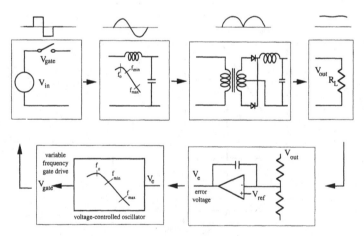

Figure 8.36 The Sequence Demonstrating How the Selectivity of an LC Filter Can Be Used to Make a Resonant Converter

POWER MANAGEMENT AND DISTRIBUTION

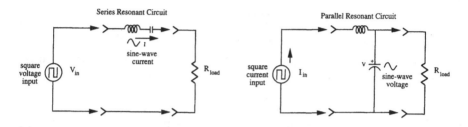

Figure 8.37 Resonant Converters

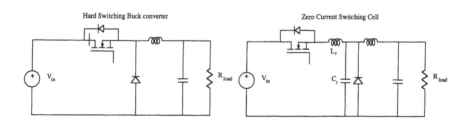

Figure 8.38 The Zero-Current Switching Quasi-Resonant Buck Converter

3. Components and packaging

Most communication satellites are now designed for 15 years of operation on orbit, and with the typical cost of a satellite in orbit approaching $500 million, the importance of the design robustness cannot be over stressed. This robustness does not end with the electrical design. The most properly designed, worst-case analyzed and tested electrical circuit can fail due to an improper solder joint, an overheated component, or one with latent ESD damage. Robust packaging design meeting the stringent requirements of launch, pyro shock, and long life is just as important as the electrical design. The packaging design includes worst-case finite element stress dynamics and thermal analyses. The designers must be mindful of issues such as coefficient of thermal expansion among components, PC boards, and associated housings; and thermal conduction and radiation paths must be analyzed and tested to predict maximum board and device junction temperatures. A packaging designer must work with the electrical designer, thermal, materials, radiation and reliability engineers, and understand the trade off in attempting to engineer the highest performance with lowest cost and weight. Simply selecting components that perform their intended function over the temperature range is not sufficient. Established reliability components with known aging effects, radiation and temperature performance, and adequate testing to the stringent space-environment standards, must be used. Lot traceability is required for each of the piece parts to ensure manufacturing process control and, if necessary, determination of the cause of failure.

In this section we will examine some of the unique requirements associated with the part selection and packaging techniques for use in space flight.

3.1 High-reliability space-grade parts

Although most civilian and defense satellites have been utilizing Class-S (space-grade) components, as manufacturing technologies have improved, new initiatives are moving towards more commercial-grade, high-reliability parts. Several operational satellites have successfully used these parts for certain applications. However, as the demand for spacecraft longevity increases (a minimum of 15 years is now expected for commercial communication satellites), the normal practice is still to use established-reliability, space-grade parts. Most spacecraft manufacturers typically establish and maintain their own approved parts list, and engineers are urged to design new products using these lists which identify readily available items from certified vendors. Specialized parts, optimized for a specific parameter, may have certain drawbacks which may make them less attractive (e.g., a high precision, extremely low offset voltage op-amplifier may need a high-power, dual-rail supply, or a low power consumption BiCMOS PWM may not tolerate a high level of total dose radiation). Spacecraft costs and cycle times can be reduced only by simplifying designs, reducing the parts count, and using standard parts. Program managers must remain aware of the importance of part specification, vendor selection, and parts management.

POWER MANAGEMENT AND DISTRIBUTION

Resistors

The very nature of switching DC-DC converters demands that dissipative components such as resistors not be part of the power train, although low value resistors are used as current sense devices. Resistors are mostly used in the control, protection, and telemetry circuits within a power converter. There are generally two types of resistor technologies in use for spacecraft: film and wirewound. Each is available in a variety of power ratings, tolerance, and package sizes. It is important to understand the temperature coefficient and aging effects for each type and, in particular, the failure modes must be understood to properly assess the impact on DC-DC converter operation. Radiation effects have typically not been a problem for resistors, but other factors such as the frequency characteristics or noise generation may be important and require special consideration.

The application guidelines for parts in general are not limited to their electrical characteristics. Other physical characteristics such as the mounting method are just as important, e.g., proper lead forming with stress. Power resistors are typically mounted near the edge of a PC board with a proper conduction path through a thermal copper layer to a chassis boss, often with a thermal compound used to assure a low thermal resistance. Larger resistors are either tied-down or mounted with solithane compound for bonding and strength. The following is a brief discussion of the specific characteristics of several resistor types typically used in spacecraft power converters.

Fixed film resistors are usually best suited for low noise and high frequency applications. Film resistors are made of nichrome, tantalum nitride, or other resistive metal films laminated inside or on an insulating material. Film resistors are well suited for precision circuits such as the feedback loop and the error amplifier in a power converter. The RNC style can be purchased in tolerances as low as 0.1% with a temperature coefficient of better than five PPM. They are also highly stable over their operational life. Typically 0.1% stability is used in the worst-case analysis for a 15-year mission. RLR style resistors are typically used for lower precision applications such as pull-up resistors. Their temperature coefficient is not as good as the RNC but RLR resistors are much less expensive.

The wirewound resistors are made of resistive wire wound on an insulated cylinder. These resistors are usually selected for their high power rating and are available in low ohmic values. The power rating ranges from 1 to 10 watts but the most popular ones for spacecraft power converters are typically one and two watts. They are available in 0.1% initial tolerance but, considering typical applications in filter damping networks and output pre-loading circuits, the 1% tolerance is more widely used. These resistors can be purchased in both inductive and non-inductively wound varieties.

Various other resistor types are used in a power converter. Thermistors are special devices with decreasing resistance with increasing temperature. They may be used in an over-temperature, shut-down circuit or a temperature-compensated output linear regulator. Often in an SSPA, the gate voltage is adjusted to temperature-compensate the RF output power by placing a thermistor in the RF section and feeding back the signal to the power converter. The gate voltage is usually generated with a (series pass) linear regulator fed from an auxiliary winding in a multi-output converter.

Thermistors are also available in positive temperature coefficient designs, sometimes referred to as sensistors. Resistor networks, available in single-inline (SIP) or dual-in-line (DIP) package styles, are also used when board space is at a premium. Variable resistors are usually not used in space applications.

Capacitors

Capacitors are available in a multitude of varieties with even more styles and specific usage characteristics than resistors. Capacitors are one of the ideally non-dissipative basic elements in a DC-DC converter power stage. As a general rule, capacitors are extremely sensitive devices so that care is required both in the process control during manufacturing and during installation in circuits.

There are three types of capacitors used for space applications: (1) multilayer ceramics that are extremely popular with power designers, (2) tantalum dielectric including both solid and wet-slug varieties and, (3) the glass, porcelain, and mica varieties.

The selection of capacitors is based on the specific circuit application requirements. Output filter capacitors, for example, require low equivalent series resistance (ESR) and low equivalent series inductance so that the output ripple is minimized. In applications requiring high temperature stability, NPO multilayer ceramic capacitors may be the most appropriate. Several parameters are often traded in the process of selecting a capacitor: the capacitor voltage and the ac current rating are first determined, and these generally determine the broad family of capacitors that are best suited. A number of other factors are then considered to narrow the selection: the capacitor tolerance and its variation with temperature, operating temperature range, equivalent series resistance and capacitance, self resonance frequency, reliability, packaging style, cost, and production lead time, among others. Once a selection has been narrowed to two or three styles, a capacitor type with the highest volumetric efficiency that meets the circuit application is selected.

Multilayer ceramic capacitors, made from a powder of various of titanate compounds, are by far the most popular capacitors in use today. There are three common types of ceramics, each distinguished by the characteristics of the dielectric used. The Z5U, typically made from barium titanate, has a very high dielectric constant. These capacitors provide a high volumetric efficiency but the initial tolerance, temperature coefficient, and long term stability are usually poor. They also demonstrate a poor voltage and frequency stability. The Z5U are best suited for bulk filter capacitors where large variations in capacitance may be tolerated. The Z5U capacitors are typically available with beginning-of-life tolerance of 5-10% and voltage ratings from 50V to 200V, but with a capacitance variation of ±10% due to swings in applied voltage and a 5% per 1000 hour variation due to aging. To obtain higher precision and better temperature and aging stability, the X7R ceramic material is typically used. These capacitors use calcium titanate. The dielectric constant is much less than the Z5U material and, therefore, smaller capacitance values are typically available given a case size. Their low ESR, good temperature, and end-of-life performance make them a suitable choice in many instances. X7R material offers very low dissipation factors (high Q), exhibit no varactor effects (capacitor does not change with applied voltage), and provide excellent temperature and long-term stability. If even more precision, lower ESR, long-term stability, and very low temperature coefficients are required, a third ceramic style alternative, the COG, is available. These capacitors use a non-ferroelectric ceramic material and are the lowest of all in their dielectric constant. Consequently, these capacitors are much larger and are only used when extreme precision is required, such as timing circuits in a PWM, a VCO for a resonant converter control, or, due to their low losses, as resonant capacitors.

Tantalum capacitors, solid and the wet slug designs, are often used for filtering applications requiring high capacitance values at low frequencies. They provide a high volumetric efficiency and good temperature stability. Wet slug capacitors are more often used as input and output filter capacitors of dc-dc converters although, with the advent of higher density multilayer ceramic capacitors, they are losing popularity. Solid tantalum capacitors are popular with commercial manufacturers of power electronics equipment but, due to a failure mechanism in high current applications, they are usually not recommended for high reliability space applications, although they can be safely used with a series limiting resistor or in low-energy circuits. This also eliminates their usefulness as an output filter capacitor; however, with a series resistor they can be used as the damping element in an output-input filter dc-dc converter application. Tantalum capacitors are polarized devices and care must be taken not to exceed their maximum rating.

Wet slug tantalum capacitors also provide high volumetric efficiencies similar to that of their solid counterparts but do not suffer from their impurity failure mechanism.

At relatively low capacitance values, mica and glass capacitors are an excellent choice. They both exhibit superior performance in temperature (as low as ±1%) and life stability, provide high self-resonance frequencies, and exhibit very low dissipation factors. Mica is a natural material and due to its crystalline structure can be cut into very thin sheets. It has a relatively low dielectric constant but provides better temperature stability and voltage performance compared to its ceramic counterpart. Glass capacitors offer even tighter tolerances and provide extremely low dissipation factors but are often more expensive. Due to better material uniformity, glass capacitors are better suited for applications with high ac voltage.

Magnetics

Second only to MOSFETs, the most expensive components used in the spacecraft power electronics equipment are transformers and inductors, principally because these parts are specifically designed and optimized for a unique application and are therefore custom devices. Magnetic devices are in the power-processing path of power electronics equipment, and optimizing their performance greatly impacts the throughput efficiency of the power equipment. Unlike other electronic components for which vendors have developed manufacturing lines with highly optimized processes producing standard components, magnetic devices are usually built-to-print from customer drawings. This means that producibility and repeatability can be major issues. The tight, low-leakage windings produced in the engineering laboratory by an expert technician might not be cost-effective or repeatable at the vendor's factory. Small process changes in the winding impregnate application method or the curing temperature can produce widely varying results. Even if every single conceivable parameter is specified, monitored, and tested to produce an electrically and mechanically consistent part, small process variations may lead to parameter out-of-specification conditions and subsequent lot failure.

Two types of magnetic devices are used in power electronics equipment: inductors, which are energy storage elements, and transformers, which are used to transfer energy, isolate the circuit grounds, or step-up or step-down the voltage or current in a circuit. Other combinations of these are also used, such as the coupled inductor in a flyback dc-dc converter which does both (i.e., it is an energy storage element with two windings).

The reader who is not familiar with the behavior of magnetic materials is referred to the Appendix for a summary of several topics critical to power system design.

Spacecraft magnetic components are specified and procured based on a set of requirements involving the design, manufacturing process, and testing procedures. Many issues that are unique to the space environment are detailed, from the smallest wire gauge that can be used to the method of splicing allowed.

Many core materials, sizes, and shapes are available to the power electronics designer. The key is to use the best materials suited for the application and to optimize for highest performance at the lowest cost. By far the most popular core material among spacecraft power electronics designers is the class of soft ferrites. Ferrites are homogenous ceramic materials made from a mixture of iron oxide, manganese, magnesium, zinc, nickel, or other rare-earth elements. They provide a high permeability with relatively high saturation flux densities (3000-5000 Gauss) and are well suited for modern high switching-frequency converters. Because the metals are used in an oxide form, the bulk resistivity of the ferrite is increased by several orders of magnitude compared to laminated iron cores, and this high resistivity lowers induced eddy current losses, thus making ferrites ideal for high frequency applications. Ferrites exhibit excellent aging and temperature stability, and the availability of different mixes allows manufacturers to offer special materials optimized for various frequency ranges.

POWER MANAGEMENT AND DISTRIBUTION

The optimum magnetic core structure is the toroid. Flux lines are nicely contained in a the toroidal core structure and flux crowding problems are minimized. The windings are placed directly on top of the core in a low-cost single step; since the windings are exposed, the toroid provides an excellent thermal path from the winding to a heat sink. Toroidal designs are the most efficient in high-frequency resonant inductor applications with a large ac component. For input and output inductors where large dc currents are present, designers typically use molypermalloy powder cores.

Since magnetic materials change from ferromagnetic to paramagnetic at their Curie temperature, thermal management is critical to guarantee that operations remain below the Curie point. Ferrites have a typical Curie temperature in the range of 150-250°C. However, the Curie point is not always the limiting factor. Some high frequency ferrites have an unusally high core loss rising exponentially at temperatures above 100°C and proper analysis must be performed to ensure thermal stability. Other limiting factors may be the maximum temperature of the insulating and binding materials used in their structure.

Semiconductors

Digital semiconductor technology is rapidly changing. Whereas power components (such as magnetics, power MOSFETs, and diodes) are in the power path of an EPS, low-power semiconductors devices (such as op-amps, PWMs, and comparators) are key to support functions such as control circuits. Microprocessors requiring 0.9V at greater than 50 amperes will demand output rectifiers for the power converter other than the traditional pn-junction or Schottky diodes. Considering the step load requirements and the magnitude of the current, placement of the power converter becomes critical. Even a 10A transient load and a rise time of 100ns from a source with only 10nH of lead inductance would produce a 1V drop. If the microprocessor were being powered by a 0.9V source voltage, this would be completely unacceptable. At 50A, a higher distribution voltage is required to the point-of-load with a subsequent down conversion to the desired voltage. This is not a new challenge in the spacecraft power technology. To meet the demands of SSPAs and TWTA's operating in a TDMA or multicarrier mode, power designers have developed innovative methods to address the transient response issue.

Several aspects of power semiconductors directly impact the power electronics designs and can dramatically improve efficiency and performance. Radiation hardened power MOSFETs with lower $R_{DS(ON)}$ and lower total gate charge are needed. Higher voltage versions can be used as the primary power switch, and lower voltage versions as the secondary synchronous rectifiers. In applications where output voltage is high enough (generally above 5V) at low to medium power levels, better pn-junction or Schottky diodes are needed. For the pn junction diodes, the forward voltage is never low enough and higher switching speeds are never fast enough. Although Schottky diodes generally provide lower forward voltage drops and much faster switching speeds, they are usually not available in higher reverse breakdown voltages or a low enough reverse leakage current.

3.2 Packaging technologies

The space environment places unique demands on electronics assemblies. Packaging is a multidisciplinary function that can be successful only when it is treated as a system. Achieving a reliable and efficient design requires trades and compromises involving electrical, mechanical, thermal, materials, radiation, reliability, and manufacturing engineering functions. Even though satellites are designed to be immune to single-point failures, redundancy does not address bad design. Since most redundant functions are carbon copies of the primary assembly, it will only be a matter of time before the secondary assembly fails in much the same way as the primary one did, thus rendering the spacecraft inoperable. The art of spacecraft electronic packaging starts with the demanding task of robust designs that are proven to last for 15 years or more. The real challenge is to do this while providing the highest density, lowest weight, and smallest size assembly to the user.

As components are brought closer to each other without leads and packages that provide stress relief, CTE mismatch, adequate thermal conduction paths, EMI, ESD, high voltage arcing, and interconnect fatigue become serious design issues.

The basic packaging technique for spacecraft electronics uses one or more printed circuit boards placed in an aluminum or magnesium housing. The boards are made from organic materials such as polyimide or glass epoxies and use the traditional multilayer plated-through-hole technology. Several standards do exist that govern various board materials, conductor thickness and plating, trace widths and clearances, solder mask and markings, board sizes, and many other parameters. Standard board sizes have been carried over from avionics and shipboard equipment into the spacecraft arena. For the most part, however, electronics assemblies within the spacecraft are highly optimized against weight and heat transfer requirements.

4. System examples

Despite numerous conflicting requirements, design optimization trades, and inter-related system issues, engineers designing PMAD systems have successfully developed and produced highly reliable spacecraft systems with on-orbit life times far exceeding their intended missions. When one considers that a communications satellite launched today is expected to still be operational in the year 2015, the high level of attention that is

POWER MANAGEMENT AND DISTRIBUTION

given to details in the spacecraft design and production process can be appreciated. What follows are a few examples of highly successful PMAD subsystems that have demonstrated a proven on-orbit track record, or in the case of ISS, hold the promise to do so.

4.1 The Lockheed Martin A2100

The Lockheed Martin A2100 spacecraft system was designed from a "clean sheet of paper" to set a new standard in commercial communication satellite technology. From the beginning, the system goals were modularity and reconfigurability to meet various customer designs. Every subsystem designer was given the task of improving performance (higher efficiency and lower mass), increasing producibility, and reducing cycle time and cost. New technologies were qualified and infused into the design over a three-year development period. Higher spacecraft power levels (up to 11 kW) with increased life expectancy to 15 years were made available serving a multitude of missions, including fixed satellite services (FSS) and direct-to-home and mobile communication services. Launch mass was reduced by 30% resulting in significant cost savings per launch. Several A2100 spacecraft are successfully operating on-orbit, and a brief description of that system is presented here.

The A2100 spacecraft generates its power from dual four-panel rigid gallium arsenide and silicon solar array structures (Figure 8.39). Solar cells are supported on Kevlar-graphite light weight panels and are driven with computer controlled stepper motors. Eclipse power is generated from dual nickel-hydrogen batteries made up of 100 and 131 Ahr individual pressure vessel (IPV) cells. Autonomous battery charge management (based on cell pressure and temperature), solar array digital sequential shunt control, bus regulation, and all of the autonomous fault management and telemetry functions are provided by a single modular and expandable (1kW-14kW) box called the power regulation unit (PRU). The PRU is fully internally redundant and provides the spacecraft with a regulated 70V bus (±2% at point-of-load, ±0.5% at the PRU). The 70V bus was chosen as an optimum point, high enough in voltage to significantly lower the spacecraft harness weight and payload DC-DC converter mass, yet low enough to utilize standard flight proven radiation-hardened MOSFETs and standard M123 multilayer ceramic capacitors. Point-of-load DC-DC converters use three standard designs, each optimized for maximum efficiency, lowest weight, and high producibility. The number of components was reduced by more than 40% compared to previous designs. Common part types were used whenever possible. Efficiencies for the payload converters were dramatically improved. TWTA high voltage electronic power conditioner efficiency is improved to 94% and SSPA EPC efficiency was improved to 92%. The power density was doubled from 75W/lb to 150W/lb.

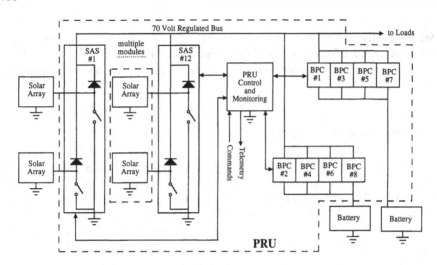

Figure 8.39 The Lockheed Martin A2100 Schematic

The PRU utilizes a novel bi-directional power converter (BPC) topology to charge and discharge the batteries. An N+1 redundancy scheme is used for each battery. BPCs are packaged as high density stand-alone subassemblies and are individually tested using automated test equipment before installation into the PRU.

Full sequential shunts are utilized to control the solar array power. This function is performed by the solar array shunt assembly (SAS). Again, these are stand-alone subassemblies and are individually tested before installation into the PRU. The control function for the SAS and BPC modules is provided by the bus control board (BCB).

The BCB is triply redundant and provides a bus voltage error signal which is selected in a voting scheme at the point of control. Other functions within the PRU include internal housekeeping EPCs, phase synchronization function, and the command and telemetry interfaces. The main power distribution is achieved with a rigid-flex assembly called the power bus assembly (PBA). The PBAs serve instead of traditional wiring harnesses and bus bars. Again PBAs are stand-alone assemblies and are individually tested before installation into the PRU.

The PRU's innovative yet robust packaging design has completely eliminated the need for internal point to point wiring. Units are consistent off the line and expensive box level troubleshooting has been virtually eliminated.

The A2100 spacecraft housekeeping and payload power converters use three standardized DC-DC converters. Low power TT&C and receiver applications use a high frequency flyback topology. Medium to high power SSPA applications utilize a Lockheed Martin proprietary high-efficiency resonant converter and the TWTA high efficiency EPC also utilizes a proprietary high-frequency topology.

POWER MANAGEMENT AND DISTRIBUTION 407

4.2 Global positioning system block IIR

The GPS block IIR spacecraft provides accurate longitudinal and latitudinal coordinates to U.S. military personnel and civilians anywhere in the world at any time. The GPS spacecraft is a three-axis stabilized satellite designed for 10 years of operational service. The spacecraft power subsystem provides 1100W at the end of life and 1700W at the beginning of life.

The power subsystem includes two solar array wings per spacecraft. Each solar array wing consists of two panels. Positioning the solar arrays normal to sun is achieved by two solar array drive (SAD) assemblies. The SADs provide power through a slip ring assembly within the SAD drive shafts. Two nickel hydrogen 40 Ah IPV batteries provide power during eclipse and when the solar power is insufficient to meet the spacecraft needs. Cell bypass circuits prevent the loss of a battery due to an open cell failure. Battery state of charge is monitored by temperature and pressure transducers. Additionally, ampere-hour integration is used to determine the exact state of charge for each battery.

The power regulation unit (PRU) provides bus-voltage regulation, battery charge and discharge functions, telemetry, and fault detection and protection functions. The PRU provides the spacecraft with a regulated 28V bus. Three basic modes of operation are possible. During excess power capacity (over and above the spacecraft need and the maximum battery charge rate), sequential shunts external to the PRU at the solar array shunt boom assembly (SBA) are enabled. This mode regulates the bus to 28V. During the battery charge mode, excess solar array power is used to charge the batteries and to regulate the bus. Three battery charge control modules are used for this purpose in a N+1 redundancy. In eclipse, or when excess power above the solar array output is required, battery discharge converters are used.

4.3 The International Space Station

The International Space Station (ISS) requires electrical power for a variety of functions: command and control, communications, lighting, heating, life support, as well as powering a host of scientific experiments. There are two electrical power systems (EPS), one each primary for the U.S. Orbital Segment (USOS) and the Russian Orbital Segment. Each EPS can provide power for its segment as well as provide shared power to the other segment and to support international partners. We will focus this discussion on the USOS EPS.

The USOSS EPS is a distributed power system consisting of three main subsystems: primary power, secondary power, and support. The primary power is obtained from two photovoltaic modules (PVM) operating at 160 Vdc. Each of these two power channels, designed to provide continuous power both during insolation and eclipse, is composed of a number of interacting elements: the solar array wing, a sequential shunt unit (SSU), the gimbal assembly (BGA), an electronics control unit (ECU), a direct current switching unit (DCSU), and three battery charge/discharge units (BCDU), and battery assemblies. Primary power generation is through the conversion of solar energy to electrical energy in four photovoltaic blankets, and primary energy storage is performed by banks of NiH_2 battery assemblies. The batteries are designed to require only a 35% depth of discharge to supply nominal ISS power requirements. If power generation were to fail, the batteries could supply power for one complete orbit at a reduced consumption rate.

As shown in Figure 8.40, much of the equipment is contained in an integrated equipment assembly that also provides thermal management, transferring heat from the Integrated Equipment Assembly (IEA) hardware to the deployable radiator. The BGA rotates the array to track the Sun while the ECU is the command and control link for power generation. The SSU maintains unregulated power in a specified voltage range, 130 to 180 Vdc. Note that the power storage function is accommodated in the primary power system since this centralized storage design results in decreased weight and cost over a decentralized one. It further provides a degree of flexibility in the event of degraded power generation capacity.

Primary power distribution is a function of the DCSU through a network of high-power switches that interconnect the arrays and batteries to the power bus. While the arrays are illuminated, the DCSU routes power to the ISS and to the battery charging units, and battery power supplies ISS requirements during eclipse.

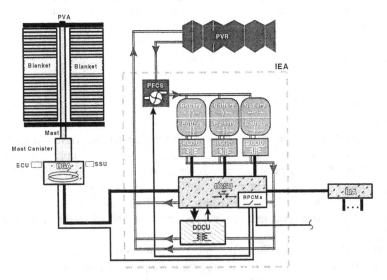

Figure 8.40 One of Two ISS USOS EPS Power Channels (Courtesy of NASA)

POWER MANAGEMENT AND DISTRIBUTION 409

The secondary power system (Figure 8.41) converts the 160V primary power to a secondary voltage of 124V. This conversion occurs not only on the IEA but throughout the ISS as well. After conversion, the secondary power is distributed to individual users who have the responsibility for any further dc-dc conversion. The key part of the secondary power distribution system is the Remote Power Controller Module (RPCM) which contains solid state or electromechanical relays that are remotely commanded to control the flow of power to individual users. The two-level EPS design allows the system to compensate for aging and line losses, and consistent with the earlier discussion on distributed systems, the higher voltage is used for transmission and the lower voltage for local distribution.

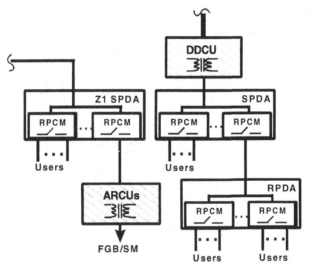

Figure 8.41 The USOS Secondary Power Conversion and Distribution System (Courtesy of NASA)

4.4 The Modular Power System

The Modular Power Subsystem (MPS), though an early vintage, is a flexible electrical power system that was designed to meet a wide range of power requirements in orbits ranging from near-Earth to GEO. The MPS controls, stores, distributes, and monitors power from mission-unique solar arrays and supplies unregulated +28Vdc power to spacecraft loads. The MPS has flown on a number of missions, including Solar Max, TOPEX, GRO, URAS, EUVE, and LANDSAT D.

The MPS can receive power from external sources during any mission phase, from pre-launch through spacecraft retrieval operations. Switching is done with relays. Either 20Ahr or 50Ahr NASA standard nickel cadmium batteries can be used to provide eclipse power, the choice dependent on the total energy storage requirements. A block diagram of the module is shown in Figure 8.42.

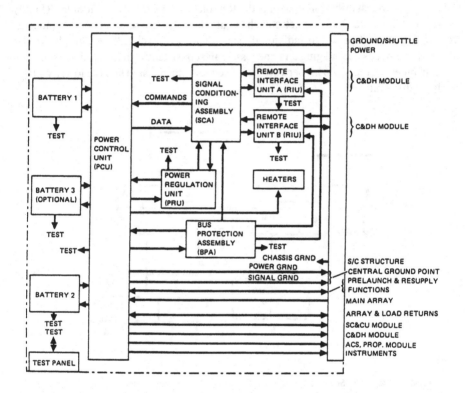

Figure 8.42 The Modular Power Subsystem (Courtesy of NASA)

The PCU serves as the distribution point for the power bus and for battery recharge power. The unit contains relays for power control and for arming and disarming the MPS as well as current sensors for charge control and solar array input, battery charge/discharge, and load current monitors.

The PRU (see Fig. 8.43) conditions the solar array power for use by the various loads and for battery charging. The input to the PRU can vary between 40 and 125 V, 72 A max, and 4500 W max, while the output is unregulated between 22 and 35 V at 108 A max and 3600 W max. It uses a non-dissipative series switching regulator consisting of six 18 A power regulator modules operating in parallel and a redundant PPT circuit, dc-dc converters and regulators, and battery charging logic. Its three modes of operation

POWER MANAGEMENT AND DISTRIBUTION

include PPT, drawing maximum power from the array; voltage limit, with eight selectable battery charge voltage levels; and standby, during which array power is not available.

The SCA provides the necessary interface circuitry between the power subsystem and the RIU coders and decoders. Among its design features are redundant dc-dc converters to supply MPS regulated power, the logic to maintain one battery on charge at all times, an independent parallel command redundancy, and automatic individual battery charge inhibit controls.

The RIU provides all command and telemetry interfaces between the power subsystem and the Communications and Data Handling subsystem.

Finally, the BPA provides redundant fusing for all non-critical MPS loads. The fuses are operated well below their normal rating to avoid failures other than those due to true circuit overload. The primary and redundant fuses are placed on separate plug-in cards for added protection against an accidental mechanical disconnect of both fuses.

The thermal design of the MPS is a passive equipment-to-baseplate radiator arrangement sized for the worst-case, hot-mode operating point. Active thermal control is provided by louvers during cooler operating times, and for maximum cold-mode operation, heaters and MLI blankets are used (see Chapter 9).

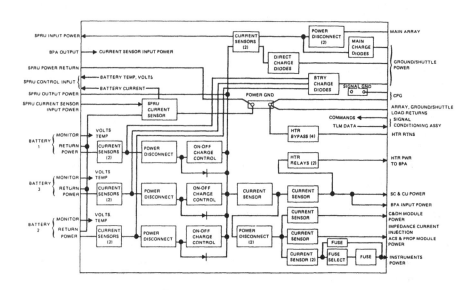

Figure 8.43 Standard PCU Functional Block Diagram (Courtesy of NASA)

5. References

Anderson, P. M. and Wolverton, R.H. Capt., *A Low Mass, Modular Power System for 300W to 1.8kW Space Craft*, internal technical memorandum, (Lockheed Martin Astronautics, Denver, CO).

Bingley, J.D., *A2100 Power Regulation Unit, Design and Predecessor Hardware*, internal memorandum, (Lockheed Martin, 1996).

Determan, W.R., *Concepts for Multi Kilowatt Space Power Applications*, internal memorandum 889241, (Rockwell International/Rocketdyne Division, Cagona Park, CA).

Dixon, Lloyd, *An Electrical Circuit Model for Magnetic Cores*, Unitrode Power Supply Design Seminar, SEM-1000, (1994) 6-9

Ebersole, T., *Electrical Power Subsystem*, GPS Orbital Operations Handbook, internal document, (Lockheed Martin, 1996).

El-Hamamsy, S. A. and Chang, E.I., *Magnetics Modeling for Computer-Aided Design of Power Electronics Circuits*, IEEE Transactions on Power electronics (1989).

Floyd, H. and Linkowski, F., *Spacecraft Applications of Lithium Ion Battery Technology*, internal memorandum (Lockheed Martin, 1998).

JPL Web Site, *Mars Rover Power Subsystem,* http://mars.jpl.nasa.gov/roverpout/power.htl.

Kassakian, J.G., Schlecht, M.F., Verghese, G.C., *Principles of Power Electronics*, (Addison-Wesley, New York,1991).

Kim, S.J. and Cho, B.H., *Analysis of Spacecraft Battery Charger Systems*, internal technical memorandum, (Virginia Power Electronics Center of the Department of Electrical Engineering, VPI, Blacksburg, VA).

Kota, Gandi K. and Owens, W., *Electric Power Generation and Conditioning for Spacecraft Dynamic Isotope Power Systems*, IEEE Transactions on Power Electronics (1987).

Krummann, W. and Ayrazian, H., *The Hughes HG601 HP Spacecraft Power Subsystem*, IECEC Proceedings (1998).

Magnetics Inc., *Data Book for Ferrite Cors for Paner and Filter Applications.*

Mammano, R., *Resonant Mode Converter Topologies*, Unitrode Power Supply Design Seminar, SEM-1000 (1994).

Mclyman, Wm. T., *Transformer and Inductor Design Handbook* (Second Edition) (Deklcer, New York, 1988) 11-12.

Department of Electrical Engineering, Massachusetts Institute of Technology, *Magnetics Circuits and Transformers*, internal memorandum (Massachusetts Institute of Technology, Boston).

Murray, W. E., *Space Station Electrical Power Distribution System Development* (Society of Automotive Engineers, 1985).

Papula, P., *Battery and Solar Array Overview*, internal technical memorandum (Lockheed Martin, 1997).

Patil, A.R., Cho B.H., and Lee, F.C., *Design Considerations for a Solar Array Switching Unit*, internal technical memorandum (Virginia Power electronics Center of the Department of Electrical Engineering, VPI, Blacksburg, VA).

Pressman, Abraham I., *Switching Power Supply Design* (McGraw Hill, New York, 1991).

Reppucci, G.M. and Sorenson, A.A., *Space Station Power System Challenges*, P-85/164 (Society of Automotive Engineers, 1985).

Schiffer, S.F., *Nickel-Hydrogen Batteries Come of Age*, RCA Engineer (RCA Corporation, 1984).

Schulten, M., Steigerwald, R.L., and Kheraluwala, M. H., *Characteristics of Load Resonant Converters Operated in a High Power Factor Mode*, IEEE Transactions on Power Electronics, Vol 7, No 2, (1992).

Severns, R.P. and Bloom, G.E., *Modern DC-DC Switchmode Power Converter Circuits*, Bloom Associates (1985) 14.

Steigerwald, Robert L., *A Comparison of Half-Bridge Resonant Converter Topologies*, IEEE Transactions on on Power Electronics, Vol 3, No 2, (1988).

Steigerwald, Robert L., *Practical Design Methodologies for Load Resonant Converters Operation Above Resonance*, private communication.

Troutman, Joe, *EOS EPS Bus Voltage Analysis*, internal technical memorandum, (Lockheed Martin Corporation, 1991).

Troutman, J., Bounds, R., and Stewart, D., *EOS Bus Voltage Trade Study*, internal technical memorandum (Lockheed Martin Corporation, 1991).

Vanduyne, J., Schysani, M., *Enhanced Power Regulation Unit for the Lockheed Martin A2100*, internal technical memorandum, (Lockheed Martin Corporation, 1998).

Walthall, E.R., *Spacecraft Design Course Notes, Communication Satellite Design*, RCA Astro Electronics (1986).

CHAPTER 9

THERMAL MANAGEMENT

1. Introduction

The thermal environment aboard a spacecraft is managed by a thermal control system (TCS) which maintains the proper operating temperature regimes for individual components and for the spacecraft as a whole. In addition to the heat loads imposed by the external environment, inefficiencies in the spacecraft systems will generate internal heat that must be accommodated. Thermal management is a multidisciplinary technology based on detailed knowledge of the orbit and its environment, materials and their long-term behavior in space, heat transfer, and spacecraft modeling. While much of the thermal load on a spacecraft comes from the external environment, e.g., the direct solar flux, the TCS is also sensitive to the design and operation of the onboard electrical power system because of the need to control, collect, transport, and reject the heat generated due to inefficiencies in the operation of the electrical subsystems and components. As important as the TCS is to proper spacecraft operation, it accounts for less than five percent of the total spacecraft cost and mass.

Two aspects of the space environment, the vacuum of space and zero gravity, combine to eliminate free convection, the most common method employed on Earth for transporting heat and operating cooling devices. During operation, the spacecraft absorbs heat from the Sun and reflected Sunlight and infrared radiation from planets, and, in addition, will generate some heat onboard from the operation of a number of spacecraft systems and subsystems. Unlike systems that operate on Earth, systems aboard spacecraft cannot easily reject the heat that continues to accumulate. Compounding the challenge of maintaining overall thermal equilibrium is the need to keep selected parts of the spacecraft operating within strict temperature limits that may be greatly outside the equilibrium temperature of the spacecraft. In some cases, the temperature control demands placed on the TCS by individual components throughout the spacecraft may range from near zero Kelvin to over 1000 Kelvin.

The thermal control of a spacecraft presents a number of difficult engineering issues. Since free convection is not a viable option in space, maintaining a proper thermal balance for both the structure and the various pieces of equipment aboard the spacecraft over long periods of time becomes a delicate balancing act between the various sources of heat and the ability of the spacecraft to radiate the excess heat. In some limited applications (as in the case of cryogenic boiloff to maintain very low temperatures for sensors), mass ejection has been used to remove heat, but this is generally not available for long-term operation of large systems.

In this chapter, we discuss the considerations that are a part of the design of the TCS, how the spacecraft mission and the thermal environment of space impact that design, how the TCS might interface with the power system, and some of the key components of a thermal control system.

1.1 Definition and purpose of a TCS

The thermal control system is designed to maintain the temperature of various elements of the spacecraft within defined bands under the varying conditions of orbital flight and over the relatively long lifetime of an operational satellite. The thermal loads on the spacecraft will vary due to changes in the external environment, over which the thermal engineer may have little control, and the internal environment, whose design characteristics often involve tradeoffs with other systems. Thermal control is more than just a case of maintaining balance between the minimum cold temperature of the spacecraft and its maximum hot temperature. Throughout the operational life of the satellite there will be specific temperature demands placed on its various components. Figure 9.1 shows several of these components and the temperature bands within which they must be maintained under varying conditions such as changing orbits, power generation and usage, and environmental degradation of the materials comprising the TCS. The necessity of maintaining the temperature within these bands varies from component to component. For example, while the solar arrays can operate over a wide temperature range, their efficiency decreases as the temperature rises so that a tradeoff between decreased efficiency and the cost of maintaining control of the temperature must be made. In other examples, cryogenic systems such as IR sensors fail to operate above a narrow temperature range, and hydrazine fuel freezes at the lower end of its storage-temperature range and decomposes at the upper end of that storage temperature. The broad range shown for typical structures can be limited because of requirements for accurate pointing and tracking systems attached to the structural member which may be expanding and contracting with changes in temperature.

The thermal control system maintains the necessary balance within the systems and across components comprising those systems. The relationship between the TCS and the spacecraft electrical power system is a particularly close one because of changing internal thermal loads due to varying demands for electrical power and the internal heat generated by all electrically powered devices. As we will discuss later, a critical issue within the spacecraft is the thermal conductivity of the spacecraft structure necessary to provide low resistivity pathways from components to radiators located on the exterior surfaces.

THERMAL MANAGEMENT 417

Although the structural components will most often provide the high conductivity pathways needed, the electrical insulation in the power system will present high thermal resistance as well. Materials research aimed at this and related problems (such as developing electrical conductors that can serve as solar array coatings) have produced solutions that, while available, remain costly. Large, cyclic temperature gradients across the craft can create structural problems as well as contributing to difficulties in pointing, tracking, and alignment of sensors and antennae. Through clever design, the thermal engineer seeks to protect domains of tight temperature control at low cost in dollars and mass and in a way that is compatible with the other operating systems of the spacecraft.

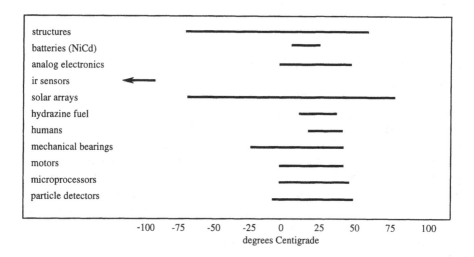

Figure 9.1 Representative Temperature Ranges for Spacecraft Components

1.2 Characterization and design of the thermal control process

Many considerations go into a complete characterization of the thermal management system, only some of which are at the discretion of the designer. Foremost is the mission, which overshadows all other issues. The mission will dictate features such as the orbit (and the attendant space environmental issues for that orbit choice), the on-station duration and design life, reliability criteria, operations requirements and timelines, configuration constraints (e.g., sensor field of view, allowable jitter), launch and ascent constraints (e.g., ground cooling, booster interfaces), safety, survivability, contamination control, spacecraft maintenance and interfaces, and programmatic constraints (cost, schedule, risk) (Vernon, 1989). In a simple case of a satellite in LEO with no significant internal heat loads, the design process may be a straightforward passive thermal control system. As the spacecraft design or operation becomes more complex, additions beyond the passive design may become necessary.

As depicted in Figure 9.2, the process starts with a definition of the mission in order to establish requirements. The thermal design requirements may be specified (as in choice of launch vehicle, orbit, design lifetime, acceptable temperature limits), derived (as in duty cycles, design techniques, temperature margins), or allocated (as in power, weight, volume). Preliminary subsystem definition is then performed on not just the TCS, but all spacecraft subsystems. This preliminary analysis defines the overall sufficiency of the TCS concept. Tradeoffs among subsystems lead to a preliminary satellite concept design and configuration. More detailed analyses, which might involve interactions of the TCS with the structure, the power system, placement and specific temperature requirements of components, among others, lead to modifications of the original concepts and result in a preliminary analytical model for the TCS. After refinements, verification of the modeling is usually done by thermal vacuum testing (Agrawal, 1986; Vernon, 1989).

There are a number of system constraints that must be accommodated within the trade-off allowed for each of the features demanded by the mission, and, in most cases, the interdependence of choices narrows options as the design progresses. As an example, in the choice of orbit, consideration must be taken for percent eclipse coverage. This, in turn, for the common case of photovoltaic power, impacts the power system design, the choice of battery configurations and operating parameters (such as depth of discharge) of the power management system, the weight allocated to the power system, and reliability, among others. This same interaction among systems and design philosophy will apply to all thermal design system constraints such as redundancy design, design margin philosophy (especially power margin), choices of an active or passive TCS, specific temperature specifications for components, weight, volume, and cost considerations, and acceptable failure modes.

The proper design of a TCS goes beyond simply maintaining overall spacecraft temperatures between some maximum and minimum values. A variety of design parameters, many related to temperature, enters into the thermal engineer's task (Wise, 1993). For example:

- Maximum and minimum non-operational temperatures– during ground operations and during flight prior to achieving final orbit, there may be temperature constraints placed on certain components. Allowing the thermal environment to stray outside of these limits may cause damage to the component or may not allow the component to begin operation when required. These limits are usually more generous than those imposed while the system is operating.
- Maximum and minimum operational temperatures– allowing the temperatures to go outside these limits may cause problems ranging from loss of calibration to component failure.

THERMAL MANAGEMENT

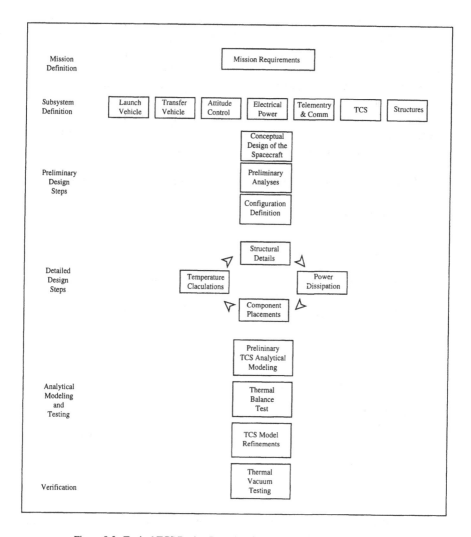

Figure 9.2. Typical TCS Design Process (after Agrawal, 1986; Vernon, 1989)

- Minimum turn-on temperatures– often a component may be stored at a much lower temperature than the norm for operation, but must be raised to a higher minimum temperature before safe operation can be assured. A simple example is a heat pipe that may be 'frozen' during dormancy but must be heated prior to operation.
- Rate-of-change limitations ($\Delta T/\Delta t$)– when bringing a system or component into its safe operational range there may be limitations imposed on the rate at which heat

may be added to or removed from the component by the TCS. Optical systems may be particularly sensitive to this rate of change of temperature.

• Controlled temperatures– there are numerous components (e.g., oscillators, ovens, sensors, and devices that operate at cryogenic temperatures) whose operating temperatures must remain within very tight limits. These devices usually require spot heaters or refrigerators to maintain tolerances.

• Temperature versus lifetime– it is usual for the temperature of a spacecraft to tend to rise with time, often because of changes in the surface properties of the radiator; overdesign of the TCS may be required to accommodate this temperature versus time trend.

The INSAT-2A, a geosynchronous communications satellite launched on an Ariane vehicle in 1992, offers a specific example of the component temperature limits. These are presented in Table 9.1.

Table 9.1 Component Temperature Limits for INSAT-2A (Kaila and Bhide, 1994) degrees C

Spacecraft Component	Non-Operating Min	Max	Operating Min	Max
Traveling Wave Tube Amplifier	-40	80	-15	75
Solid State Power Amplifier	-40	65	-15	60
Battery (NiCd)	0	35	0	35
Very High Resolution Radiometer	-30	55	-15	50
Gyros	-15	70	-10	65
Momentum Wheels	-15	55	-5	50
Earth Sensor	-40	60	-15	55
HgCdTe IR Sensor	-168	-158	-168	-158
Propulsion Tanks	5	45	5	45
Reflectors	-175	90	-175	90
Solar Panel	-120	115	-120	115
Spacecraft Structure	-50	65	-50	65

Thermal design considerations also involve compromises with the structural, attitude control, and power systems, as well as meeting cost, weight, and volume constraints. The issue of mass constraints in spacecraft design is particularly critical to the TCS. Lightweight spacecraft tend to be inherently poor thermal conductors so that the task of conducting internally-generated heat to the surface radiators becomes an even more difficult problem. As the availability of prime geostationary orbits decreases, the need to increase the number of channels and power per channel on these communications satellites will become an option of choice. This, in turn, will increase the internally generated heat loads and make thermal management even more challenging.

THERMAL MANAGEMENT

2. The thermal environment

The thermal loads with which the TCS must deal come from several sources, as shown in Figure 9.3. Although each of the sources contributes to the overall thermal environment, the dominant source is the Sun which dictates much of the TCS design. At the Earth's average distance from the Sun (1 AU), about 1370 W/m² of thermal load is delivered to the surface of a spacecraft. In addition to this direct illumination, the Sun's radiation affects the spacecraft in three other ways: the reflection of the Sun's radiation from the surface of the Earth, the albedo, presents a thermal load to the spacecraft which varies with the latitude and the orbit; the Sun's radiation scattered from one part of the spacecraft to another; and the infrared radiation from the Earth that is present even on the nighttime side of the orbit. Finally, during launch and orbital positioning there are aerodynamic and rocket plume thermal loads that must be accommodated by the TCS.

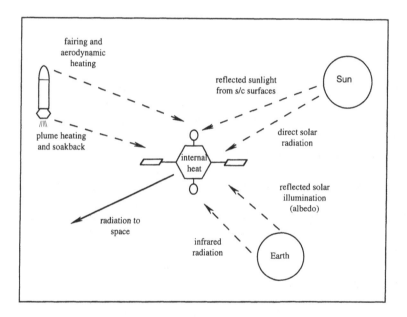

Figure 9.3 Sources of Heat Affecting a Spacecraft

Prior to launch, the thermal management of the spacecraft is controlled by ground units either in a clean room or on the launch vehicle itself. It is common for a satellite to spend weeks or months in a vehicle while preparing for launch. The TCS requirements during this phase are not as demanding as those on-orbit because of the greater freedom available to terrestrial thermal management systems. Once the launch preparation phase is completed, however, and thermal management is under the spacecraft's TCS, the thermal environment of the spacecraft will never again be static. There are several distinct thermal environments that must be accommodated: launch and ascent, orbit transfer, and final orbit parking. During launch and ascent, aerodynamic heating can generate loads of several hundred W/m^2 on the fairings. Once the vehicle has reached an altitude of greater than about 100 km, the aerodynamic heating is no longer an issue, but is replaced by the several other sources of heat with which the spacecraft must deal throughout its operational lifetime. While there may be little distinction between the environmental characteristics of the transfer orbit and the final operational orbit, there is a significant distinction between them in the internal heat generated by systems operating in the final orbit that are not active during orbit transfer. (Pisacane and Moore, 1994). The thermal loads will continuously change due to orbital dynamics, the operation of onboard systems, and the interactions of the spacecraft with the environment. As we will see later in this section, these changes occur with time constants varying from minutes to years.

2.1 Solar radiation

The primary source of heat for a typical spacecraft is radiation heating from the Sun, about 1370 W/m^2. This number will vary by about 5% during the course of the year as the Earth's distance from the Sun increases and decreases. The 'blackbody' temperature of the Sun is about 5860 K (see Section 3.2), and the energy radiated as a function of wavelength for a blackbody of that temperature is shown in Figure 9.4.

At the Sun's blackbody temperature, virtually all of the solar radiation occurs at wavelengths between 0.2 and 2.6 micrometers so it is important that materials used in spacecraft construction have their absorptivity fully characterized for these wavelengths. The behavior of materials exposed to this wavelength band, especially the degree to which the energy is absorbed, is of paramount importance to the thermal design engineer. The actual heating that this solar radiation causes is dependent on the characteristics of the materials covering the spacecraft, particularly the solar absorptivity, α, and thermal emissivity, ε, which are discussed in detail in Section 3.3.

The extent to which solar heating occurs is also dependent on the orbit of the spacecraft and the method by which the satellite is stabilized. The amount of time the spacecraft spends in eclipse, and thus out of the Sun's direct radiation, may vary from more than 40 minutes of every 90-minute low-Earth orbit to as little as zero for much of the year for satellites in geosynchronous orbits. The solar heating load will change dramatically during eclipse and vary during the year as the Earth proceeds in its orbit.

The method by which the satellite is stabilized will also influence the thermal balance. The thermal design for spin-stabilized spacecraft uses the cylindrical solar array surrounding the satellite to radiate internally-generated heat. An α/ε ratio of unity for

THERMAL MANAGEMENT

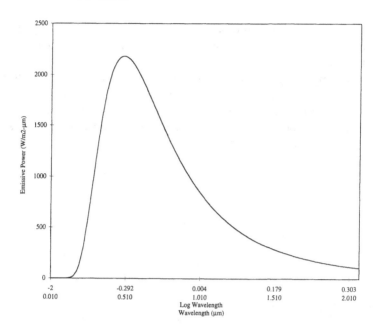

Figure 9.4 Radiated Power versus Wavelength for a Blackbody at 5870 K

solar panels makes them a useful heat sink and radiator. Heat is conducted either by direct contact or via a structural member of the satellite to the solar array as it spins about an axis normal to the direction of the Sun (Gilmore, 1994).

The most common method of stabilization, especially for geosynchronous satellites, is three-axis, and generally all use the same thermal management techniques, namely a combination of multilayer insulation and efficient radiators. Electronic boxes generating large amounts of heat are usually mounted directly on the walls of the structural members supporting the outward-facing radiators. Boxes internal to the spacecraft transfer their heat to the radiators by radiation or conduction. As we will see, radiators are designed to be low solar absorbers. This low absorbing surface, combined with the multilayer insulation used on the other parts of the satellite, makes the three-axis stabilized spacecraft less sensitive to variations in the external heat loading than those that are spin-stabilized (Gilmore, 1994).

Variations in the solar flux do occur, however, and must be a part of the thermal design. As Agrawal (1986) points out, in the case of geosynchronous satellites the variations occur on two different time scales. The geosynchronous satellite makes one complete revolution every 24 hours, and this gives rise to a diurnal (i.e., daily) variation that is depicted in Figure 9.5. In this case, the solar flux is a maximum on the various faces of the satellite as the day progresses: anti-Earth facing at noon, west at dawn, etc. Because the north and south facing sides are never in direct illumination, there is no significant variation in flux loading for these faces. They are the faces most likely to show the best temperature stability. In addition to the diurnal variations, there is a seasonal variation as the Sun rises and falls 23 degrees from the equinox direction. As seen in the figure below, and as discussed in Chapter 2, it is only at equinox that eclipses can occur for geosynchronous satellites. The maximum eclipse period during these semi-annual eclipse seasons is about 75 minutes.

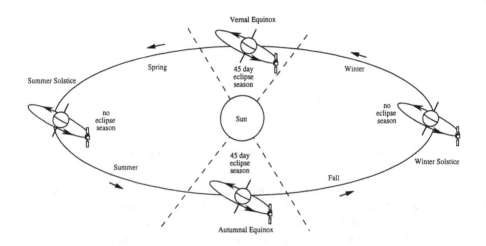

Figure 9.5 Solar Flux Variations at Geosynchronous Orbit (Sabripour, 1999)

2.2 Planetary radiation

Although this section will be restricted to a discussion of the radiation from the Earth, all of what is presented can be generalized to the thermal load imposed on a spacecraft in orbit around any planet.

The Earth contributes to the overall thermal load of a satellite in two ways: direct reflection of Sunlight (albedo) and infrared emissions defined by the equilibrium temperature of the planet. These effects are important in low Earth orbits and are only minor additions to the thermal models at geosynchronous altitudes.

The Earth albedo

The albedo characterizes the Sun's reflected radiation from the surface of the Earth and is usually expressed as a percentage of the Sun's radiation reflected back into space. While a nominal value of about 0.3 is often used in calculations, the actual albedo is not a constant. The amount of Sunlight reflected varies depending on whether the reflection occurs from continents or oceans, forests or ice packs, by direct illumination or grazing incidence, etc. There are much data available on the details of albedo calculations, some of which is nicely summarized by Gilmore (1994). One should also note that the albedo is independent of any particular orbit. Orbital averages are usually sufficient in thermal calculations because of the relatively short time constants of most periods and the thermal capacity of most satellites. When using the albedo to estimate the thermal flux reaching the satellite, it is important to remember that the flux will decrease as the satellite passes the subsolar point of the orbit. The solar energy reaching the Earth, and so being reflected from the surface, varies as the cosine of the angle from the subsolar point.

The albedo of the Earth is also a function of the latitude with the largest value occurring at the poles and the smallest at the equator. This variation is to be expected since reflection from snow is greater than that from dense vegetation.

The Earth IR Emission

The Earth itself absorbs some of the solar radiation. The equilibrium temperature of the Earth is above zero Kelvin and so emits electromagnetic radiation approximated by a blackbody at its equilibrium temperature. This equilibrium temperature can be estimated simply by equating the solar energy absorbed by the Earth to the energy radiated by the Earth:

$$S\pi R_e^2 (1-\rho) = \sigma \varepsilon (4\pi R_e^2) T^4$$

$$T = 290 K$$

In the expression above, S is the solar radiance (1370 W/m^2), R_e is the radius of the Earth, ρ is the albedo (0.3), σ is the Stephan-Boltzmann constant, and ε is an assumed emissivity of 0.9 for the Earth.

At this low temperature, the peak amplitude of this radiation occurs at a wavelength of about 10 micrometers, far above the visible portion of the electromagnetic spectrum, but in the infrared portion. It is common, therefore, to refer to this radiation as the Earth

IR emission. Near the Earth's surface, the average value of the thermal radiation is about 240 W/m^2. The IR emission varies with latitude, with the largest emission, as might be expected, taking place near the equator. The average IR emitted at the equator is about 50% greater than that at the poles, as shown in Table 9.2.

Table 9.2 Variations in Earth IR Emission and Earth Albedo as a Function of Latitude (used with permission, Gilmore, 1994)

Latitude Range (degrees)		Average IR (W/m^2)	Average Albedo (percent)
90	80	177	69
80	70	177	68
70	60	189	53
60	50	202	44
50	40	218	37
40	30	240	31
30	20	259	26
20	10	256	24
10	0	240	25
0	-10	252	23
-10	-20	256	24
-20	-30	259	24
-30	-40	240	28
-40	-50	218	35
-50	-60	202	45
-60	-70	183	56
-70	-80	161	74
-80	-90	136	74

At all spacecraft altitudes where the flux is significant, the Earth's IR is greater than the albedo flux. Also, recall that the albedo contribution to spacecraft heating will disappear while the vehicle is in the Earth's shadow. The IR flux is present at all times, however, independent of whether or not the satellite is being shielded by the Earth from the solar radiation. A graphical comparison of the relative values of IR and albedo flux averaged over the surface of a spherical satellite for low-to-mid altitudes is shown in Figure 9.6. Both sources of heating are negligible at geosynchronous orbits, the IR flux being of order 0.5 W/m^2 and the albedo about half that value.

2.3 Spacecraft-generated heat

The launch vehicle and the spacecraft itself will be the source of some heat that must be collected, transported, and then radiated. In addition to the planetary radiation, there is a thermal load of about 0.5 W/m^2 throughout the solar system due to cosmic rays interacting with the spacecraft. While these heat loads will be significantly less than those

THERMAL MANAGEMENT

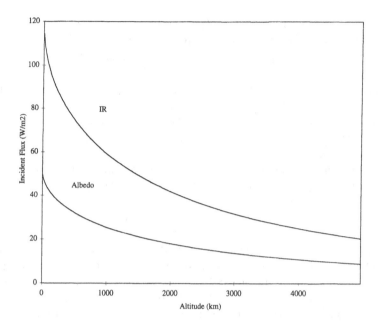

Figure 9.6 Incident IR and Albedo Flux on a Spherical Satellite versus Altitude

imposed by solar radiation or the Earth's IR, they may be sufficient to interfere with the performance of sensitive components, especially those operating at cryogenic temperatures.

On-board systems

Electrical power components on satellites are designed to operate at the highest efficiency to reduce the heat generated. Generally these components function at efficiencies between 0.15 (e.g., solar arrays) and 0.97 (inverters) and at power levels from a few watts to hundreds of watts. Any inefficiency will produce some localized heat, usually low-grade (i.e., low temperature) heat that the TCS must accommodate, and low-grade heat is the most difficult to manage in space.

Aerodynamic heating

As we saw in Chapter 2, even at altitudes of 150 km there remains some measurable atmosphere. As the launch vehicle leaves the denser atmosphere, the fairings protecting the payload are discarded to minimize dead weight. This exposes the satellite to the ambient atmosphere, and the collisions of these molecules with the satellite surface generates heat. The heating rate is proportional to the product of the ambient density (decreasing with increasing altitude) and the cube of the velocity (increasing with increasing altitude) and although this heating source may be present only for a short time (tens of minutes) during the ascent, it plays a role in the thermal analysis of any orbit whose perigee is below about 150 to 200 km (Gilmore, 1994).

Above the altitudes where there is any sensible neutral atmosphere lie the Van Allen belts of charged particles (see Chapter 2). These trapped electrons are found at altitudes between several hundred km and more than 50,000 km while the trapped protons are found primarily in the inner belt between altitudes of several hundred km to about 5500 km. Satellites whose orbits cross these regions are subject to some modest surface heating which may be neglected for spacecraft whose equilibrium temperature is above the cryogenic range, perhaps greater than 200 K (Figure 9.7). However, a significant local temperature rise can be seen as these charged particles are stopped in the first tens of micrometers, and this local heating can adversely impact systems designed to operate at cryogenic temperatures (Gilmore, 1994).

Note in this example that at an altitude of 2.0 R_e, a radiation surface at 20 K can experience a potentially unacceptable temperature increase of almost 40 K.

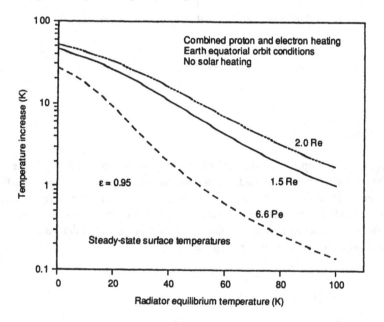

Figure 9.7 Temperature Increase in an Aluminum Radiator Due to Charged-Particle Heating (used with permission, Gilmore, 1994)

3. Heat transfer mechanisms

Except for the operation of closed loop systems (see Section 9.5.2), convection does not play a role in heat transfer onboard spacecraft. Conduction inside the spacecraft and radiation both inside and outside the spacecraft are the dominate heat transfer mechanisms.

3.1 Heat transfer by conduction

Conduction is the primary method by which heat is transferred from the interior of the spacecraft to the external surface. Conductive heat transfer is described by Fourier's law, which for the steady-state, one-dimensional case is

$$Q = \frac{kA}{d}(T_2 - T_1) \qquad (9.1)$$

where Q is the rate at which energy is being transferred (W), k is the thermal conductivity (W/m-K) of the material involved, A is the area across which the heat is being transported (m^2), d is the distance along which the heat is conducted (m), T_2 is the hot-side temperature (K), and T_1 is the cold-side temperature (K). The key design issue governing this method of heat transfer is the thermal conductivity of the materials involved. The thermal conductivity of several materials useful in TCS design is given in Table 9.3 below.

Table 9.3 Densities and Thermal Conductivities of Selected Spacecraft Materials (Eden, 1991; Juhaz, 1997)

Material	Thermal Conductivity (W/m-K)
Aluminum alloys	120-210
Diamond (natural)	2000
Diamond (VD)	500-1600
Aluminum nitride	170-230
Gallium arsenide	45
Molybdenum	145
Aluminum	237
Copper	395
Silver	428
Stainless steel (304)	17
Carbon steel	52
Phenolic	.03

Once cannot help but notice the extremely large thermal conductivity of diamond. While not yet a realistic material to be considered for large-scale applications, the use of thin-film diamond in conductive heat transfer from component-dense electronics packages is becoming more practical.

When possible, components generating large heat loads, such as the collectors of travelling wave tube (TWT) amplifiers, are mounted directly to the surface of the spacecraft to minimize the conduction path. When these components are mounted further inside the spacecraft, care must be taken to ensure adequate heat transport. Transport by conduction alone may impose mass requirements on the structure that cannot be accommodated, so the component is placed to allow for a combination of conduction and radiative heat transport.

3.2 Heat transfer by radiation

Radiation is the principal mechanism by which the environment transfers heat to the satellite, and the method by which the satellite rejects its waste heat into the space environment. In many designs, it is also the primary method of heat transfer onboard the spacecraft from individual systems to the radiators.

All bodies with a temperature above zero Kelvin emit and absorb electromagnetic energy, i.e., radiation. The most efficient radiator of this electromagnetic radiation is the 'blackbody,' a surface that emits the maximum amount of energy for a given equilibrium temperature. Such a blackbody also is the most efficient absorber of thermal radiation, a fact that will be useful in our later discussion of properties of materials used in a TCS. The concept of a blackbody is more than just a theoretical nicety; in practical calculations, real bodies are often approximated by these blackbodies at a fixed temperature.

In 1900, Planck proposed an empirical expression, based on quantum theory, to fit experiments measuring spectral distribution of radiation emitted by blackbodies at different temperatures. This expression, Planck's Law, relates the blackbody absolute temperature to the rate at which energy is emitted:

$$E(\lambda, T) = \frac{2\pi h c^2}{\lambda^5 \left(e^{hc/\lambda kT} - 1\right)} \tag{9.2}$$

where E is the energy emitted by a blackbody at temperature T (K) and at wavelength λ, h is Planck's constant (6.6252×10^{-34} J-s), c is the speed of light (3×10^8 m/s), and k is Boltzmann's constant (1.38×10^{-23} J/K). The spectral radiancy is defined such that E dλ

THERMAL MANAGEMENT

is the power radiated per unit area for wavelengths in the wavelength interval λ to $\lambda + d\lambda$. The units of E are then watts per unit area per unit wavelength (W/m^2/μm).

The remarkable feature of the equation at the time it was initially proposed by Planck is that it contained only the single unknown constant, h, and yet produced with great accuracy a fit to the emission data then available. Eq. (9.2), of itself, has limited direct value to the thermal engineer. It is, however, the basis for several expressions that are of great practical value. A plot of Eq. (9.2) is shown for several temperatures (Figure 9.8). Note that the amplitude of the emission increases with increasing temperature, and that the wavelength of the peak amplitude decreases (hence the energy of the corresponding photons increases) with increasing temperature. Recall from Chapter 2 that the solar spectrum can be approximated using a blackbody temperature of 5780 K, and the Earth's IR spectrum corresponds to a temperature of about 290 K.

By differentiating Eq. (9.2) and equating the result to zero, one can derive Wien's displacement law which defines the wavelength λ_{max} at which the spectral radiancy has its maximum value for a given temperature T. The Wien displacement law

$$\lambda_{max} \, T = 2.898 \times 10^{-3} \, m \, K \tag{9.3}$$

relates the temperature of a blackbody to the peak in the spectral intensity, i.e., the 'color' of the body. The Earth's IR spectrum peaks at about 10 μm, consistent with the Earth's equilibrium temperature of about 290 K, while the solar spectrum is a maximum around 0.5 μm in agreement with its blackbody temperature of 5780 K.

An even more useful expression for thermal management is obtained by integrating Planck's expression over all possible wavelengths, that is, from zero to infinity. This integrated result, Q, the radiancy, is the power per unit surface area radiated into the forward hemisphere from all wavelengths. This produces an expression for the rate at which energy is emitted by a blackbody in equilibrium at temperature T (Kelvin). The resulting expression

$$Q = \sigma T^4 \tag{9.4}$$

is known as the Stefan-Boltzmann equation and is fundamental to maintaining thermal balance onboard spacecraft. The constant, σ, is the Stephan-Boltzmann constant (5.67 x 10^{-8} W/m^2/K^4), and T (Kelvin) is the temperature of the radiating body. As will be discussed later, the radiative behavior of various materials is different, so that in practical use Eq. 9.4 is written as

$$Q = \varepsilon \, \sigma T^4 \tag{9.5}$$

where ε is the emissivity of the material. As one might expect at this point, the absorptivity and emissivity of the materials covering the outside of spacecraft are critical to

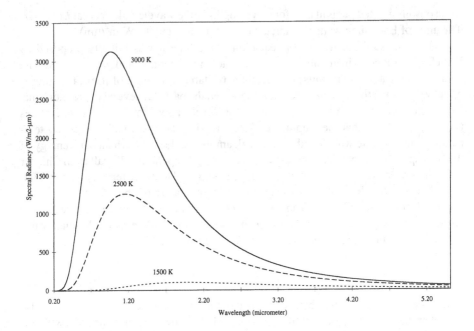

Figure 9.8 Blackbody Radiation as a Function of Wavelength and Temperature

thermal management. In the following section, a more detailed discussion of these properties is presented.

3.3 Absorptivity and emissivity

The external surfaces of the spacecraft radiatively couple the craft to the space environment. These surfaces are also exposed to all external sources of energy, and so both their absorptive and radiative properties are critical to the thermal management process. All external loads must pass through these surfaces, and all excess heat onboard must be radiated by these surfaces.

The energy incident on the surface of an object can be described by Kirchhoff's law for solid bodies:

$$E_r + E_t + E_a = E_i$$

that is, the sum of the reflected, transmitted, and absorbed energy is equal to the total incident energy. Because of the opacity of the materials used in spacecraft design, the

THERMAL MANAGEMENT

solar radiation incident on the surface of the spacecraft is not usually transmitted so that Kirchhoff's law becomes

$$E_r + E_a = E_{inc}$$

$$\frac{E_r}{E_{inc}} + \frac{E_a}{E_{inc}} = 1$$

or

$$\rho + \alpha = 1$$

from which it is customary to define the solar absorptivity, α, and the solar reflectivity, ρ.

The energy absorbed will not raise the temperature indefinitely, but is balanced at some point by the energy radiated as the body reaches an equilibrium temperature. This radiated energy is described by Planck's law (Eq. 9.2) and the Stephan-Boltzmann equation (Eq. 9.5).

We have seen how the energy spectrum radiated from a body depends strongly on the temperature. Figure 9.9 compares the two radiation spectra of a body at the temperature of the Sun to a body at a nominal temperature of an operating spacecraft. Note that these curves are not plotted to the same vertical scale. At the Sun's temperature, all of the radiated energy will be between 0.2 μm and 3 μm with a peak value at about 0.5 μm. Quite a different situation occurs at 300 K where all the radiation is emitted at wavelengths greater than about 4 μm with a peak occurring well into the long wavelength infrared at about 10 μm.

The important point from the thermal design perspective is that the solar energy impinging on the spacecraft surface and the energy emitted by a spacecraft surface occur in non-overlapping portions of the electromagnetic spectrum, and that the responses of materials to these two portions of the spectrum can be quite different.

The percentage of incident radiation absorbed by the surface is a function of the type of surface, the wavelength of the radiation, the angle of incidence, and the temperature of the surface. This percentage is the absorptivity. Likewise, the ratio of energy emitted from the surface to that emitted from an ideal black surface, at the same wavelength, direction, and temperature, is called the emissivity.

Kirchhoff also showed that at a given wavelength the absorptivity is equal to the emissivity:

$$\alpha(\lambda) = \varepsilon(\lambda)$$

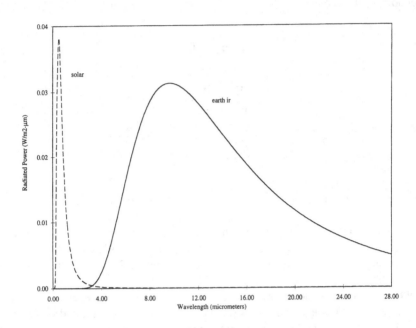

Figure 9.9 Power Emitted by a Blackbody at 5800 K and 300 K

Although the absorptivity and emissivity are equal at a given wavelength, the spacecraft will absorb solar radiation primarily at the short wavelengths (the uv through short wavelength IR) but will emit radiation primarily at much longer wavelengths (long wavelength IR and above).

Most surfaces will not display a flat absorptivity or emissivity across all wavelengths. As an example, Figure 9.10 shows a plot of $\alpha(\lambda)$ (or equivalently $\varepsilon(\lambda)$) versus the wavelength for white paint.

Notice that the absorptivity is small at the short wavelengths corresponding to the incident solar radiation but that the emissivity is large at the longer wavelengths at which the surface will emit IR radiation. This allows a surface covered with white paint to operate at a relatively cool temperature since it is a 'poor' absorber of energy and a 'good' emitter of energy. In practice, the averages of absorptivity and emissivity across some band of wavelengths are used to describe surfaces. The average of absorptivity across the solar band is referred to as the solar absorptivity, and similarly, the average of emissivity across the longer wavelength IR band is called the IR emissivity. Materials similar to the white paint shown in the figure above, which may display very different values of solar absorptivity and IR emissivity, offer the thermal designer a means of

THERMAL MANAGEMENT

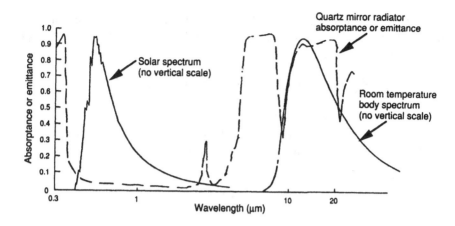

Figure 9.10 Absorptivity (Emissivity) versus Wavelength for White Paint
(used with permission, Gilmore, 1994)

passively controlling the spacecraft temperature. As we shall see later, the ratio of α/ε is a determining factor in the design of a TCS.

The absorptance of a surface will generally increase with extended exposure to the space environment, while the emittance will remain constant. The implications of this are discussed in Section 5.1 of this Chapter.

4. The basics of thermal analysis

The calculation of spacecraft temperatures occurs at two levels, an equilibrium temperature for the spacecraft as a whole and a more precise calculation of minimum and maximum temperatures, and temperature variations with time at specific sites and for specific components. The overall equilibrium temperature may be approximated from a simple conservation of energy expression:

$$Q_{stored} = Q_{incident} + Q_{generated} - Q_{radiated} \qquad (9.6)$$

At a more microscopic level, the temperature bands of specific components within the spacecraft will require a detailed knowledge of the design and construction of the satellite. The computer codes used at this level of modeling are often contractor specific and always quite complex. A discussion of these codes is beyond the purpose of this book; the reader is referred to the work of Ormsby (1994) for a background on such models.

As a simple example of the more macroscopic analysis, let us estimate the equilibrium temperature of a conductive sphere in orbit around the Earth. We assume no internal heat generation and since it is conductive, we also assume it is isothermal. If the satellite is in geosynchronous orbit, we can neglect the eclipse periods in this example. For this case

$$Q_{incident} = \text{solar radiation} + \text{albedo} + \text{Earth IR}$$
$$= S \alpha \pi R^2 + \text{negligible amounts at GEO}$$

$$Q_{radiated} = \varepsilon \sigma (4 \pi R^2) T^4$$
$$Q_{generated} = Q_{stored} = 0$$

So,

$$S \alpha \pi R^2 = \varepsilon \sigma (4 \pi R^2) T^4$$

and

$$T = 278 \left(\alpha/\varepsilon\right)^{\frac{1}{4}} K \qquad (9.7)$$

Here we can assume several values for the ratio of absorptivity to emissivity with dramatically different results for the equilibrium temperature of the satellite:

Table 9.4 Equilibrium Temperature as a Function of α/ε

Surface	α/ε Ratio	$T_{equilibrium}$
white paint	0.1	156 K
black paint	1.0	278 K
gold	10.0	494 K

For a small α/ε ratio, the IR emissivity is enhanced over the solar energy that is absorbed, and the sphere achieves a relatively low equilibrium temperature. At the other extreme, for an α/ε ratio of 10, the solar absorptivity is enhanced over the IR emissivity and the equilibrium temperature is sharply increased.

As an example, consider a geosynchronous satellite with a photovoltaic panel that remains oriented toward the Sun (Wise, 1997). While in the Sunlight, the panel will reach an equilibrium temperature that is a function of the effective absorptance of the front of the solar panel, the angle of incidence of the solar radiation, and the effective emissivity of the front and back of the panel:

$$\alpha_{eff} = \alpha_s + f \eta$$

where α_{eff} is the effective absorptance of the solar panel, α_s is the average solar panel absorptance, f is the solar cell packing factor, and η is the cell operating efficiency.

The equilibrium operating temperature in Sunlight then becomes

$$T_{eq} = \left(\frac{\alpha_{eff} A_f S \cos\theta}{(\varepsilon_f A_f + \varepsilon_b A_b)\sigma} \right)^{\frac{1}{4}}$$

where A_f and A_b are the areas of the front and back of the array, S is the solar constant, θ is the angle of incidence of the solar radiation, ε_f and ε_b are the emissivities of the front and back, and σ is the Stephan-Boltzmann constant.

With the assumption that the front and back areas are equal, the solar panel absorptivity is 0.8 and has a packing ratio of 0.9 with a cell efficiency of 0.1, a front panel emissivity of 0.6, and a rear panel emissivity of 0.85, one arrives at an equilibrium temperature

$$T = 58\ °C$$

When the satellite goes into eclipse, the situation changes significantly. The governing expression for this transient period is

$$m c_p \frac{dT}{dt} = -\varepsilon A \sigma T^4$$

where m is the mass of the solar panel and c_p is the effective specific heat. In this example, the emissivity is assumed to be 0.8, and the mass is replaced by an average density (1.38 g/cm^3) and thickness (0.13 cm.) The result, Figure 9.11, is a plot of the temperature of the solar array during the maximum eclipse period (about 75 minutes). During the eclipse, the temperature is seen to drop to about -175 $°C$ before quickly recovering once it is exposed to the solar radiation.

5. Thermal management techniques

Spacecraft thermal control systems are generally classified into two types: active and passive. Active systems are more complex and expensive and are used in satellites that have particularly stringent requirements on component temperatures or the need to dissipate large quantities of internally-generated heat. These systems employ components such as refrigerators, open and closed loop pumped systems, louvers with thermostats, and heaters. By comparison, passive systems can be much simpler and are the first choice when they can be used. These systems, for example, rely on conduction and

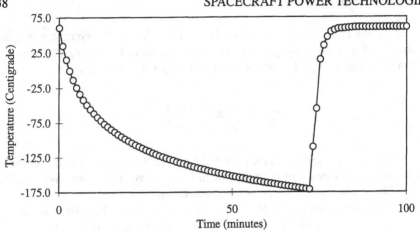

Figure 9.11 Temperature Variation of a Solar Panel During GEO Eclipse

internal radiative coupling to transport the heat to an external surface where it is radiated into space. The hardware techniques used in satellite thermal control depend to a large extent on the spacecraft geometry and attitude control. The concept of spinning the platform for passive thermal control has been used for more than two decades. The non-spinning vehicle usually has north or south faces that receive little incident solar energy and are natural locations for heat dissipating radiators. Several of the components used in these systems are discussed below.

5.1 Passive thermal management

Figure 9.12 shows an example of a passive TCS designed to reject the heat generated in the collector of a TWT amplifier. These amplifiers are commonly found in communications satellites, and the heat generated in the collector can be several hundred watts. In this design, the collector is mounted directly to the outside of the satellite to reduce the conduction path to the radiator. A thermal doubler (a heat spreader with very high thermal conductivity) is used to increase the conduction to a larger area for transfer to the radiator. Several components often found in a passive TCS are discussed in this section.

Coatings and paints

The most important part of both active and passive systems is the satellite coating. In both cases the coating is used to control the solar energy absorbed and to improve the radiation of IR energy.

THERMAL MANAGEMENT

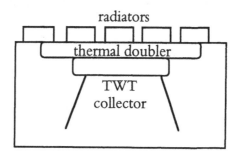

Figure 9.12 Example of Passive Thermal Contol

Coatings may be selected to enhance or decrease absorptivity and emissivity throughout any wavelength band. Figure 9.13 compares the four fundamental thermal control surfaces, and it is through a combination of these surfaces that the overall balance between solar energy absorbed and excess spacecraft heat radiated is maintained. As the names imply, the solar absorber absorbs at the short wavelengths but has low emissivity at the longer wavelengths, while the solar reflector has just the opposite characteristics. For that reason, the solar reflector is the surface that is used for radiator design. The flat absorber has high solar absorptance and IR emittance at all wavelengths while the flat reflector has low values at all wavelengths. Shown in each figure is a solid line representing the ideal surface and a dotted line showing the characteristics of a real surface approximating the ideal.

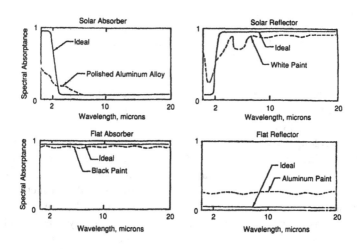

Figure 9.13 The Four Fundamental Thermal Control Surfaces
(used with permission, Gilmore, 1994)

Figure 9.14 compares the solar absorptivity and IR emissivity for several surfaces commonly used in the construction of spacecraft. The proper combination of these or other surfaces will produce an overall average α and ε needed for a given application and will, in many cases, be sufficient to maintain the proper spacecraft equilibrium temperature.

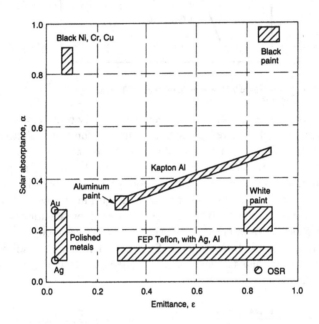

Figure 9.14 Surface Properties of Selected Material Finishes
(used with permission, Wingate, 1994)

The solar absorptivity and IR emissivity of some representative materials used in spacecraft construction are given in Table 9.5. The values for absorptivity indicate a BOL (beginning of life) value because the solar absorptivity of many of these materials increases with prolonged exposure to the space environment. There are several causes for these changes, including the choice of orbit, the materials used in the construction of the spacecraft, and the nature of the surface coating itself. The orbit will determine the exposure to solar ultraviolet radiation and charged particles, both of which can affect the coatings, and the materials within the spacecraft may outgas which may cause material to be deposited on surfaces, again changing surface absorptivity, Since all materials are not equally susceptible to interactions with the environment, the materials that are resistant to changes in their radiation properties are normally the surfaces of choice.

THERMAL MANAGEMENT

Table 9.5 Absorptance and Emissivity of Selected Finishes
(after Gilmore, 1994)

Material	Absorptivity	Emissivity	α/ε
Optical Solar Reflectors			
quartz mirrors	0.07	0.80	0.01
aluminized Teflon	0.14	0.70	0.02
Vapor Deposited Metals			
aluminum 0.08	0.02	4.0	
gold	0.19	0.02	9.5
nickel	0.38	0.04	9.5
silver	0.04	0.02	2.0
White Paints			
Z93	0.19	0.92	0.2
Chemglaze A276	0.23	0.88	0.3
Black Paints			
Catalac	0.96	0.85	1.1
Chemglaze Z306	0.94	0.89	1.1
Films			
beta cloth 0.40	0.86	0.5	
Kapton (0.25 mil on Al backing)	0.31	0.43	0.7
Mylar (0.25 mil on Al backing)	0.15	0.34	0.4
Metals			
black copper	0.98	0.63	1.6
buffed aluminum	0.16	0.03	5.3
polished aluminum	0.15	0.05	3.0
electroplated gold	0.23	0.03	7.7
sandblasted gold	0.48	0.14	3.4

The changes that do occur will affect absorptivity more than emissivity, and the change will always increase α rather than decrease it. This change is most troublesome in the demands it places on the radiators. The radiators must be sized to accommodate the increase in thermal load due to the increase in absorbed solar radiation. As Gilmore and Stuckey (1994) point out, oversizing the radiators to accommodate the end-of-life heat loads will cause the satellite to operate cooler at the start of its mission. This may necessitate the use of heaters to maintain a high enough temperature early on.

White paints, useful in the coating of antennae reflectors and as radiator surfaces because of their large emissivity and low α/ε ratio, are among the most susceptible to environmental degradation, some tripling their value of absorptivity in only a few years. The same can be said of anodized aluminum used in the interior structure of many spacecraft (Agrawal, 1986).

Radiators

Radiators, the surfaces that radiate excess heat from the spacecraft into space, play a critical role in both active and passive TCS designs. The radiator material is selected for its low absorptivity and high emissivity, and must maintain as low an α/ε ratio as possible throughout the time in orbit. Often, the radiators are placed to minimize exposure to the solar flux, for example, on the north and south surfaces of three-axes stabilized GEO spacecraft. The radiators are often second surface mirrors (also called optical solar reflectors, OSR) as pictured in Figure 9.15. A typical design uses a 6-mil fused silica surface with a silver film on the second surface. The highly reflective surface (silver in this example) is protected by a thin transparent cover that serves several purposes: to protect the reflecting surface from environmental effects while allowing incoming radiation access to the reflecting surface (i.e., have a high transmissivity), and to provide for an efficient IR emission from the topmost surface (i.e., have a large value of ε). The OSR has typical values of $\alpha \sim 0.15$ and $\varepsilon \sim 0.8$. From Eq. (9.7), a Sunlit sphere with this α/ε ratio would have an equilibrium temperature of about 180 K.

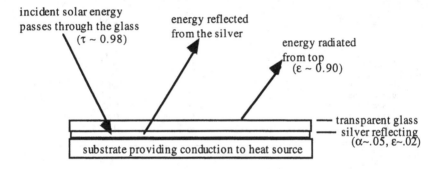

Figure 9.15 The Optical Solar Reflector Radiator

The OSR remains susceptible to surface contamination which will give rise to an increase in absorptivity and thus an increase in equilibrium temperature. Wingate (1994) points out that teflon-based OSRs are also susceptible to degradation from charged particles. The degradation profiles of several TCS surfaces are shown in Table 9.6.

Table 9.6 Surfaces in the Space Environment
(after Wingate, 1994)

Environment	Material	Environmental Effect
solar ultraviolet	white paint	5% to 100% change in α
local outgassing	thermal coatings	varying increase in α
		low ε surfaces may increase ε
charged particles	teflon-based single surface mirrors	increased α for moderate doses- >10^{15}/cc with E = 5keV-1MeV
atomic oxygen	kapton, mylar	loss of strength, disintegration

THERMAL MANAGEMENT

Phase change materials

Often, high-power components do not operate continuously, but rather cycle on and off as the mission demands. The heat load such components might generate is illustrated in Figure 9.16. The use of phase-change materials (PCM) allows the TCS to be scaled for an average thermal load (perhaps tens of watts) rather than the peak load (perhaps hundreds of watts). PCM systems can also be useful in dampening temperature variations as satellites move in and out of the Earth's shadow. In the example shown in Figure 9.17, a PCM is selected based on three characteristics: the temperature at which it will change phase, typically solid to liquid, will fix the maximum operating temperature of the system; the latent heat of fusion which determines the total amount of heat the PCM can absorb; and the thermal conductivity of the PCM which will limit the power density of the system.

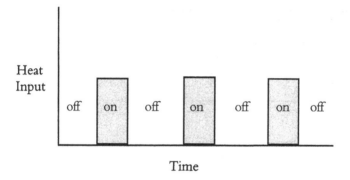

Figure 9.16 Time-Varying Heat Loads

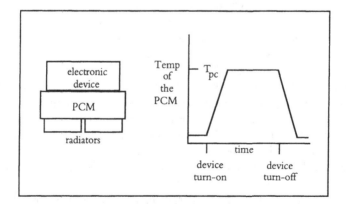

Figure 9.17 Typical Operation of a PCM Heat Sink

As the temperature of the PCM rises, it approaches the phase change temperature (T_{pc}) which is the maximum temperature the system can reach as long as there remains some material with its phase unchanged. Once the phase change has been completed, the continued input of heat will give rise to an unwanted temperature increase. When the electronic device has been turned off, the heat input will stop and the heat stored in the PCM can flow to the radiator and allow for phase reversal (liquid to solid) to occur. The PCM can also be incorporated into the radiator itself allowing the radiator to be sized for average rather than peak heat load (Bledjian et al., 1994). Several examples of PCM in various temperature operating ranges are given in Table 9.7.

Table 9.7 Representative Materials used in PCM Systems
(after Hale et al., 1994; Rose et al., 1991)

Material	Melting Point (K)	Heat of Fusion (J/kg)
Lithium hydride	960	3.0×10^6
Water	273	3.3×10^5
Calcium chloride	302	1.7×10^5
Nitrogen pentoxide	303	3.2×10^5
N-Heptane	182	1.4×10^5
Ethane	101	9.3×10^4

Insulation

Insulation is used to minimize the heat transfer between adjacent regions which must be held at different temperatures. Many modern satellites are also wrapped in insulation blankets with areas cut out to accommodate radiator surfaces. Rigid high temperature ceramic insulation is used in engine nozzle and re-entry shields, and plastic foams are used for short duration cryogenic tankage insulation. Insulation blankets are also used to isolate thermal radiator surfaces, long-duration cryogenic tanks, and IR detector systems from spacecraft structural members and external environments.

Multilayer insulation (MLI) is composed of multiple alternating layers of low-emittance surfaces and low-conductance separators. The simplest construction is a layered blanket assembled from crinkled thin mylar (0.6×10^{-3} mm thickness) aluminized on one surface (5×10^{-5} mm thickness). The crinkling results in the sheets touching only at a few points, thus eliminating the need for a low conductance separator. Heat transfer in the MLI blanket is by a combination of gaseous conduction, solid conduction, and radiation. Gas trapped between the layers and the outgassing of the MLI materials contribute to the conduction, but with time on orbit, this trapped-gas pressure will decrease and with it the conductivity.

THERMAL MANAGEMENT

The theoretical performance for crinkled one-surface aluminized mylar insulation shows that the equivalent effective emittance for a multilayer blanket varies as $1/(1+N)$, where N is the number of layers. Simply increasing the number of layers past a certain value will not improve performance because, as the density of layers increases the contact between layers also increases with the result that the decreasing radiation value is countered by the rising solid conductivity. The optimum value appears to be about 25 layers.

5.2 Active thermal management

In many cases, the thermal loads generated by onboard components, or special temperature requirements of individual components, may go beyond the capabilities of passive control systems. Passive control is always preferred because of the penalties that accompany active systems: increased mass, electrical power demands, and complexity of the system. But as Hager *et al.* (1993) point out, active control may be mandated in instances of large temperature extremes as might accompany the operation of NaS batteries (625K), power conditioning subsystems (300K), and IR sensors (10K) all operating within the same spacecraft.

Active systems differ greatly and thus do not lend themselves to a universal example. A descriptive example, shown in Figure 9.18, is a single-phase pumped loop system used to carry excess heat from the collector of a TWT amplifier to a radiating surface. The increased complexity of the piping and the electrical power requirements of the pump are representative of the active systems. This is in contrast to the passive design that was described in Figure 9.12

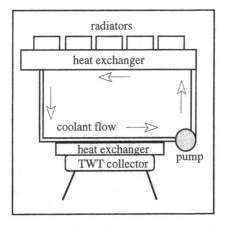

Figure 9.18 Example of Active Thermal Control

Several components of active systems that are often related to electrical power system design and operation are described briefly in this section. For a complete discussion of these and many other active components, the reader is referred to the excellent text by Gilmore (1994.)

Heaters

Heat generation from simple I^2R heaters aboard spacecraft is often used to maintain proper temperatures in cold-case extremes. Some heat may be available from the normal operation of electrical power system components, but these heat sources may not be available or suitable when the spot heating of a specific component is required. The demand for heat can occur, for example, when the design of the TCS is based on the anticipated heat load from the operation of some system that may not be active for some period of the orbit, or when additional heating is required during eclipse. Heater electrical power can be a significant factor in sizing the overall power demands such as the batteries of a photovoltaic system operating during eclipse. Additional heating may also be necessary early in the orbit lifetime because of the overdesign of radiators to accommodate EOL absorptivity degradation.

The heaters may be simple ohmic strips, wires, or patches, usually configured to allow for redundancy. The heaters may be controlled from a ground station or internally through thermostats, solid-state controllers, or onboard computers connected to distributed sensors. Again, however, the heaters and their control units will place demands on the spacecraft power system which passive systems avoid.

Thermoelectric coolers

The thermoelectric coolers (TE) are solid state devices that provide spot cooling for modest heat loads, generally in the 20-30 K range. Larger heat loading may require mechanical refrigeration systems which suffer from lower reliability, the need for vibration isolation, and generally higher electrical power requirements. The TE devices are based on the Peltier effect in which the passage of electric current through a junction of dissimilar metals produces a localized cooling at the junction.

In the simplest example, passing a small electric current through the p-n junction of bismuth telluride creates a warm and cold side of the junction with a temperature difference of several to tens of Kelvin. The larger the temperature difference required, the larger will be the ratio of input power to the junction (watts of electrical power) to power of refrigeration (watts of cooling). Typically the efficiency of practical TE units is much less than 1%, and so the use of these devices is restricted to the spot cooling of very sensitive electronic components. For example, a TE device may be integrated into

THERMAL MANAGEMENT

the packaging of a low-temperature preamplifier, with the warm side of the TE conducting heat directly to a radiator surface.

Heat pipes

A heat pipe is a thermal device that is capable of transferring large amounts of heat between two points with almost no temperature difference between them. The key to the design of a heat pipe is the selection of the proper working fluid for the application at hand. In practical terms, a heat pipe can be viewed as an extremely high thermal conductivity material constructed as a closed container with a capillary wick and a small amount of vaporizable fluid, as shown in Figure 9.19.

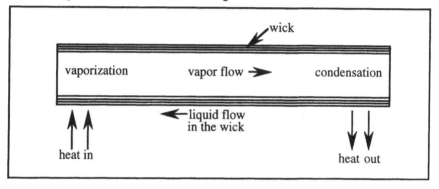

Figure 9.19 Operation of a Heat Pipe

As the fluid is heated, it vaporizes and is conducted by vapor convection to the cool end of the tube (heat out). Since the pressure drop is slight across the length of the heat pipe, the operation is essentially isothermal, and since the latent heat of vaporization of the working fluid is usually high, only a small amount of the material is needed to transport large quantities of heat. The condensed vapor is returned to the warm end by capillary action. The reliability of the system is very high since the heat pipe has no moving parts except for the motion of the fluid/vapor mixture and requires no power input except for the heat that is to be conducted. The key design option is the choice of the working fluid based on the temperature range expected. Several candidate fluids, shown in Figure 9.20, span the temperature range from cryogenic to very-high temperature operations.

Modifications can be made to the basic heat pipe design to allow it to function as more than a conductivity medium. One such variation is the heat diode in which heat is conducted from the 'normally' hot end to the 'normally' cool end, but the conductivity

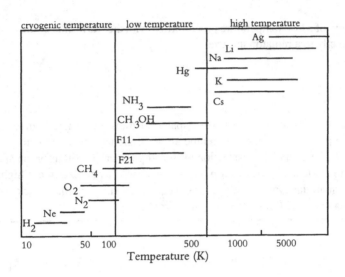

Figure 9.20 Heat Pipe Working Fluids (after Wise, 1985)

is reduced to essentially nil if the temperatures of the two ends are reversed. The leakage conduction along the walls of the pipe and through the wick can be significant in the case of cryogenic systems. An excellent discussion of the details of these diode designs is given by Prager (1994).

Variable conductance heat pipes (VCHP) will maintain a constant temperature of the heat input side over a large range of input levels. The VCHP operate with the addition of a non-condensable gas reservoir to the cold side of the pipe, with the gas chosen to have a pressure about equal to the saturation vapor pressure of the working fluid. This reservoir, which is several times larger than the volume of the pipe itself, allows the pipe to operate at a constant pressure independent of the heat input in the following way: as the heat flux increases at the hot end, the volume of the non-condensable gas in the pipe decreases and the active area of the pipe condenser is increased. As the heat input at the hot end decreases, the volume of the non-condensable gas increases, and the active area of the condenser also decreases. Wise (1985) points out the advantages that the VCHP offers in satellite TCS design. This version of the heat pipe operates as a variable radiator area in the cold case and opens to full unconstrained heat rejection under full thermal load, and many of the problems of radiator sizing under varying thermal loading are reduced, specifically the need for electrical power for heaters for those periods of reduced thermal loads during which radiators may be oversized.

Pumped systems

Pumped loop systems have been proven to be reliable for spacecraft with moderate ($<10^3$ W) heat loads. These systems provide efficient transfer of large amounts of heat through forced liquid convective cooling. In Figure 9.18, a system is shown which consists of a heat exchanger from the TWT amplifier, a pump to force the flow of liquid, and a second heat exchanger to transfer the heat to the radiators. The flow can be either laminar or turbulent. The liquid used is chosen based on several characteristics, including vapor pressure, specific heat, dynamic viscosity, and thermal conductivity. Among the more commonly used working fluids are Freon, water/methanol solutions, water/glycol solutions, and carbon tetrachloride (Lam, 1994).

These single-phase systems are not, however, without their disadvantages, the primary one being the temperature gradient along the loop. This gradient can be minimized with larger pipes and higher flow rates, but at the cost of increased mass and electrical power. To overcome these limitations, two-phase systems have been developed during the past twenty years. These two-phase systems take advantage of the latent heat of vaporization which is often several orders of magnitude higher than corresponding heat capacities of liquids. This translates into lower mass, higher heat transfer rates, lower power requirements for pumps, and a significantly lower temperature gradient over the loop than the corresponding single-phase designs.

6. References

Agrawal, B. N., *Design of Geosynchronous Spacecraft,* (Prentice Hall, Englewood Cliffs, NJ, 1986), 265-322.

Anderson, E.C. and Clark, L. G., *Geometric Shape Factors for Planetary-thermal and Planetary-reflected Radiation Incident upon Spinning and Nonspinning Spacecraft*, NASA Technical Report TN-D-2835 (1965).

Bledjian, L., Hale, D. V., Hoover, M. T., and O'Neill, M. J., *PCMs and Heat Sinks*, in *Satellite Thermal Control Handbook* , Gilmore, David G., (ed.), (The Aerospace Corporation Press, El Segundo, California, 1994), 4-147:4-157.

Eden, R. C., in *Applications of Diamond Films and Related Materials*, Tzeng, Y., Yoshikawa, M., Murakawa, M., and Feldman, A., (eds.), (Elsevier Science Publishers, Amsterdam, 1991), 259-268.

ESA, *Spacecraft Thermal Control Design Data*, ESA PSS-03-018, Iss. 1 Rev.1 (1989).

ESA, *Data for the Selection of Space Materials*, ESA PSS-01-701, Iss. 1, Rev. 1 (1990).

Gilmore, David G., (ed.), *Satellite Thermal Control Handbook* (The Aerospace Corporation Press, El Segundo, California, 1994).

Gilmore, D. G. and Stuckey, W. K., in *Satellite Thermal Control Handbook*, Gilmore, David G., (ed.), (The Aerospace Corporation Press, El Segundo, California, 1994), 4:1-4:16.

Griffin, Michael D. and French, James R., *Space Vehicle Design*, Prezemieniecki, J. R., (ed.), (AIAA Education Series, Washington, D.C., 1991), 371-394.

Hale, D.V., Hoover, M.J., and O'Neill, M.J., in *Satellite Thermal Control Handbook*, Gilmore, David G., (ed.), (The Aerospace Corporation Press, El Segundo, California, 1994), 4:149-4:157.

Hager, B.G. et al., *Effects of Payload Heat Flux on Space Radiator Area, Journal of Spacecraft and Rockets*, 30, No. 2 (1993), 225-226.

Henderson, R.A., in *Spacecraft Systems Engineering*, Second Edition, Fortescue, P. and Stark, J., (eds.), (Wiley and Sons, Chichester, 1995), 337-368.

Juhaz, A. NASA Lewis Research Center. (private communication, 1998).

Kaila, V. K. and Bhide, R. S., *Design of Thermal Control System of INSAT-2A and Its Initial In-orbit Performance, Journal of Spacecraft Technology*, 4, No. 1 (1994), 132-143.

Lam, T. T., in *Satellite Thermal Control Handbook*, Gilmore, David G., (ed.), (The Aerospace Corporation Press, El Segundo, California, 1994), 4:161-4:210.

McMordie, R.K., in *Space Mission Analysis and Design*, Second Edition, Larson, W. J. and Wertz, J. R., (eds.), (Microcosm, Inc., Torrence California and Kluwer Academic Publishers, Dordrecht, 1992), 409-430.

NASA Space Materials Handbook, NASA SP-3051, 3rd Edition (1969).

Ormsby, Rachel A., *Reference Manual for the Thermal Analyst's Help Desk Expert System*, NASA Contractor Report 194941 (1994).

Pisacane, Vincent L. and Moore, Robert C., (eds.), *Fundamentals of Space Systems*, (Oxford University Press, Oxford, 1994).

Prager, R.C., *Heat Pipes and Capillary Pumped Loops*, in *Satellite Thermal Control Handbook*, Gilmore, David G., (ed.), (The Aerospace Corporation Press, El Segundo, California, 1994), 7:1-7:22.

Rose, M. F., et al., *Novel Techniques for the Thermal Management of Space-Based, High-Power Microwave Tubes*, IEEE Transactions on Electron Devices, 38, No. 10 (1991), 2252-2263.

Sabripour, Shey S., Lockheed Martin Missiles and Space Communications and Power Center, private communication (1999).

Soop, E. M., *Handbook of Geostationary Orbits*, (Kluwer Academic Publishers, Dordrecht, 1994).

Vernon, Robert, *Spacecraft Design: Section 5-Thermal Control*, internal lecture notes, Lockheed Martin Corporation (1989).

Wingate, Clarence A., in *Fundamentals of Space Systems*,. Pisacane, Vincent L. and Moore, Robert C., (eds.), (Oxford University Press, Oxford, 1994), 433-468.

Wise, Peter C., *Spacecraft Thermal Control Technology: Design Challenges into the*

1990s, Proceedings of the 36th Congress of the International Astronautical Federation, Stockholm, 1985 (Pergamon Press, Oxford, 1985).

Wise, Peter C., *Spacecraft Systems Engineering: Thermal Control Subsystem*, internal lecture notes, Lockheed Martin Corporation (1993).

Wise, Peter C., private communication, 1997.

Wolverton, R. W., *Flight Performance Handbook for Orbital Operations*, (John Wiley and Sons, Inc., New York, 1961).

APPENDIX

MAGNETIC MATERIALS IN POWER MANAGEMENT

The physical laws of electromagnetic circuits are quite analogous to their electrical counterparts. What makes it seem challenging at times may be the many trades that need to be made every step of the way to obtain an optimized design. It is best to use an iterative process for this purpose (i.e., to complete the design to the end and then optimize it with several iterations). Although a detailed discussion of magnetics theory is beyond the scope of this review, to familiarize the reader with the general concepts in this field, a brief discussion is presented here. The object is not to present the material as a design and analysis tool but rather to establish basic intuitive concepts in this area. To this end, units and conversion constants are omitted to simplify the flow of this discussion.

When a current is passed through a wire, a magnetic field **H** is established normal to the wire in a circular path (Figure A.1). The direction of the field follows the right hand rule with the thumb in the direction of the current and the other fingers curled around the wire in the direction of the magnetic field. To increase this magnetic field, the wire can be formed into a coil. The intensity of this magnetic field or the resulting flux density (number of flux lines per cross sectional area) is proportional to the product of the current in the circuit and the number of turns (Figure A.2). This is called the magnetomotive force (mmf) or **F**. A comparison of the expressions for a uniform magnetic field and an electric field is shown:

$$H = \frac{F}{l_m} \qquad \mathcal{E} = \frac{V}{l_e}$$

H = magnetic field (oersted)   **E** = electric field
F = magnetomotive force   **V** = electromotive force
l_m = magnetic path length l_e = electric field path length

It can also be shown that, in a uniform field, magnetic flux density is given as

$$B = \frac{\Phi}{A_C} \qquad J = \frac{I}{A_C}$$

B = flux density (gauss)   **J** = current density
Φ = flux   **I** = current
A_c = cross sectional area A_c = cross sectional area

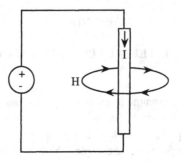

Figure A.1 The Magnetic Field

For the circuit shown in Figure A.2, the magnetic field is plotted against the flux density and is shown in Figure A.3. A linear relationship, valid only for free space, exists between the magnetic field and the flux density. The constant of proportionality, or the slope of the curve, is the permeability of free space, μ_0.

$$B = \mu_0 H$$

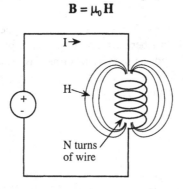

Figure A.2 The Magnetic Field of a Coil

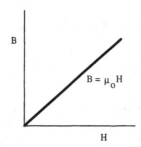

Figure A.3 The Relationship between **B** and **H** in Free Space

APPENDIX: MAGNETIC MATERIALS   455

This can be interpreted as indicating that air is a poor conductor of magnetic flux. To increase the flux density, a core material with higher permeability can be inserted into the coil as shown in the circuit in Figure A.4. The permeability of magnetic material is expressed in terms of their permeability relative to that of free space. Therefore, with

$$\mu = \mu_r \mu_0$$

where μ_r is the relative permeability of the magnetic material, the new relationship for a core using a magnetic material is then

$$\mathbf{B} = \mu \mathbf{H}$$

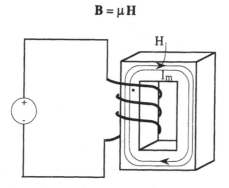

Figure A.4 the Increased Flux Density through the Use of a High Permeability Core

The relative permeability of magnetic materials can be very high (in the order of thousands), but unlike free space, the relationship between **B** and **H** is no longer linear. Figure A.5 shows the same plot as Figure A.3, except that a magnetic core is inserted in the coil.

When the core is completely demagnetized, as the magnetic field increases, the flux density increases slowly and then follows a slope close to its relative permeability until it reaches a point beyond which increasing the magnetic force will not result in increased flux density. The core is said to be in saturation (B_{sat}). In a real circuit application, the core is usually not driven to this point since at saturation the magnetic element cannot sustain a voltage and behaves like a short circuit. As the magnetic field H is decreased, the curve does not follow the original path during magnetization. When the magnetic field is now reduced to zero, there is a finite flux density, the remanence flux, shown as flux density B_r. To drive this flux to zero and demagnetize the core, the magnetizing field must now be driven negative. The field required to achieve this (intersection of the curve at B=0) is called the coercive force H_c. The magnitude of the remanence flux and the coercive force indicated provide a figure of merit describing the quality of the material as a permanent magnet (not that this property is desirable in power electronics materials). On the contrary, usually soft magnetic material with minimum remanent flux and coercive force are preferred for power electronics applications.

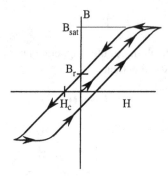

Figure A.5 A Hysteresis Plot of **B** versus **H** for a Magnetic Core

The energy that is used by the magnetomotive force to magnetize and demagnetize the core is not fully recovered. When a magnetic material is exposed to a changing flux, there are two types of losses that occur. One is due to the hysteresis property of the material to retain magnetism as discussed in the BH loop hysteresis curve. The other, I^2R losses in the core material result from an induced current (eddy currents) arising from the changing flux in accordance with the Faraday induction law. The combination of hysteresis and eddy current losses makes up the total core loss in the magnetic material.

Two more relations are important to complete the basic magnetic circuit understanding- those attributed to Ampere and Faraday. Referring back to Figure A.4, Ampere's law states that the line integral of the magnetic field around a closed magnetic path length is equal to the total current going through the path. Again assuming a uniform field, this is shown as:

$$\oint H \cdot dl_m = NI = H\, l_m$$

where N is the number of turns in the coil, I is the current, and l_m is the magnetic path length.

Since the magnetomotive force **F** is proportional to **NI**, then

$$F = NI = Hlm = \frac{Blm}{\mu} = \frac{\Phi lm}{\mu A_c} = \Phi \mathcal{R}$$

APPENDIX: MAGNETIC MATERIALS

where $\mathcal{R} = \frac{l_m}{\mu A_c}$ is the reluctance of the magnetic core material. Again, the analogy with electrical circuits is clear:

$$\mathcal{R} = \frac{l_m}{\mu A_c} \qquad\qquad R = \frac{l}{\sigma A_c}$$

R = reluctance of the core
l_m = magnetic path length
μ = permeability of the core
A_c = cross sectional area

R = resistance
l = wire length
σ = conductivity of the wire
A_c = cross sectional area

The equivalent magnetic circuit is compared to its electrical analog in Figure A.6:

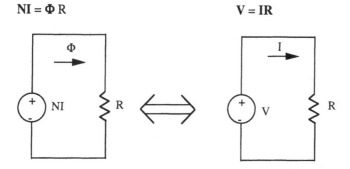

$$NI = \Phi R \qquad\qquad V = IR$$

Figure A.6 Equivalent Magnetic and Electric Circuits

Turning our attention now to time-varying situations, Faraday's law states that when a time varying flux is place through an N-turn coil of wire, a voltage (electromotive force) is induced and is given by

$$v = N\frac{d\Phi}{dt}$$

But, since

$$\Phi = \frac{NI}{\mathcal{R}}$$

then

$$v = N\frac{d}{dt}\left(\frac{Ni}{\mathcal{R}}\right)$$

or

$$V = \left(\frac{N^2}{\mathcal{R}}\right)\frac{di}{dt}$$

This can be compared to the familiar expression

$$V = L\frac{di}{dt}$$

and since

$$\Phi = \frac{NI}{\mathcal{R}}$$

and

$$\mathcal{R} = \frac{l_m}{\mu A_c}$$

then

$$L = \frac{N^2}{\mathcal{R}} = \frac{N\Phi}{i} = \frac{N^2 \mu A_c}{l_m}$$

Inductance is directly proportional to square of turns and permeability and inversely proportional to reluctance which should be intuitive. Having discussed Faraday's and Ampere's laws, we can also develop the relationship for the hysteresis loss as discussed previously. Assuming the magnetic core shown in Figure A.7, the energy required to traverse the BH curve per cycle is:

$$E = \int v(t)i(t)dt$$

and the power dissipated is

$$P_H = f\int v(t)i(t)dt$$

where E is the core magnetizing energy per cycle, P_H the hysteresis core loss, f the core excitation frequency, $v(t)$ the excitation electromotive force, and $i(t)$ the excitation current.

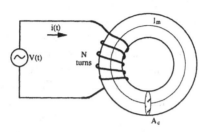

Figure A.7 The Torroidal Core

Substituting Faraday's and Ampere's relationships instead of $v(t)$ and $i(t)$:

or

$$P_H = f\int (N\frac{d\Phi}{dt})(\frac{Hl_m}{N})dt$$

$$P_H = f\int \left(NA_c \frac{dB}{dt}\right)\left(\frac{Hl_m}{N}\right)dt$$

Simplifying yields

or

$$P_H = f\int (A_c dB)(Hl_m)$$

$$P_H = f(l_m A_c)\int HdB$$

Recognizing that $l_m A_c$ is the magnetic core volume V_c, the expression for the hysteresis loss becomes

$$P_H = fV_c \int HdB$$

The integral of magnetic field **H** with respect to flux density **B** is the area enclosed in the **BH** hysteresis loop. This discussion helps to gain an understanding of the core loss within a magnetic core. In practice, however, the hysteresis curve for the given core geometry and material is not available although core manufacturers do provide core loss.

Again note the frequency term in above expressions. There are no hysteresis core losses associated with a DC biased core. We now have the tools to discuss the second part of a magnetic core loss, the eddy current losses.

If the material used for the magnetic core exhibits low electrical conductivity, the changing flux in the core can induce a current in the core itself. These eddy currents result in I^2R losses and heat up the core (Figure A.8). According to Lenz's law, the direction of this current is such that it will produce a flux to oppose the original flux that produced this current. In order for current to be induced, a closed circuit path must be present, and this is consistent with Lenz's law which only applies to a closed circuit. It is for this reason that iron cores are laminated and powder cores consist of individually insulated magnetic powder material. The magnitude of eddy currents can be greatly reduced by cutting the core material so that the path length for the induced current is minimized.

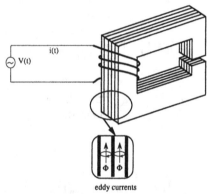

Figure A.8. The Introduction of Laminated Cores to Reduce Eddy Currents

The basic concepts of a transformer follow the same basic principles as those discussed above. A transformer is a magnetic element which has a minimum of two windings wound on a common magnetic core material. Based on Faraday's law, and referring to Figure A.9, we have:

$$v_p = N_p \frac{d\Phi}{dt}$$

and

$$v_s = N_s \frac{d\Phi}{dt}$$

APPENDIX: MAGNETIC MATERIALS

Since both windings are wound on the same core, they are exposed to the same varying flux. Thus, substituting one equation into the other gives

$$\frac{v_s}{v_p} = \frac{N_s}{N_p}$$

which gives the elegant voltage relationship in a transformer. The current relationship in the windings is rather intuitive and can be found assuming an ideal transformer for which the instantaneous power into the transformer primary must equal the power delivered out of the transformer secondary:

Then,
$$P_p = P_s$$

$$v_p \cdot i_p = v_s \cdot i_s$$

or
$$\frac{v_s}{v_p} = \frac{i_p}{i_s}$$

$$\frac{v_s}{v_p} = \frac{N_s}{N_p} = \frac{i_p}{i_s}$$

To develop the primary to secondary impedance relationship, we follow the same logic with the result

$$Z_s = \frac{v_s}{i_s} = \frac{v_p N_s}{N_p} \frac{N_s}{i_p N_p} = \frac{v_p}{i_p}\left(\frac{N_s}{N_p}\right)^2$$

$$Z_s = Z_p\left(\frac{N_s}{N_p}\right)^2$$

To establish the transformer equivalent circuit, we refer to Figure A.9. Note that just as with its electrical counterpart in which the sum of the voltages around a loop is zero, the sum of the megnetomotive forces around a loop must also equal zero. For the transformer core shown in Figure A.9, we have for the magnetomotive force

$$F = NI = \Phi R$$

$$\Phi R = N_p i_p + N_s i_s$$

$$\Phi = \frac{N_p i_p + N_s i_s}{\mathcal{R}}$$

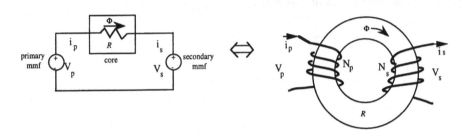

Figure A.9 The Transformer and Its Equivalent Circuit

We again made use of Faraday's law to write

$$v_p = N_p \frac{d\Phi}{dt}$$

and substitute

$$v_p = N_p \frac{d}{dt}\left(\frac{N_p i_p + N_s i_s}{\mathcal{R}}\right)$$

$$v_p = \frac{N_p^2}{\mathcal{R}} \frac{d}{dt}\left(i_p + \frac{N_s}{N_p} i_s\right)$$

Comparing this expression with the now familiar

$$V = L \frac{di}{dt}$$

APPENDIX: MAGNETIC MATERIALS

we conclude that

$$L_m = \frac{N_p^2}{\mathcal{R}}$$

where L_m is defined as the transformer magnetizing inductance and is shown in Figure A.10 as a lumped element inductor in parallel with the primary of the transformer. The second part of the equation is identified as the transformer magnetizing current and is defined from the expression above as

$$I_m = \left(i_p + \frac{N_s}{N_p} i_s \right)$$

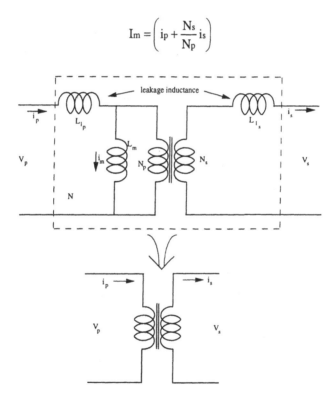

Figure A.10 A Detailed Model of the Transformer Showing the Leakage Inductance

This brief discussion is intended to present only the basic concepts underlying magnetics design and analysis. The reader is referred to any of several basic texts on electromagnetic fields for a more thorough discussion. Several of them are listed among the references.

The designer must make trade-offs and optimize the design for many other important parameters that go into a high efficiency inductor or transformer component. These include the skin effect and proximity effect losses and the effects of fringing flux in a gapped core (both by inducing eddy currents in the windings and creating hot spots in the core at the point of entrance to the core). Other issues include the opposing requirements to achieve a low leakage, tightly wound, interleaved windings while maintaining a low inter-winding capacitance. Despite this, gaining an intuitive understanding of the basic electromagnetic circuits and their physical properties allows the power electronics designer to develop and analyze a highly efficient, predictable, and producible magnetic component.

467

A

Activation polarization, 178
AFC, 223, 224, 225, 235
Air mass one, 48
Air mass zero, 48, 80, 83
Albedo, Earth, 29, 318, 421-426, 436
Apollo
  telescope mount, 37
ARPS, Pluto Express, 253
Atmosphere, near Earth, 1, 15, 17, 33-34, 41, 47-51, 74, 82, 197, 238, 244, 249, 257, 281, 427-428
Atomic oxygen, 24-29, 34, 36, 47, 50, 63, 135
Atomic oxygen, erosion, 36
Attitude control system, 6, 12, 16, 40, 124-125, 161, 259-260, 420, 438

B

Back surface fields, 88-89, 97-98
Balance-of-system (BOS), 74
Band edge efficiencies, photocell, 143
Base layer, 76
Battery, 1, 13, 14, 33, 127, 136-137, 144, 157-166, 170-205, 211- 226, 235-241, 271, 357-374, 405-409, 418
  capacity, 360
  cathode, 162-170, 172, 180, 188-190, 193-197, 201-203, 216, 219, 224, 226, 231- 238, 296, 307-308, 310
  charging, V/T limit, 368
  common pressure vessel, 162, 211-213
  CPV, 162, 211, 212
  depth of discharge (DOD), 14, 33, 136, 137, 159-162, 174-181, 193, 209-213, 222, 237, 360-369, 408, 418
  individual pressure vessel (IPV), 161, 211- 213, 405-407
  IPV, 161, 211- 213, 405-407
  Li-BCX, 162, 193-199
  Li-MnO$_2$, 202
  LiSO$_2$, 12
  Li-SOCl$_2$, 162, 193-199
  Lithium-iodine, 193, 203
  Lithium-ion, 360, 370
  Lithium-sulfur dioxide, 193- 200
  Lithium-thionyl chloride, 193- 202, 236, 359
  Li-TiS$_2$, 162, 219
  Nickel-Cadmium (NiCd), 17, 137, 159-168, 177-183, 203-217, 221, 238, 360-370, 420
  Nickel-Hydrogen, 159, 211
  Ni-H$_2$, 162, 177, 204-205, 211-221
  Ni-MH, 162, 205
  overcharging, 178, 239, 368-370
  primary, 1, 9, 12, 15, 159, 176, 185-191, 359
  primary, Ag-Zn, 159
  reconditioning, 211, 353-374
  regulation and charge control, 161, 172-80, 211-218, 370, 407, 410
  secondary, 9, 157, 218, 362
  Silver-cadmium, 204, 213, 217
  Silver-Zinc, 1, 12, 159, 204, 213, 217
  Silver-Zinc primary, 1
Battery cell, 157, 174, 226, 271, 368, 374
  hermetically sealed, 211
  Nickel-Cadmium (Ni-Cd), 159
  Sodium, 222
Beginning-of-life (BOL), 24, 73, 109, 122-125, 258, 440

Bismuth telluride, 292, 446
Blackbody temperature, 422, 431
Blackbody, radiation curve, 48
Brayton thermodynamic cycle, 11, 140, 250, 265, 323, 332-334, 341
Bus bars, 357, 373, 406

C

Capillary action, 447
Cell bypass diodes, 132, 370
Centre National d'Estudes Spatiales, 16
CMOS, 60
Coatings, 134, 135, 299, 309, 315, 330, 337, 417, 440
Coatings, anti-reflective, 76
Common pressure vessel, 213
Component test power converter (CTPC), 327, 329
Composite materials, 34, 264, 337
Concentration polarization, 178, 227
Convective heat transfer, 142, 347, 415, 429, 447
CPV, 162, 211, 212

D

Debris, man-made, 23, 24
Debye length, 44, 45
Depth of discharge (DOD), 14, 33, 136-137, 159-181, 193, 209-213, 222, 237, 360-369, 408, 418
Differential absorbed dose, 102
Differential charging, 45, 130
Differential fluence spectrum, 102
Diffusion length, 78-79, 91-104, 148
Diode saturation current, 78, 92
DIPS, 334-341
Direct-write e-beam lithography (DEBL), 316
Displacement damage dose, 95-107, 115-116, 148
Dowtherm A, 342- 345
Dynamic conversion, 20, 250, 323
Dynamic conversion, chemical, 9
Dynamic isotope power system (DIPS), 334-341

E

Eagle-Picher, 161-162, 212, 213
Earth, axis of rotation, 46, 53
Eclipse, 25
   duration, 1, 11, 14, 25- 33, 45, 54, 125, 131-137, 146, 159, 180, 204, 213, 226, 232, 241, 353-363, 372, 407-409, 418, 422-424, 436-437, 446
   fraction of orbit, 31
   geosynchronous orbit, 30
Effective isotropically radiated power, 354, 355
EIRP, 354, 355
Electrical breakdown of dielectrics, 42
Electrical potential, 42
Electrical power system, 3, 5, 7, 10, 12, 18, 24, 25, 33, 41, 58, 407, 409, 415-416, 446
   Cassini, 12, 196-197, 250, 278, 359
   Envisat, 17
   Galileo, 15, 196, 250, 259, 278, 294, 359
   ISS, 13, 14, 120-123, 362, 405-409
   Lockheed Martin A2100, 405-408
   Magellan, 12
   Modular power system, 16, 161, 181, 409-410
   PMAD, 10, 361
   Spartan, 12
   TOPEX, 16, 17, 161, 181, 409
Electrochemical cell, 157, 163-177, 191-194, 204, 235, 360

INDEX

Electrode polarization, 368
Electromagnetic radiation, solar, 23, 29, 46-52, 315, 425
Electronics, CMOS, 60
Electronics, radiation-resistant devices, 61
Emission bandwidth, 147
Emissivity, 141-142, 145, 261, 264, 303, 309, 318, 422-441
Emitter efficiency, 142-146
Emitter layer, 76, 88
End-of-life (EOL) design, 73, 98, 108, 120, 126, 353-357, 363, 446
End-of-life (EOL) performance, 10, 13-17, 73, 98, 138, 334, 353, 362, 377-378, 401, 441
Energetic proton fluxes, 54
Energy conversion
   dynamic, 20, 250, 323
      Brayton Isotope Power System (BIPS), 334, 340
      Brayton Rotating Unit (BRU), 334
      Brayton, BIPS, 334, 340
      Brayton, BSF, 88, 89, 98
      chemical, 9
      kilowatt isotope power system (KIPS), 341
      Rankine, Rotary Fluid Management Device (RFMD), 345
      Stirling rhombic drive, 324
      Stirling, CTPC, 327, 329
      Stirling, FPSE, 324-330
      Stirling, SSPC, 327-329
   dynamic system
      Brayton, DIPS, 334-341
      Rankine cycle, ORC, 341-346
      Rankine, Rotating Jet Condenser (RJC), 345
      solar, SDGTD, 334, 339
      Stirling, Space Power Demonstrator Engine (SPDE), 327
      Stirling, Space Power Research Engine (SPRE), 327
   dynamic Brayton Rotating Unit (BRU), 334-335
   static
      AMTEC, 11, 253, 287-288, 307-310
      AMTEC, Nernst equation, 165, 309
      RTG, ALSEP, 249
      thermionic, arc drop, 299-302
      thermionic, arcing, 24, 38, 39, 42, 130-136, 404
      thermionic, back voltage, 300-301
      thermionic, barrier index, 300-302
      thermionic, Electron cooling, 302
      thermionic, retarding region, 298
      TPV, ytterbium oxide, 142
   static system
      general purpose heat source, 249-261, 293, 310-318, 343
      thermionic, Knudsen mode, 306
      thermionic, TECMDL, 302
      thermoelectric, ISOTEC, 244
   static, AMTEC, BASE, 307-311
   TPV
      antenna filter, 315-316
Energy sources, 149, 324, 340
Energy storage, 11, 136-146, 157-176, 185-193, 223, 237-241, 357, 390, 402-409
Environmental coatings, 438-439
Environmental factors of space, 10, 23, 29
Equilibrium temperature of a spacecraft, 29, 415, 424-442
Equinox, autumnal and vernal, 30
European Retrievable Carrier, 63

EWB, 136

F

F10.7 index, 47, 51
Faraday constant, 163-166, 228, 309
Fill factor, 77, 92, 313-316
Forward bias, 77, 380- 388
Forward mode, 77
Front surface fields, 76
Fuel cell, 9, 18, 157-166, 212, 223-233
  alkaline electrolyte H_2-O_2, 223
  alkaline fuel, 229
  Gemini, 231
  regenerative, 9, 11, 232
Fuel cell system, 157-158, 223-230

G

Galactic cosmic ray, 52, 55
General purpose heat source (GPHS), 249-261, 293, 310-318, 343
Geomagnetic field, 44, 53-56
Geomagnetic field, magnetic dipole, 53
Geomagnetic field, South Atlantic Anomaly, 23, 29, 53, 61
Geomagnetic substorm activity, 42
Geosynchronous orbit, 3, 23, 28-30, 62-65, 95, 120, 136-137, 204, 232, 420-426, 436
GPHS, 249-253, 258-261, 293, 310-311, 316-318, 343
Gravity, 223, 225, 231, 415
Grounding, 4, 372-373

H

Half-cell reactions, 164
Heat pipe, 266, 308, 329-330, 419, 447-448
Heat transfer mechanisms, 429

Heaters, 29, 174, 225, 231, 271-273, 280, 420, 437, 441, 446-448
High voltage vs. low voltage systems, 361
Huygens, 12

I

In-band emissivity, 142
Individual pressure vessel (IPV), 161, 211-213, 405-407
Insulation, 45, 66, 134, 166, 225, 230, 258-261, 291-294, 304, 308-313, 326, 330, 372, 374, 417, 423, 444-445
Interagency Nuclear Safety Review Panel (INSRP), 278
Internal series resistance, 77, 368
International Commission on Radiological Protection (ICRP), 275
International Space Station, 13, 14, 74, 120-123, 135, 362, 405-409
IR emission, 29, 426, 442
IR emission, Earth, 426
IR emissivity, 434, 436, 440
ISOTEC, 244

K

Kilowatt isotope power system (KIPS), 341
Kinetic temperature, 39, 41-44
KIPS, 343-345
Knudsen mode, 306
Kurchatov Institute of Atomic Energy, 255-257

L

Latent heat of vaporization, 447-449

INDEX

Launch vehicles
  Ariane, 420
  Lockheed Martin A2100, 372, 405

M

Magnetic dipole, 53
Manganese oxides, 162
Masked ion beam lithography (MIBL), 316
Materials degradation, absorptivity at EOL, 446
MCD, 161
MCFC, 224
Meteoroids, 62, 63, 64
Micrometeoroid environment, 23, 64
Molniya orbit, 29, 65
MPS, 16, 161, 181, 409-411
Multistage sequential topology, 365

N

NASCAP GEO, 136
NASCAP LEO, 136
Nernst equation, 165, 309
NIEL, 101-104, 115
Nuclear reactor
  Bouk, 255-257, 270, 281, 287
  Krasnaya Zvezda, 255-257
  Medium Power Reactor Experiment (MPRE), 341-342, 347
  Multimegawatt Space Nuclear Power Program, 252
  Nuclear energy reactor for vehicular applications (NERVA), 252
  Pied Piper program, 242
  Romashka, 255, 265, 270
Nuclear reactors, 9, 241-242, 250-257, 264-287, 341
Nuclear safety
  ICRP, 275
  INRSP, 278
  United Nations, 282
Nuclear systems, 9, 18, 242, 250, 255, 281, 303-304

O

Obstructed mode, 301-302
Obstructed region, 301-302
Open circuit voltage, 77, 80, 88, 91, 93, 100, 143, 172, 222, 289, 309, 313, 366
Open-cell failure, 370
Optical solar reflector (OSR), 442
Orbit
  GEO, 28-30, 42, 59, 62-65, 74-75, 95, 120-137, 162, 176-181, 204-205, 211-213, 223, 249, 357-369, 409, 436, 442
  geosynchronous, 23-30, 95, 204, 232, 422, 426, 436
  LEO, 16, 28-30, 34, 41, 65-66, 75, 89, 98, 111, 118, 132-137, 176, 180, 204-213, 223, 357, 363, 368, 417
  low-Earth, 12, 17, 28, 359, 422
  Sun-synchronous, 29
Orbital elements, 27, 28
Outgassing of materials, 37, 39, 57, 444
Outgassing rate, temperature dependence, 37
Out-of-band emissivities, 142

P

PACF, 224, 226
Parasitic currents, 42
Paschen electrical breakdown of gases, 38
Paschen minimum, 39
Peak power tracker, 366

Peak power tracking, maximum power, 77
Peltier coefficient, 290
Peltier effect, 290, 446
PEMFC, 223-225, 230
Phase change material, 443-444
Phase change temperature, 444
Photocurrent, 45
Photoelectrons, low-energy, 45
Photoemission, 45
Photoionization, 41-43
Photovoltaics, 89, 123
Photovoltaics, infrared, 311
PMAD, harness routing, 373
PMAD, voltage choice, 361
Polarization, 172-183, 226, 238
Polymers, 34
Power converter
   series, 372
Power converters
   boost, 381
   buck, 379-387, 396
   buck derived, 387
   buck-boost, 383-388
   DC-DC, 363-366, 372-375, 384-386, 390, 396-406
   pulse width modulated, 364, 384-385, 390-396
   regenerative DC-DC, 396
   regulated vs unregulated, 361
Power distribution
   AC vs DC, 362
   high voltage vs low voltage, 361
   regulated vs unregulated, 362
Power generators
   radioisotope thermoelectric, 9, 12
Power management
   buck regulators, 379
   buck-derived, 383-388
   buck-derived converters, 387
   DC-DC converters, 363-366, 372-375, 384-390, 396-400, 405-406
   dissipative, 362, 364
   linear regulators, 372-378
   pulse width modulated converters, 364, 384-396
   resonant DC-DC converters, 396
   series converters, 372
Power management and control, 363-364
Power management and distribution, 14, 33, 57, 137, 363
   magnetic components, 390, 403, 453, 463
   main bus, 373-375
Power management and distribution (PMAD), 11, 14, 20, 353-363, 372, 404-405
Power management and distribution, efficiency, 5, 71-77, 85-90, 95-127, 138-148, 176, 183, 188, 221-236, 250, 265, 287, 302-394, 400-406, 416, 427, 437, 446, 464
Power output, 47, 79, 126, 146, 190, 253, 258, 266, 270, 299, 302, 387, 389
Power regulator
   boost, 381
   buck, 379
   buck-boost derived, 383
   switching, 372, 381
Power source
   prime, 4, 7, 9
Power system
   advanced radioisotope (ARPS), 253
   Cassini, 12, 196, 197, 250, 278, 359
   Envisat, 17
   Galileo, 15, 196, 250, 259, 278, 294, 359
   GPS (IIR), 81, 407
   ideal, 353
   ISS, 407
   ISS USOS, 407
   LES 8, 242, 249, 343

LES 9, 242, 249
Lockheed Martin A2100, 372, 405, 406
Magellan, 12
nuclear reactor, SNAP-50, 341
RTG, SNAP, 242-252, 262-265, 287-293, 341
Spartan, 12
TOPAZ, 253-257, 265, 270-272, 294
TOPEX, 16, 17, 161, 181, 409
Power systems
  non-dissipative, 366
Primary batteries, 359
Primary cell, 157, 162, 170-177, 185-193, 203-205, 359
  Zinc anode, 214
Prismatic cell, 221
Prismatic Ni-Cd cells, 161
Project Feedback, 242

R

Radiated power, total equivalent isotropic, 354
Radiation
  displacement damage dose, 95-98, 101-107, 115-116, 148
Radiation resistance, 75, 88, 97-98, 107-118, 148
Radiation, absorbed dose, 101-104
Radiation, differential absorbed dose, 102
Radiation, differential fluence spectrum, 102
Radiators, 316, 339-340, 416-423, 430, 438-449
Radio flux at 10.7 cm wavelength, 47
Radioisotope thermoelectric generators (RTG), 9, 12, 144, 242-278, 291-294, 359
Rankine cycle, 11, 140, 250, 323, 340-348
Rankine fluids, Dowtherm A, 342-345
Rechargeable cells, 176-178, 205, 236-237
Reversed bias, 77
RTG, 144, 242-278, 291-294, 359
RTG, Cassini, 12, 196-197, 250, 278, 359
RTG, General Purpose Heat Source (GPHS), 293

S

Schottky effect, 301
Second surface mirrors, 442
Seebeck coefficients, 288-290
Seebeck, Thomas, 288
Selective emitter, 140-146
Semiconductors, 96, 103, 112, 313, 403
Semiconductors, CMOS, 60
Shadowing factor, 138
Shielding effects, 64
Short circuit conditions, 79
Short circuit current, 77-92
Shunts
  linear and switching, 364
Single event upset, latch up, 60
Solar absorptivity, 422, 433-440
Solar activity, F10.7 index, 47, 51
Solar array, 6, 9, 12, 15, 29, 34, 55, 71-75, 120-139, 144-148, 255, 359, 363, 407-416, 427
  area power density, 72
  aspect ratio, 125
  body-mounted, 120, 358
  control, 363-366
  fill factor, 77, 91, 313
  lifetime, 75
  packing factor, 127, 129, 315
  shadowing factor, 138
  shielding effects, 64

slip rings, 357, 372
Solar cells
 MBG cells, mechanically stacked, 111
 MBG cells, monolithically grown, 111
Solar constant, 18, 48, 84, 437
Solar cycle
 Solar maximum, 39, 47-55
 Solar minimum, 47-55
Solar electromagnetic radiation, 23, 29, 46-52, 315, 425
Solar power, 77
Solar proton flares, 52
Solar radiation, 29, 80, 422-437, 441
Solar radiation, absorptivity, 50, 140-141, 313-315, 422, 432-442
Solar spectrum, air mass zero, 48
Solar spectrum, EUV, 23, 47, 50
Solar sunspot cycle, 47
Solid state power amplifiers, 353
Solstice, winter and summer, 30
South Atlantic Anomaly (SAA), 23, 29, 53, 61-62
Space environment
 atomic oxygen, 24-29, 34-36, 47-50, 63, 135
 atomic oxygen erosion, 36
 charged particle, 29, 33, 41-55, 61, 428, 440, 442
 cosmic ray, single-event upset, 58, 60
 cosmic ray, single-particle effects, 58
 cosmic ray, single-particle event, 60
 debris, 29, 62-65, 129, 257, 281
 galactic cosmic rays, 23, 29, 55, 60, 426
 magnetosphere, 15, 41, 54-55, 79, 90, 139
 meteoroids, 62-64
 micrometeoroids, 23, 64

 near-Earth, 23, 33, 46
 near-Earth, 18, 23, 33, 52-53, 63
 neutral atmosphere, 23, 33, 41, 47, 135
 neutral atmosphere, 29, 33-34, 39, 42, 428
 outer radiation belts, 23
 particulate environment, 62-63
 plasma, 41-47, 52
 plasma, high-energy, 23
 plasma, ionospheric, 29, 42, 45
 plasmas, 33, 42, 44
 pressure, ambient, 37
 pressure, residual, 37
 radiation damage, 53, 56, 74-79, 88-108, 114, 119, 126, 132-138, 148
 radiation damage in electronics, 53
 radiation degradation factor, 138
 radiation environment, 28, 41, 46, 75, 98, 111, 127, 303
 radiation, ionizing, 46, 58, 60
 radiation, planetary, 426
 radiation, SEU and latch up, 60
 radiation, total absorbed dose, 102-103
 radiation, UV, 23, 41-50, 62, 130
 solar activity, F10.7 index, 47, 51
 solar flare, 29, 47-60, 74, 79, 90, 95, 130-133
 solar proton flares, 52
 solar radiation, 29, 80, 422-427, 433-437, 441
 solar spectrum, air mass zero, 48
 solar spectrum, EUV, 23, 47, 50
 solar sunspot cycle, 47
 solar wind, 41, 53-55
 space solar spectrum, 80
 thermospheric temperature, 44, 51
Space power demonstrator engine (SPDE), 327-329
Space power research engine (SPRE), 327-329

INDEX

Space-charge effects, 45
Spacecraft
   Cassini, 12, 196-197, 250, 278, 359
   Clementine, 162, 213
   Cosmos 1402, 257, 281
   Cosmos 1818, 257
   Cosmos 1867, 257
   Cosmos 1900, 281
   Cosmos 954, 257, 281-282
   CRESS, 57
   Envisat, 17
   EUVE, 409
   Explorer 23, 161
   Explorer 6, 119, 159
   Explorer I, 33
   Galaxy IV, 6
   Galileo, 15, 196, 250, 259, 278, 294, 359
   geosynchronous communications, 3
   GRO, 161, 409
   Hubble telescope, 63
   INSAT-2a, 420
   International Space Station, 13, 74, 120-122, 135, 362, 407
   ISS, Russian Orbital Segment, 407
   LANDSAT D, 409
   LDEF, 63, 196
   LES, 242, 249, 343
   LES 8, 242, 249, 343
   LES 9, 242, 249
   Lincoln Experimental Satellites (LES), 249
   Magellan, 12
   Mariner 2, 159
   Mars Global Surveyor Mission, 162
   Pioneer, 159, 248-249, 359
   Pioneer 10, 248-249
   Pioneer 11, 249
   Pluto Express, 253
   Poseidon, 16, 17
   Radar Ocean Reconnaissance Satellite (RORSAT), 255
   Ranger 3, 159
   RORSAT, 255, 270, 281, 287
   SMM, 16, 161, 190, 211
   Solar Max Mission (SMM), 161, 190, 218
   Spartan, 12
   Sputnik I, 1, 12, 159
   TIROS, 159
   TOPEX, 16-17
   Ulysses, 250, 259
   UOSAT2, 62
   URAS, 409
   Vanguard I, 1, 88, 119, 139
   Vanguard Test Vehicle 3, 159
   Viking, 249
   Voyager, 249, 343, 359
Spacecraft charging, 41, 54, 75, 129, 132
Spacecraft differential charging, 45, 130
Spacecraft materials properties, 291
Spacecraft stabilization, methods, 423
Spacecraft stabilization, three-axis, 358, 373, 407, 423
Spacecraft, architecture of, 5
Specific area, 73
Specific energy, 159-166, 174-176, 186, 194, 199, 204, 205, 211-222, 359-360, 370
Specific mass, 72
Specific power, 3, 72, 98, 123-126, 137, 213, 293
Spectral response, 84-87, 114, 140
Spin-stabilized satellites, 120, 358
SPV, 162, 211-213
SSPA, 353-355, 375, 400, 405-406
Standard cells, 161
Stirling cycle, 11, 140, 250-253, 265-266, 323-330
Sun synchronous orbit, 29
Sunspots, F10.7 index, 47, 51
Surface absorptivity, 440

T

Temperature coefficients, 86, 401
Temperature degradation factor, 138
Temperature margin, AMTEC, 310, 418
Temperature-compensated voltage, 161
Themionics, unignited mode, 299
Thermal analysis of a spacecraft, 428, 435
Thermal control system, 13, 415-418, 437
Thermal cycling, 33, 73, 75, 130, 138, 345
Thermal environment, 20, 415-418, 421
Thermal management, 4, 10, 20, 25-37, 48-50, 65, 222, 223, 235, 340, 356, 358, 403, 408, 417-445
   active control, 445
   aerodynamic heating, 261, 281, 421, 422
   heat pipe, 447-448
   heat pipe working fluids, 307, 323-327, 335-449
   heaters, 446
   multilayer insulation, 423, 444
   paints, 62, 438, 441
   passive, 445
   phase change materials, 443
   pumped systems, 437
   radiative efficiency, 142
   surface absorptivity, 440
   thermal control system, 415-448
Thermionic
   diode saturation current, 78, 92
   saturation region, 298-302
   Schock instability, 302
Thermionic conversion, 11, 245-257, 265, 271, 272, 287, 288, 294-306
Thermionic fuel element verification program (TFEVP), 253
Thermionic, ignited mode, 299, 300, 302
Thermodynamic cycle, Brayton, 11, 140, 250, 265, 323, 332-341
Thermoelectric conversion, 11, 144, 244-292, 334, 359, 446
Thermoelectric coolers, 446
Thermoelectric materials, figure of merit, 237, 456
Thermoelectric Seebeck coefficients, 288-290
Thermonuclear fusion, 47
Thermospheric temperature, 44, 51
Thomson coefficient, 291
Three-band model, 142
TPV material, Erbia, 143
TPV rare-Earth emitters
   Erbia, 143
   Holmia, 143
   Neodymia, 143
TPV, Ytterbia, 142
TPV, ytterbium oxide, 142
TPV. direct-write e-beam lithography, 316
Transition point, thermionic conversion, 299-302
Traveling wave tube amplifiers, 353
TWTA, 353-355, 403-406

U

Universal gas constant, 309

W

Wien displacement law, 431
Wiring loss factor in PMAD, 138
Work function, thermionic, 297-306
Working fluid, heat pipe, 307, 323-327, 335-449

Y

YTP, 161